International RILEM Conference on Early-Age and Long-Term Cracking in RC Structures

RILEM BOOKSERIES

Volume 31

RILEM, The International Union of Laboratories and Experts in Construction Materials, Systems and Structures, founded in 1947, is a non-governmental scientific association whose goal is to contribute to progress in the construction sciences, techniques and industries, essentially by means of the communication it fosters between research and practice. RILEM's focus is on construction materials and their use in building and civil engineering structures, covering all phases of the building process from manufacture to use and recycling of materials. More information on RILEM and its previous publications can be found on www.RILEM.net.

Indexed in SCOPUS, Google Scholar and SpringerLink.

More information about this series at http://www.springer.com/series/8781

Fragkoulis Kanavaris · Farid Benboudjema ·
Miguel Azenha
Editors

International RILEM Conference on Early-Age and Long-Term Cracking in RC Structures

CRC 2021

Springer

Editors
Fragkoulis Kanavaris
Advanced Digital Engineering
Specialist Technology and Research
ARUP
London, UK

Farid Benboudjema
Laboratoire de Mécanique et Technologie
ENS Paris-Saclay
Gif-sur-Yvette, France

Miguel Azenha
ISISE, Department of Civil Engineering
University of Minho
Guimarães, Portugal

ISSN 2211-0844 ISSN 2211-0852 (electronic)
RILEM Bookseries
ISBN 978-3-030-72920-2 ISBN 978-3-030-72921-9 (eBook)
https://doi.org/10.1007/978-3-030-72921-9

This Springer imprint is published by the registered company Springer Nature Switzerland AG
The registered company address is: Gewerbestrasse 11, 6330 Cham, Switzerland

Organizing Committees

Conference Chairs

Azenha, Miguel
Benboudjema, Farid
Kanavaris, Fragkoulis

Scientific Committee

Kanavaris, Fragkoulis (Chair)
Adia, Jean-Luc
Alam, Sayed Yasir
Asamoto, Shingo
Azenha, Miguel
Benboudjema, Farid
Bourdot, Alexandra
Briffaut, Matthieu
Courtois, Alexis
Fairbairn, Eduardo
Faria, Rui
Fau, Amélie
Ferreira, Nuno
Forth, John
Funchal de Faria
Hamilton, David
Honorio, Tulio
Jedrzejewska, Agnieszka
Jyosyula, Sri Kalyana Rama

Kaklauskas, Gintaris
Kang, Shao-Bo
Kanstad, Terje
Khadka, Binod
Klausen, Anja
Kuperman, Selmo
Lacarriere, Laurie
Oliver-Leblond, Cecile
Pacheco, Jose
Pérez, Alejandro
Pesavento, Francesco
Phillip, Nivin
Sadowski, Lukasz
Sánchez Montero, Javier
Schlicke, Dirk
Seetharam, Suresh
Serra, Carlos
Sfikas, Ioannis
Smilauer, Vit
Sousa, Carlos
Sriskandarajah, Sanchayan
Tan, Reignard
Torrenti, Jean Michel
Torres Martín, Julio
Ulfkjaer, Jens Peder
Usui, Tatsuya
van den Bos, Ab Alexander
Vijay, Er. A.
Vollum, Robert
Wang, ChenFei
Zych, Mariusz.

Preface

Even though Reinforced Concrete (RC) is one of the most used manmade materials in the world, and several mature models exist for the prediction and design of the Ultimate Limit State (ULS) capacity of RC structures, the prediction of Serviceability Limit State (SLS) is still not mature enough for actual design purposes. This may lead to immediate repair needs, inadequate service life behavior even when regulatory prescriptions for design are strictly followed, or at other times to unnecessary overdesign of reinforcement, which has negative consequences on both sustainability and economic aspects. This is amplified by development of new more sustainable cements and supplementary materials since the range of material properties needed for SLS design is considerably wider than for ULS-design.

One of the most common issues relevant to many RC structures which is encountered on-site is that of cracking. Understanding the interplay between fundamental mechanisms involved in the processes of hydration-induced temperature, shrinkage, creep, loading and herewith caused cracking, requires interdisciplinary integration of materials science and structural engineering. A comprehensive approach to this problem requires the design of robust experimental techniques, development of multiscale models (micro-, meso-, and mainly, macro-scale), and evaluation of code-based and other analytical approaches relevant to crack control in concrete which is also the aim of RILEM TC 287-CCS: Early-age and long-term crack width analysis in RC structures.

The CRC2021 conference aimed to bring together experts to share the state-of-the-art developments on the aforementioned fields and the progress within RILEM TC 287-CCS. The conference had contributions by researchers and other professionals from a variety of scientific disciplines in works targeted to crack control in concrete or similar materials and crack healing. These proceedings are considered of interest for researchers and practitioners at any career level who have an interest in deepening knowledge on the targeted subjects, and are divided into the following 5 sections:

- Structural design and specification
- Experimental and on-site evaluation
- Modelling and simulation
- Binders, additions, and admixtures
- Crack healing

The event was hosted by ENS Paris-Saclay, France, on the 9 April 2021 with a significant share of online participants due to restraints to travelling caused by COVID-19.

As a final remark, the editors would like to acknowledge the voluntary and competent job made by the members of the Scientific Committee of CRC2021, that ensured that every single contribution shown in this publication has been checked, commented and approved by a set of two independent reviews.

London, UK — Fragkoulis Kanavaris
Gif-sur-Yvette, France — Farid Benboudjema
Guimarães, Portugal — Miguel Azenha

Contents

Structural Design and Specification

Age-Adjusted Effective Elastic Modulus of High-Performance Concrete at Early Age 3
Arosha Dabarera, Liang Li, Jiahang Li, Vishvendra Singh Jamwal, Qifan Yang, and Vinh Dao

Plain Geopolymer Concrete Cross-Section Surface Analysis After Creep and Shrinkage Tests in Compression and Tension 13
Rihards Gailitis, Andina Sprince, Leonids Pakrastins, Kinga Korniejenko, and Tomass Kozlovskis

Crack Width Estimation in Steel Fibre Reinforced Concrete (SFRC): Design and Limitations 25
Ned Gould, Fragkoulis Kanavaris, and Nuno Ferreira

Cracking Analysis of Partially Prestressed Concrete Tie Under the Effect of Primary and Secondary Cracks 37
Maurizio Taliano

Design, Evaluation and Specification for Concrete Mixtures with Low Potential for Cracking 51
Jose Pacheco

Experimentation and On-Site Monitoring

Crack Initiation and Propagation in Fiber-Glass Reinforced Mortars 63
Pascale Saba, Tulio Honorio, Omar-Ateeq Mahmood, and Farid Benboudjema

Distributed Optical Fibre Sensors for Strain and Temperature Monitoring of Early-Age Concrete: Laboratory and In-situ Examples 77
Rafał Sieńko, Łukasz Bednarski, and Tomasz Howiacki

Experimental Investigations of Cracking in Reinforced Concrete Beams of Different Depth 89
Aleksandr Sokolov, Deividas Rumsys, Karolis Sakalauskas, Darius Bacinskas, and Gintaris Kaklauskas

Imposed Deformation Reduction in Semimassive Walls of RC Tanks by Internal Cooling 101
Mariusz Zych and Trong-Chuc Nguyen

Initiation Period of Corrosion by Chloride Ion in Cracked Concrete Structures 115
Julio Torres and Javier Sánchez

Investigation on Crack Control and Crack Pattern Analysis of Self-compacting Concrete Exposed to Standard Fire Exposure 127
Mervin Ealiyas Mathews, N. Anand, A. Diana Andrushia, and Tattukolla Kiran

Long-Term Experimental Campaign on RC Shrinkage Cracking: Conceptualization, Planning and Experimental Procedures 141
José Gomes, José Granja, Carlos Sousa, Cláudio Ferreira, Dirk Schlicke, Rui Faria, and Miguel Azenha

Strength Behaviour of a High-Performance Concrete Under Drying 155
Ismail Yurtdas, Nicolas Burlion, Jianfu Shao, and Alex Li

The Cracking Sensitivity of a Na-Geopolymer 165
Heng Sounean, Hannawi Kinda, and Darquennes Aveline

The Effect of Curing Duration on Restrained Shrinkage Cracking of Concrete Containing Macro Polymeric Fibers 175
Alireza Bagheri, Armin Jamali, Mina Pourmir, and Hamed Zanganeh

Modelling and Simulation

Crack Control of Upstream Polder Face Using Calibrated Thermo-mechanical Simulations 189
Vít Šmilauer, Petr Havlásek, Pavel Reiterman, and Petr Huňka

CSI (Cement Science Investigation): Using Machine Learning to Guess the OPC Pastes Composition from the Elastic Response 201
Tulio Honorio, Sofiane Ait-Hamadouche, and Amélie Fau

Development of Expansive Agent Model in a Multi-scale Thermodynamic Framework Based on Hydration and Microstructure Formation 211
Kolneath Pen, Tetsuya Ishida, Igarashi Go, Yuya Takahashi, and Shinya Ito

Double Wall Containment Building Leak-Tightness Prediction: Strategy and Application 227
Laurent Charpin, Charles Toulemonde, Jean-Luc Adia, Florian Escoffier, Sylvie Michel-Ponnelle, Guillaume Boulant, Benoît Masson, and Julien Niepceron

In-situ X-CT Test on Mortar Micro-specimen Coupled with Mesoscale Numerical Simulations of Fracture 239
Aliaksandra Tsitova, Fabien Bernachy-Barbe, Benoît Bary, and François Hild

Numerical Analysis of Size Effects on the Stabilized Cracking Stage in RC Members 253
David E. M. Bouhjiti, Noam Demri, Miquel A. Huguet, Silvano Erlicher, and Julien Baroth

Numerical Estimation of Crack Openings in Concrete Structures Under Multi-physic Loadings 265
François Soleilhet, Farid Benboudjema, Xavier Jourdain, and Fabrice Gatuingt

Sustainability of Internally-Cured Concrete for Mitigating Shrinkage Cracking Using Service Life Prediction Models 277
Sara Kalantari, Mohammad Ali Dastan Diznab, and Fariborz M. Tehrani

Validation of Thermo-hygro-Mechanical FEM Analysis of Thick Restrained RC Members by Comparison with Experiments 291
Cláudio Ferreira, Dirk Schlicke, Carlos Sousa, José Gomes, Rui Faria, and Miguel Azenha

Binders, Additions, and Admixtures

Bleeding of Cement Pastes Modified with the Addition of Granite Powder 307
Adrian Chajec and Łukasz Sadowski

Effect of Expansive Additives on the Early Age Elastic Modulus Development of Cement Paste by Ambient Response Method (ARM) 319
Mayank Gupta, Igarashi Go, Jose Granja, Miguel Azenha, and Tetsuya Ishida

Effect of MgO-Based Expansive Agent on the Cement-Based Mortar Behavior 329
Lina Ammar, Kinda Hannawi, and Aveline Darquennes

Plasticizers as Dispersants for Carbon Nanotubes in Self-sensing Cement Composites 341
Pedro de Almeida Carísio, Oscar Aurelio Mendoza Reales, Eduardo de Morares Rego Fairbairn, and Romildo Dias Toledo Filho

Crack Healing

Autogenous Healing of High Performance Concrete Using Metakaolin 355
Reeja P. Soman and Nivin Philip

Bacteria Based Self-healing of Later-Age Cracks in Concrete 367
Lorena Skevi, Bianca Reeksting, Susanne Gebhard, and Kevin Paine

Effects of *Bacillus subtilis* on Crack Remediation in Thermally Degraded Limestone Calcined Clay Cement Mortars 377
Joseph Mwiti Marangu and Mark Bediako

Effect of Binary-Use Mineral Admixtures for the Advanced Autogenous Self-healing Behavior of Fiber-Reinforced Cementitious Composites 389
Oğuzhan Öztürk and Marta Roig-Flores

RILEM Publications

The following list is presenting the global offer of RILEM Publications, sorted by series. Each publication is available in printed version and/or in online version.

RILEM PROCEEDINGS (PRO)

PRO 1: Durability of High Performance Concrete (ISBN: 2-912143-03-9; e-ISBN: 2-351580-12-5; e-ISBN: 2351580125); *Ed. H. Sommer*

PRO 2: Chloride Penetration into Concrete (ISBN: 2-912143-00-04; e-ISBN: 2912143454); *Eds. L.-O. Nilsson and J.-P. Ollivier*

PRO 3: Evaluation and Strengthening of Existing Masonry Structures (ISBN: 2-912143-02-0; e-ISBN: 2351580141); *Eds. L. Binda and C. Modena*

PRO 4: Concrete: From Material to Structure (ISBN: 2-912143-04-7; e-ISBN: 2351580206); *Eds. J.-P. Bournazel and Y. Malier*

PRO 5: The Role of Admixtures in High Performance Concrete (ISBN: 2-912143-05-5; e-ISBN: 2351580214); *Eds. J. G. Cabrera and R. Rivera-Villarreal*

PRO 6: High Performance Fiber Reinforced Cement Composites - HPFRCC 3 (ISBN: 2-912143-06-3; e-ISBN: 2351580222); *Eds. H. W. Reinhardt and A. E. Naaman*

PRO 7: 1st International RILEM Symposium on Self-Compacting Concrete (ISBN: 2-912143-09-8; e-ISBN: 2912143721); *Eds. Å. Skarendahl and Ö. Petersson*

PRO 8: International RILEM Symposium on Timber Engineering (ISBN: 2-912143-10-1; e-ISBN: 2351580230); *Ed. L. Boström*

PRO 9: 2nd International RILEM Symposium on Adhesion between Polymers and Concrete ISAP '99 (ISBN: 2-912143-11-X; e-ISBN: 2351580249); *Eds. Y. Ohama and M. Puterman*

PRO 10: 3rd International RILEM Symposium on Durability of Building and Construction Sealants (ISBN: 2-912143-13-6; e-ISBN: 2351580257); *Eds. A. T. Wolf*

PRO 11: 4th International RILEM Conference on Reflective Cracking in Pavements (ISBN: 2-912143-14-4; e-ISBN: 2351580265); *Eds. A. O. Abd El Halim, D. A. Taylor and El H. H. Mohamed*

PRO 12: International RILEM Workshop on Historic Mortars: Characteristics and Tests (ISBN: 2-912143-15-2; e-ISBN: 2351580273); *Eds. P. Bartos, C. Groot and J. J. Hughes*

PRO 13: 2nd International RILEM Symposium on Hydration and Setting (ISBN: 2-912143-16-0; e-ISBN: 2351580281); *Ed. A. Nonat*

PRO 14: Integrated Life-Cycle Design of Materials and Structures - ILCDES 2000 (ISBN: 951-758-408-3; e-ISBN: 235158029X); (ISSN: 0356-9403); *Ed. S. Sarja*

PRO 15: Fifth RILEM Symposium on Fibre-Reinforced Concretes (FRC) - BEFIB'2000 (ISBN: 2-912143-18-7; e-ISBN: 291214373X); *Eds. P. Rossi and G. Chanvillard*

PRO 16: Life Prediction and Management of Concrete Structures (ISBN: 2-912143-19-5; e-ISBN: 2351580303); *Ed. D. Naus*

PRO 17: Shrinkage of Concrete – Shrinkage 2000 (ISBN: 2-912143-20-9; e-ISBN: 2351580311); *Eds. V. Baroghel-Bouny and P.-C. Aïtcin*

PRO 18: Measurement and Interpretation of the On-Site Corrosion Rate (ISBN: 2-912143-21-7; e-ISBN: 235158032X); *Eds. C. Andrade, C. Alonso, J. Fullea, J. Polimon and J. Rodriguez*

PRO 19: Testing and Modelling the Chloride Ingress into Concrete (ISBN: 2-912143-22-5; e-ISBN: 2351580338); *Eds. C. Andrade and J. Kropp*

PRO 20: 1st International RILEM Workshop on Microbial Impacts on Building Materials (CD 02) (e-ISBN 978-2-35158-013-4); *Ed. M. Ribas Silva*

PRO 21: International RILEM Symposium on Connections between Steel and Concrete (ISBN: 2-912143-25-X; e-ISBN: 2351580346); *Ed. R. Eligehausen*

PRO 22: International RILEM Symposium on Joints in Timber Structures (ISBN: 2-912143-28-4; e-ISBN: 2351580354); *Eds. S. Aicher and H.-W. Reinhardt*

PRO 23: International RILEM Conference on Early Age Cracking in Cementitious Systems (ISBN: 2-912143-29-2; e-ISBN: 2351580362); *Eds. K. Kovler and A. Bentur*

PRO 24: 2nd International RILEM Workshop on Frost Resistance of Concrete (ISBN: 2-912143-30-6; e-ISBN: 2351580370); *Eds. M. J. Setzer, R. Auberg and H.-J. Keck*

PRO 25: International RILEM Workshop on Frost Damage in Concrete (ISBN: 2-912143-31-4; e-ISBN: 2351580389); *Eds. D. J. Janssen, M. J. Setzer and M. B. Snyder*

PRO 26: International RILEM Workshop on On-Site Control and Evaluation of Masonry Structures (ISBN: 2-912143-34-9; e-ISBN: 2351580141); *Eds. L. Binda and R. C. de Vekey*

PRO 27: International RILEM Symposium on Building Joint Sealants (CD03; e-ISBN: 235158015X); *Ed. A. T. Wolf*

PRO 28: 6th International RILEM Symposium on Performance Testing and Evaluation of Bituminous Materials - PTEBM'03 (ISBN: 2-912143-35-7; e-ISBN: 978-2-912143-77-8); *Ed. M. N. Partl*

PRO 29: 2nd International RILEM Workshop on Life Prediction and Ageing Management of Concrete Structures (ISBN: 2-912143-36-5; e-ISBN: 2912143780); *Ed. D. J. Naus*

PRO 30: 4th International RILEM Workshop on High Performance Fiber Reinforced Cement Composites - HPFRCC 4 (ISBN: 2-912143-37-3; e-ISBN: 2912143799); *Eds. A. E. Naaman and H. W. Reinhardt*

PRO 31: International RILEM Workshop on Test and Design Methods for Steel Fibre Reinforced Concrete: Background and Experiences (ISBN: 2-912143-38-1; e-ISBN: 2351580168); *Eds. B. Schnütgen and L. Vandewalle*

PRO 32: International Conference on Advances in Concrete and Structures 2 vol. (ISBN (set): 2-912143-41-1; e-ISBN: 2351580176); *Eds. Ying-shu Yuan, Surendra P. Shah and Heng-lin Lü*

PRO 33: 3rd International Symposium on Self-Compacting Concrete (ISBN: 2-912143-42-X; e-ISBN: 2912143713); *Eds. Ó. Wallevik and I. Níelsson*

PRO 34: International RILEM Conference on Microbial Impact on Building Materials (ISBN: 2-912143-43-8; e-ISBN: 2351580184); *Ed. M. Ribas Silva*

PRO 35: International RILEM TC 186-ISA on Internal Sulfate Attack and Delayed Ettringite Formation (ISBN: 2-912143-44-6; e-ISBN: 2912143802); *Eds. K. Scrivener and J. Skalny*

PRO 36: International RILEM Symposium on Concrete Science and Engineering – A Tribute to Arnon Bentur (ISBN: 2-912143-46-2; e-ISBN: 2912143586); *Eds. K. Kovler, J. Marchand, S. Mindess and J. Weiss*

PRO 37: 5th International RILEM Conference on Cracking in Pavements – Mitigation, Risk Assessment and Prevention (ISBN: 2-912143-47-0; e-ISBN: 2912143764); *Eds. C. Petit, I. Al-Qadi and A. Millien*

PRO 38: 3rd International RILEM Workshop on Testing and Modelling the Chloride Ingress into Concrete (ISBN: 2-912143-48-9; e-ISBN: 2912143578); *Eds. C. Andrade and J. Kropp*

PRO 39: 6th International RILEM Symposium on Fibre-Reinforced Concretes - BEFIB 2004 (ISBN: 2-912143-51-9; e-ISBN: 2912143748); *Eds. M. Di Prisco, R. Felicetti and G. A. Plizzari*

PRO 40: International RILEM Conference on the Use of Recycled Materials in Buildings and Structures (ISBN: 2-912143-52-7; e-ISBN: 2912143756); *Eds. E. Vázquez, Ch. F. Hendriks and G. M. T. Janssen*

PRO 41: RILEM International Symposium on Environment-Conscious Materials and Systems for Sustainable Development (ISBN: 2-912143-55-1; e-ISBN: 2912143640); *Eds. N. Kashino and Y. Ohama*

PRO 42: SCC'2005 - China: 1st International Symposium on Design, Performance and Use of Self-Consolidating Concrete (ISBN: 2-912143-61-6; e-ISBN: 2912143624); *Eds. Zhiwu Yu, Caijun Shi, Kamal Henri Khayat and Youjun Xie*

PRO 43: International RILEM Workshop on Bonded Concrete Overlays (e-ISBN: 2-912143-83-7); *Eds. J. L. Granju and J. Silfwerbrand*

PRO 44: 2nd International RILEM Workshop on Microbial Impacts on Building Materials (CD11) (e-ISBN: 2-912143-84-5); *Ed. M. Ribas Silva*

PRO 45: 2nd International Symposium on Nanotechnology in Construction, Bilbao (ISBN: 2-912143-87-X; e-ISBN: 2912143888); *Eds. Peter J. M. Bartos, Yolanda de Miguel and Antonio Porro*

PRO 46: ConcreteLife'06 - International RILEM-JCI Seminar on Concrete Durability and Service Life Planning: Curing, Crack Control, Performance in Harsh Environments (ISBN: 2-912143-89-6; e-ISBN: 291214390X); *Ed. K. Kovler*

PRO 47: International RILEM Workshop on Performance Based Evaluation and Indicators for Concrete Durability (ISBN: 978-2-912143-95-2; e-ISBN: 9782912143969); *Eds. V. Baroghel-Bouny, C. Andrade, R. Torrent and K. Scrivener*

PRO 48: 1st International RILEM Symposium on Advances in Concrete through Science and Engineering (e-ISBN: 2-912143-92-6); *Eds. J. Weiss, K. Kovler, J. Marchand, and S. Mindess*

PRO 49: International RILEM Workshop on High Performance Fiber Reinforced Cementitious Composites in Structural Applications (ISBN: 2-912143-93-4; e-ISBN: 2912143942); *Eds. G. Fischer and V.C. Li*

PRO 50: 1st International RILEM Symposium on Textile Reinforced Concrete (ISBN: 2-912143-97-7; e-ISBN: 2351580087); *Eds. Josef Hegger, Wolfgang Brameshuber and Norbert Will*

PRO 51: 2nd International Symposium on Advances in Concrete through Science and Engineering (ISBN: 2-35158-003-6; e-ISBN: 2-35158-002-8); *Eds. J. Marchand, B. Bissonnette, R. Gagné, M. Jolin and F. Paradis*

PRO 52: Volume Changes of Hardening Concrete: Testing and Mitigation (ISBN: 2-35158-004-4; e-ISBN: 2-35158-005-2); *Eds. O. M. Jensen, P. Lura and K. Kovler*

PRO 53: High Performance Fiber Reinforced Cement Composites - HPFRCC5 (ISBN: 978-2-35158-046-2; e-ISBN: 978-2-35158-089-9); *Eds. H. W. Reinhardt and A. E. Naaman*

PRO 54: 5th International RILEM Symposium on Self-Compacting Concrete (ISBN: 978-2-35158-047-9; e-ISBN: 978-2-35158-088-2); *Eds. G. De Schutter and V. Boel*

PRO 55: International RILEM Symposium Photocatalysis, Environment and Construction Materials (ISBN: 978-2-35158-056-1; e-ISBN: 978-2-35158-057-8); *Eds. P. Baglioni and L. Cassar*

PRO 56: International RILEM Workshop on Integral Service Life Modelling of Concrete Structures (ISBN 978-2-35158-058-5; e-ISBN: 978-2-35158-090-5); *Eds. R. M. Ferreira, J. Gulikers and C. Andrade*

PRO 57: RILEM Workshop on Performance of cement-based materials in aggressive aqueous environments (e-ISBN: 978-2-35158-059-2); *Ed. N. De Belie*

PRO 58: International RILEM Symposium on Concrete Modelling - CONMOD'08 (ISBN: 978-2-35158-060-8; e-ISBN: 978-2-35158-076-9); *Eds. E. Schlangen and G. De Schutter*

PRO 59: International RILEM Conference on On Site Assessment of Concrete, Masonry and Timber Structures - SACoMaTiS 2008 (ISBN set: 978-2-35158-061-5; e-ISBN: 978-2-35158-075-2); *Eds. L. Binda, M. di Prisco and R. Felicetti*

PRO 60: Seventh RILEM International Symposium on Fibre Reinforced Concrete: Design and Applications - BEFIB 2008 (ISBN: 978-2-35158-064-6; e-ISBN: 978-2-35158-086-8); *Ed. R. Gettu*

PRO 61: 1st International Conference on Microstructure Related Durability of Cementitious Composites 2 vol., (ISBN: 978-2-35158-065-3; e-ISBN: 978-2-35158-084-4); *Eds. W. Sun, K. van Breugel, C. Miao, G. Ye and H. Chen*

PRO 62: NSF/ RILEM Workshop: In-situ Evaluation of Historic Wood and Masonry Structures (e-ISBN: 978-2-35158-068-4); *Eds. B. Kasal, R. Anthony and M. Drdácký*

PRO 63: Concrete in Aggressive Aqueous Environments: Performance, Testing and Modelling, 2 vol., (ISBN: 978-2-35158-071-4; e-ISBN: 978-2-35158-082-0); *Eds. M. G. Alexander and A. Bertron*

PRO 64: Long Term Performance of Cementitious Barriers and Reinforced Concrete in Nuclear Power Plants and Waste Management - NUCPERF 2009 (ISBN: 978-2-35158-072-1; e-ISBN: 978-2-35158-087-5); *Eds. V. L'Hostis, R. Gens, C. Gallé*

PRO 65: Design Performance and Use of Self-consolidating Concrete - SCC'2009 (ISBN: 978-2-35158-073-8; e-ISBN: 978-2-35158-093-6); *Eds. C. Shi, Z. Yu, K. H. Khayat and P. Yan*

PRO 66: 2nd International RILEM Workshop on Concrete Durability and Service Life Planning - ConcreteLife'09 (ISBN: 978-2-35158-074-5; ISBN: 978-2-35158-074-5); *Ed. K. Kovler*

PRO 67: Repairs Mortars for Historic Masonry (e-ISBN: 978-2-35158-083-7); *Ed. C. Groot*

PRO 68: Proceedings of the 3rd International RILEM Symposium on 'Rheology of Cement Suspensions such as Fresh Concrete (ISBN 978-2-35158-091-2; e-ISBN: 978-2-35158-092-9); *Eds. O. H. Wallevik, S. Kubens and S. Oesterheld*

PRO 69: 3rd International PhD Student Workshop on 'Modelling the Durability of Reinforced Concrete (ISBN: 978-2-35158-095-0); *Eds. R. M. Ferreira, J. Gulikers and C. Andrade*

PRO 70: 2nd International Conference on 'Service Life Design for Infrastructure' (ISBN set: 978-2-35158-096-7, e-ISBN: 978-2-35158-097-4); Ed. K. van Breugel, G. Ye and Y. Yuan

PRO 71: Advances in Civil Engineering Materials - The 50-year Teaching Anniversary of Prof. Sun Wei' (ISBN: 978-2-35158-098-1; e-ISBN: 978-2-35158-099-8); *Eds. C. Miao, G. Ye, and H. Chen*

PRO 72: First International Conference on 'Advances in Chemically-Activated Materials – CAM'2010' (2010), 264 pp, ISBN: 978-2-35158-101-8; e-ISBN: 978-2-35158-115-5, *Eds. Caijun Shi and Xiaodong Shen*

PRO 73: 2nd International Conference on 'Waste Engineering and Management - ICWEM 2010' (2010), 894 pp, ISBN: 978-2-35158-102-5; e-ISBN: 978-2-35158-103-2, *Eds. J. Zh. Xiao, Y. Zhang, M. S. Cheung and R. Chu*

PRO 74: International RILEM Conference on 'Use of Superabsorsorbent Polymers and Other New Addditives in Concrete' (2010) 374 pp., ISBN: 978-2-35158-104-9; e-ISBN: 978-2-35158-105-6; *Eds. O.M. Jensen, M.T. Hasholt, and S. Laustsen*

PRO 75: International Conference on 'Material Science - 2nd ICTRC - Textile Reinforced Concrete - Theme 1' (2010) 436 pp., ISBN: 978-2-35158-106-3; e-ISBN: 978-2-35158-107-0; *Ed. W. Brameshuber*

PRO 76: International Conference on 'Material Science - HetMat - Modelling of Heterogeneous Materials - Theme 2' (2010) 255 pp., ISBN: 978-2-35158-108-7; e-ISBN: 978-2-35158-109-4; *Ed. W. Brameshuber*

PRO 77: International Conference on 'Material Science - AdIPoC - Additions Improving Properties of Concrete - Theme 3' (2010) 459 pp., ISBN: 978-2-35158-110-0; e-ISBN: 978-2-35158-111-7; *Ed. W. Brameshuber*

PRO 78: 2nd Historic Mortars Conference and RILEM TC 203-RHM Final Workshop – HMC2010 (2010) 1416 pp., e-ISBN: 978-2-35158-112-4; *Eds J. Válek, C. Groot, and J. J. Hughes*

PRO 79: International RILEM Conference on Advances in Construction Materials Through Science and Engineering (2011) 213 pp., ISBN: 978-2-35158-116-2, e-ISBN: 978-2-35158-117-9; *Eds Christopher Leung and K.T. Wan*

PRO 80: 2nd International RILEM Conference on Concrete Spalling due to Fire Exposure (2011) 453 pp., ISBN: 978-2-35158-118-6, e-ISBN: 978-2-35158-119-3; *Eds E.A.B. Koenders and F. Dehn*

PRO 81: 2nd International RILEM Conference on Strain Hardening Cementitious Composites (SHCC2-Rio) (2011) 451 pp., ISBN: 978-2-35158-120-9, e-ISBN: 978-2-35158-121-6; *Eds R.D. Toledo Filho, F.A. Silva, E.A.B. Koenders and E.M. R. Fairbairn*

PRO 82: 2nd International RILEM Conference on Progress of Recycling in the Built Environment (2011) 507 pp., e-ISBN: 978-2-35158-122-3; *Eds V.M. John, E. Vazquez, S.C. Angulo and C. Ulsen*

PRO 83: 2nd International Conference on Microstructural-related Durability of Cementitious Composites (2012) 250 pp., ISBN: 978-2-35158-129-2; e-ISBN: 978-2-35158-123-0; *Eds G. Ye, K. van Breugel, W. Sun and C. Miao*

PRO 84: CONSEC13 - Seventh International Conference on Concrete under Severe Conditions – Environment and Loading (2013) 1930 pp., ISBN: 978-2-35158-124-7; e-ISBN: 978-2- 35158-134-6; *Eds Z.J. Li, W. Sun, C.W. Miao, K. Sakai, O.E. Gjorv & N. Banthia*

PRO 85: RILEM-JCI International Workshop on Crack Control of Mass Concrete and Related issues concerning Early-Age of Concrete Structures – ConCrack 3 – Control of Cracking in Concrete Structures 3 (2012) 237 pp., ISBN: 978-2-35158-125-4; e-ISBN: 978-2-35158-126-1; *Eds F. Toutlemonde and J.-M. Torrenti*

PRO 86: International Symposium on Life Cycle Assessment and Construction (2012) 414 pp., ISBN: 978-2-35158-127-8, e-ISBN: 978-2-35158-128-5; *Eds A. Ventura and C. de la Roche*

PRO 87: UHPFRC 2013 – RILEM-fib-AFGC International Symposium on Ultra-High Performance Fibre-Reinforced Concrete (2013), ISBN: 978-2-35158-130-8, e-ISBN: 978-2-35158-131-5; *Eds F. Toutlemonde*

PRO 88: 8th RILEM International Symposium on Fibre Reinforced Concrete (2012) 344 pp., ISBN: 978-2-35158-132-2, e-ISBN: 978-2-35158-133-9; *Eds Joaquim A.O. Barros*

PRO 89: RILEM International workshop on performance-based specification and control of concrete durability (2014) 678 pp, ISBN: 978-2-35158-135-3, e-ISBN: 978-2-35158-136-0; *Eds. D. Bjegović, H. Beushausen and M. Serdar*

PRO 90: 7th RILEM International Conference on Self-Compacting Concrete and of the 1st RILEM International Conference on Rheology and Processing of Construction Materials (2013) 396 pp, ISBN: 978-2-35158-137-7, e-ISBN: 978-2-35158-138-4; *Eds. Nicolas Roussel and Hela Bessaies-Bey*

PRO 91: CONMOD 2014 - RILEM International Symposium on Concrete Modelling (2014), ISBN: 978-2-35158-139-1; e-ISBN: 978-2-35158-140-7; *Eds. Kefei Li, Peiyu Yan and Rongwei Yang*

PRO 92: CAM 2014 - 2nd International Conference on advances in chemically-activated materials (2014) 392 pp., ISBN: 978-2-35158-141-4; e-ISBN: 978-2-35158-142-1; *Eds. Caijun Shi and Xiadong Shen*

PRO 93: SCC 2014 - 3rd International Symposium on Design, Performance and Use of Self-Consolidating Concrete (2014) 438 pp., ISBN: 978-2-35158-143-8; e-ISBN: 978-2-35158-144-5; *Eds. Caijun Shi, Zhihua Ou, Kamal H. Khayat*

PRO 94 (online version): HPFRCC-7 - 7th RILEM conference on High performance fiber reinforced cement composites (2015), e-ISBN: 978-2-35158-146-9; *Eds. H.W. Reinhardt, G.J. Parra-Montesinos, H. Garrecht*

PRO 95: International RILEM Conference on Application of superabsorbent polymers and other new admixtures in concrete construction (2014), ISBN: 978-2-35158-147-6; e-ISBN: 978-2-35158-148-3; *Eds. Viktor Mechtcherine, Christof Schroefl*

PRO 96 (online version): XIII DBMC: XIII International Conference on Durability of Building Materials and Components(2015), e-ISBN: 978-2-35158-149-0; *Eds. M. Quattrone, V.M. John*

PRO 97: SHCC3 – 3rd International RILEM Conference on Strain Hardening Cementitious Composites (2014), ISBN: 978-2-35158-150-6; e-ISBN: 978-2-35158-151-3; *Eds. E. Schlangen, M.G. Sierra Beltran, M. Lukovic, G. Ye*

PRO 98: FERRO-11 – 11th International Symposium on Ferrocement and 3rd ICTRC - International Conference on Textile Reinforced Concrete (2015), ISBN: 978-2-35158-152-0; e-ISBN: 978-2-35158-153-7; *Ed. W. Brameshuber*

PRO 99 (online version): ICBBM 2015 - 1st International Conference on Bio-Based Building Materials (2015), e-ISBN: 978-2-35158-154-4; *Eds. S. Amziane, M. Sonebi*

PRO 100: SCC16 - RILEM Self-Consolidating Concrete Conference (2016), ISBN: 978-2-35158-156-8; e-ISBN: 978-2-35158-157-5; *Ed. Kamal H. Kayat*

PRO 101 (online version): III Progress of Recycling in the Built Environment (2015), e-ISBN: 978-2-35158-158-2; *Eds I. Martins, C. Ulsen and S. C. Angulo*

PRO 102 (online version): RILEM Conference on Microorganisms-Cementitious Materials Interactions (2016), e-ISBN: 978-2-35158-160-5; *Eds. Alexandra Bertron, Henk Jonkers, Virginie Wiktor*

PRO 103 (online version): ACESC'16 - Advances in Civil Engineering and Sustainable Construction (2016), e-ISBN: 978-2-35158-161-2; *Eds. T.Ch. Madhavi, G. Prabhakar, Santhosh Ram and P.M. Rameshwaran*

PRO 104 (online version): SSCS'2015 - Numerical Modeling - Strategies for Sustainable Concrete Structures (2015), e-ISBN: 978-2-35158-162-9

PRO 105: 1st International Conference on UHPC Materials and Structures (2016), ISBN: 978-2-35158-164-3, e-ISBN: 978-2-35158-165-0

PRO 106: AFGC-ACI-fib-RILEM International Conference on Ultra-High-Performance Fibre-Reinforced Concrete – UHPFRC 2017 (2017), ISBN: 978-2-35158-166-7, e-ISBN: 978-2-35158-167-4; *Eds. François Toutlemonde & Jacques Resplendino*

PRO 107 (online version): XIV DBMC – 14th International Conference on Durability of Building Materials and Components (2017), e-ISBN: 978-2-35158-159-9; *Eds. Geert De Schutter, Nele De Belie, Arnold Janssens, Nathan Van Den Bossche*

PRO 108: MSSCE 2016 - Innovation of Teaching in Materials and Structures (2016), ISBN: 978-2-35158-178-0, e-ISBN: 978-2-35158-179-7; *Ed. Per Goltermann*

PRO 109 (2 volumes): MSSCE 2016 - Service Life of Cement-Based Materials and Structures (2016), ISBN Vol. 1: 978-2-35158-170-4, Vol. 2: 978-2-35158-171-4, Set Vol. 1&2: 978-2-35158-172-8, e-ISBN : 978-2-35158-173-5; *Eds. Miguel Azenha, Ivan Gabrijel, Dirk Schlicke, Terje Kanstad and Ole Mejlhede Jensen*

PRO 110: MSSCE 2016 - Historical Masonry (2016), ISBN: 978-2-35158-178-0, e-ISBN: 978-2-35158-179-7; *Eds. Inge Rörig-Dalgaard and Ioannis Ioannou*

PRO 111: MSSCE 2016 - Electrochemistry in Civil Engineering (2016), ISBN: 978-2-35158-176-6, e-ISBN: 978-2-35158-177-3; *Ed. Lisbeth M. Ottosen*

PRO 112: MSSCE 2016 - Moisture in Materials and Structures (2016), ISBN: 978-2-35158-178-0, e-ISBN: 978-2-35158-179-7; *Eds. Kurt Kielsgaard Hansen, Carsten Rode and Lars-Olof Nilsson*

PRO 113: MSSCE 2016 - Concrete with Supplementary Cementitious Materials (2016), ISBN: 978-2-35158-178-0, e-ISBN: 978-2-35158-179-7; *Eds. Ole Mejlhede Jensen, Konstantin Kovler and Nele De Belie*

PRO 114: MSSCE 2016 - Frost Action in Concrete (2016), ISBN: 978-2-35158-182-7, e-ISBN: 978-2-35158-183-4; *Eds. Marianne Tange Hasholt, Katja Fridh and R. Doug Hooton*

PRO 115: MSSCE 2016 - Fresh Concrete (2016), ISBN: 978-2-35158-184-1, e-ISBN: 978-2-35158-185-8; *Eds. Lars N. Thrane, Claus Pade, Oldrich Svec and Nicolas Roussel*

PRO 116: BEFIB 2016 – 9th RILEM International Symposium on Fiber Reinforced Concrete (2016), ISBN: 978-2-35158-187-2, e-ISBN: 978-2-35158-186-5; *Eds. N. Banthia, M. di Prisco and S. Soleimani-Dashtaki*

PRO 117: 3rd International RILEM Conference on Microstructure Related Durability of Cementitious Composites (2016), ISBN: 978-2-35158-188-9, e-ISBN: 978-2-35158-189-6; *Eds. Changwen Miao, Wei Sun, Jiaping Liu, Huisu Chen, Guang Ye and Klaas van Breugel*

PRO 118 (4 volumes): International Conference on Advances in Construction Materials and Systems (2017), ISBN Set: 978-2-35158-190-2, Vol. 1: 978-2-35158-193-3, Vol. 2: 978-2-35158-194-0, Vol. 3: ISBN:978-2-35158-195-7, Vol. 4: ISBN:978-2-35158-196-4, e-ISBN: 978-2-35158-191-9; *Eds. Manu Santhanam, Ravindra Gettu, Radhakrishna G. Pillai and Sunitha K. Nayar*

PRO 119 (online version): ICBBM 2017 - Second International RILEM Conference on Bio-based Building Materials, (2017), e-ISBN: 978-2-35158-192-6; *Eds. Sofiane Amziane, Mohammed Sonebi*

PRO 120 (2 volumes): EAC-02 - 2nd International RILEM/COST Conference on Early Age Cracking and Serviceability in Cement-based Materials and Structures, (2017), Vol. 1: 978-2-35158-199-5, Vol. 2: 978-2-35158-200-8, Set: 978-2-35158-197-1, e-ISBN: 978-2-35158-198-8; *Eds. Stéphanie Staquet and Dimitrios Aggelis*

PRO 121 (2 volumes): SynerCrete18: Interdisciplinary Approaches for Cement-based Materials and Structural Concrete: Synergizing Expertise and Bridging Scales of Space and Time, (2018), Set: 978-2-35158-202-2, Vol.1: 978-2-35158-211-4, Vol.2: 978-2-35158-212-1, e-ISBN: 978-2-35158-203-9; *Eds. Miguel Azenha, Dirk Schlicke, Farid Benboudjema, Agnieszka Knoppik*

PRO 122: SCC'2018 China - Fourth International Symposium on Design, Performance and Use of Self-Consolidating Concrete, (2018), ISBN:

978-2-35158-204-6, e-ISBN: 978-2-35158-205-3; *Eds. C. Shi, Z. Zhang, K. H. Khayat*

PRO 123: Final Conference of RILEM TC 253-MCI: Microorganisms-Cementitious Materials Interactions (2018), Set: 978-2-35158-207-7, Vol.1: 978-2-35158-209-1, Vol.2: 978-2-35158-210-7, e-ISBN: 978-2-35158-206-0; *Ed. Alexandra Bertron*

PRO 124 (online version): Fourth International Conference Progress of Recycling in the Built Environment (2018), e-ISBN: 978-2-35158-208-4; *Eds. Isabel M. Martins, Carina Ulsen, Yury Villagran*

PRO 125 (online version): SLD4 - 4th International Conference on Service Life Design for Infrastructures (2018), e-ISBN: 978-2-35158-213-8; *Eds. Guang Ye, Yong Yuan, Claudia Romero Rodriguez, Hongzhi Zhang, Branko Savija*

PRO 126: Workshop on Concrete Modelling and Material Behaviour in honor of Professor Klaas van Breugel (2018), ISBN: 978-2-35158-214-5, e-ISBN: 978-2-35158-215-2; *Ed. Guang Ye*

PRO 127 (online version): CONMOD2018 - Symposium on Concrete Modelling (2018), e-ISBN: 978-2-35158-216-9; *Eds. Erik Schlangen, Geert de Schutter, Branko Savija, Hongzhi Zhang, Claudia Romero Rodriguez*

PRO 128: SMSS2019 - International Conference on Sustainable Materials, Systems and Structures (2019), ISBN: 978-2-35158-217-6, e-ISBN: 978-2-35158-218-3

PRO 129: 2nd International Conference on UHPC Materials and Structures (UHPC2018-China), ISBN: 978-2-35158-219-0, e-ISBN: 978-2-35158-220-6;

PRO 130: 5th Historic Mortars Conference (2019), ISBN: 978-2-35158-221-3, e-ISBN: 978-2-35158-222-0; *Eds. José Ignacio Álvarez, José María Fernández, Íñigo Navarro, Adrián Durán, Rafael Sirera*

PRO 131 (online version): 3rd International Conference on Bio-Based Building Materials (ICBBM2019), e-ISBN: 978-2-35158-229-9; *Eds. Mohammed Sonebi, Sofiane Amziane, Jonathan Page*

PRO 132: IRWRMC'18 - International RILEM Workshop on Rheological Measurements of Cement-based Materials (2018), ISBN: 978-2-35158-230-5, e-ISBN: 978-2-35158-231-2; *Eds. Chafika Djelal, Yannick Vanhove*

PRO 133 (online version): CO2STO2019 - International Workshop CO2 Storage in Concrete (2019), e-ISBN: 978-2-35158-232-9; *Eds. Assia Djerbi, Othman Omikrine-Metalssi, Teddy Fen-Chong*

PRO 134: 3rd ACF/HNU International Conference on UHPC Materials and Structures - UHPC'2020, ISBN: 978-2-35158-233-6, e-ISBN: 978-2-35158-234-3; *Eds. Caijun Shi & Jiaping Liu*

RILEM REPORTS (REP)

Report 19: Considerations for Use in Managing the Aging of Nuclear Power Plant Concrete Structures (ISBN: 2-912143-07-1); *Ed. D. J. Naus*

Report 20: Engineering and Transport Properties of the Interfacial Transition Zone in Cementitious Composites (ISBN: 2-912143-08-X); *Eds. M. G. Alexander, G. Arliguie, G. Ballivy, A. Bentur and J. Marchand*

Report 21: Durability of Building Sealants (ISBN: 2-912143-12-8); *Ed. A. T. Wolf*

Report 22: Sustainable Raw Materials - Construction and Demolition Waste (ISBN: 2-912143-17-9); *Eds. C. F. Hendriks and H. S. Pietersen*

Report 23: Self-Compacting Concrete state-of-the-art report (ISBN: 2-912143-23-3); *Eds. Å. Skarendahl and Ö. Petersson*

Report 24: Workability and Rheology of Fresh Concrete: Compendium of Tests (ISBN: 2-912143-32-2); *Eds. P. J. M. Bartos, M. Sonebi and A. K. Tamimi*

Report 25: Early Age Cracking in Cementitious Systems (ISBN: 2-912143-33-0); *Ed. A. Bentur*

Report 26: Towards Sustainable Roofing (Joint Committee CIB/RILEM) (CD 07) (e-ISBN 978-2-912143-65-5); *Eds. Thomas W. Hutchinson and Keith Roberts*

Report 27: Condition Assessment of Roofs (Joint Committee CIB/RILEM) (CD 08) (e-ISBN 978-2-912143-66-2); *Ed. CIB W 83/RILEM TC166-RMS*

Report 28: Final report of RILEM TC 167-COM 'Characterisation of Old Mortars with Respect to Their Repair (ISBN: 978-2-912143-56-3); *Eds. C. Groot, G. Ashall and J. Hughes*

Report 29: Pavement Performance Prediction and Evaluation (PPPE): Interlaboratory Tests (e-ISBN: 2-912143-68-3); *Eds. M. Partl and H. Piber*

Report 30: Final Report of RILEM TC 198-URM 'Use of Recycled Materials' (ISBN: 2-912143-82-9; e-ISBN: 2-912143-69-1); *Eds. Ch. F. Hendriks, G. M. T. Janssen and E. Vázquez*

Report 31: Final Report of RILEM TC 185-ATC 'Advanced testing of cement-based materials during setting and hardening' (ISBN: 2-912143-81-0; e-ISBN: 2-912143-70-5); *Eds. H. W. Reinhardt and C. U. Grosse*

Report 32: Probabilistic Assessment of Existing Structures. A JCSS publication (ISBN 2-912143-24-1); *Ed. D. Diamantidis*

Report 33: State-of-the-Art Report of RILEM Technical Committee TC 184-IFE 'Industrial Floors' (ISBN 2-35158-006-0); *Ed. P. Seidler*

Report 34: Report of RILEM Technical Committee TC 147-FMB 'Fracture mechanics applications to anchorage and bond' Tension of Reinforced Concrete Prisms – Round Robin Analysis and Tests on Bond (e-ISBN 2-912143-91-8); *Eds. L. Elfgren and K. Noghabai*

Report 35: Final Report of RILEM Technical Committee TC 188-CSC 'Casting of Self Compacting Concrete' (ISBN 2-35158-001-X; e-ISBN: 2-912143-98-5); *Eds. Å. Skarendahl and P. Billberg*

Report 36: State-of-the-Art Report of RILEM Technical Committee TC 201-TRC 'Textile Reinforced Concrete' (ISBN 2-912143-99-3); *Ed. W. Brameshuber*

Report 37: State-of-the-Art Report of RILEM Technical Committee TC 192-ECM 'Environment-conscious construction materials and systems' (ISBN: 978-2-35158-053-0); *Eds. N. Kashino, D. Van Gemert and K. Imamoto*

Report 38: State-of-the-Art Report of RILEM Technical Committee TC 205-DSC 'Durability of Self-Compacting Concrete' (ISBN: 978-2-35158-048-6); *Eds. G. De Schutter and K. Audenaert*

Report 39: Final Report of RILEM Technical Committee TC 187-SOC 'Experimental determination of the stress-crack opening curve for concrete in tension' (ISBN 978-2-35158-049-3); *Ed. J. Planas*

Report 40: State-of-the-Art Report of RILEM Technical Committee TC 189-NEC 'Non-Destructive Evaluation of the Penetrability and Thickness of the Concrete Cover' (ISBN 978-2-35158-054-7); *Eds. R. Torrent and L. Fernández Luco*

Report 41: State-of-the-Art Report of RILEM Technical Committee TC 196-ICC 'Internal Curing of Concrete' (ISBN 978-2-35158-009-7); *Eds. K. Kovler and O. M. Jensen*

Report 42: 'Acoustic Emission and Related Non-destructive Evaluation Techniques for Crack Detection and Damage Evaluation in Concrete' - Final Report of RILEM Technical Committee 212-ACD (e-ISBN: 978-2-35158-100-1); *Ed. M. Ohtsu*

Report 45: Repair Mortars for Historic Masonry - State-of-the-Art Report of RILEM Technical Committee TC 203-RHM (e-ISBN: 978-2-35158-163-6); *Eds. Paul Maurenbrecher and Caspar Groot*

Report 46: Surface delamination of concrete industrial floors and other durability related aspects guide - Report of RILEM Technical Committee TC 268-SIF (e-ISBN: 978-2-35158-201-5); *Ed. Valerie Pollet*

Structural Design and Specification

Age-Adjusted Effective Elastic Modulus of High-Performance Concrete at Early Age

Arosha Dabarera, Liang Li, Jiahang Li, Vishvendra Singh Jamwal, Qifan Yang, and Vinh Dao

Abstract One of the key parameters governing the stresses due to restrained deformation, and thereby the associated risk of cracking, is the elastic modulus of concrete. The early-age evolution of the elastic modulus is intrinsically a function of the development of hydration degree. For concrete subjected to sustained loads, the elastic modulus evolves with creep and relaxation properties. Many design methods use the concept of effective elastic modulus or age-adjusted effective elastic modulus to account for such effects. The challenge with the existing approaches is the lack of in-depth knowledge and reliable test data to accurately determine creep and aging effects for predicting elastic modulus evolution under sustained loading conditions. This paper first provides a brief review of the current approaches to determine the elastic modulus evolution under sustained loading at early age. Second, a method is proposed to directly obtain the age-adjusted effective elastic modulus ($E_{aaef}(t, t_0)$) evolution experimentally by utilizing an advanced Temperature Stress Testing Machine. Such evolution of $E_{aaef}(t, t_0)$ in a high-performance concrete subjected to a sustained tensile stress of 30% of the 3-day tensile strength is experimentally obtained. The test data are compared with the pure elastic modulus evolution. The creep and aging coefficients are derived based on the newly-measured test data. Differences between obtained early-age creep coefficients and that predicted by existing approaches are comparatively investigated. In addition, on the basis of newly-measured test data and analytical investigations, a convenient approach to determine $E_{aaef}(t, t_0)$ using time-dependent profiles of reduction factors is proposed.

Keywords Age-adjusted effective elastic modulus · Creep coefficient · Early age · High-performance concrete · Reduction factor

A. Dabarera (✉) · L. Li · J. Li · V. S. Jamwal · Q. Yang · V. Dao (✉)
School of Civil Engineering, The University of Queensland, Brisbane, QLD 4072, Australia
e-mail: a.dabarera@uq.net.au

V. Dao
e-mail: v.dao@uq.edu.au

F. Kanavaris et al. (eds.), *International RILEM Conference on Early-Age and Long-Term Cracking in RC Structures*, RILEM Bookseries 31,
https://doi.org/10.1007/978-3-030-72921-9_1

1 Introduction

Elastic modulus is an extremely important parameter for evaluating the risk of cracking in hardening concrete structures. In essence, the evolution of elastic modulus is intrinsically a function of the development of hydration degree [1]. At an early age until setting, elastic modulus development is negligible due to the low hydration degree. This development is initiated with a gradual establishment of a percolated solid network in the microstructure around setting time. Then onwards, the elastic modulus evolves almost linearly with hydration degree and its development rate gradually diminish over time [2]. In addition to the well-known theoretical perspectives on basic mechanisms, creep and relaxation properties play a key role in the actual elastic modulus of concrete subjected to sustained load.

The influence of time-varying effect caused by creep is typically interpreted as a decrease of elastic modulus of concrete. It is reported that early-age creep can reduce elastic modulus by about 35–40% [3, 4].To consider such effects, effective modulus method and age-adjusted modulus method are generally used in practice. The effective modulus method combines both instantaneous and creep components of the strain in computing the elastic modulus. In this method, the creep strain at a given time is independent of the previous stress history. Thus, the effective modulus method is only suitable for non-ageing concrete subjected to constant stress history, therefore, its applicability is quite limited in practice [5]. Trost [6] proposed an "age-adjusted effective modulus" by taking the ageing effect of concrete into account. This method was further improved by Bazant [7], who determined the ageing coefficient in terms of relaxation function. The ageing coefficient considers the fact that gradually applied stress will cause smaller strain than that caused by instantly applied stress with the similar magnitude within similar time period. In this modified approach, the age-adjusted effective modulus of concrete ($E_{\mathrm{aaef}}(t, t_0)$) at time t can be computed by;

$$E_{aaef}(t, t_0) = \frac{1}{1 + \chi(t, t_0)\phi(t, t_0)} E(t, t_0) = k(t, t_0) \times E(t, t_0) \tag{1}$$

where t_0 is the time of first loading, $E(t, t_0)$ is the pure elastic modulus at considered age, $\chi(t, t_0)$ is the ageing coefficient and $\phi(t, t_0)$ is the creep coefficient. $E_{\mathrm{aaef}}(t, t_0)$ can be alternatively interpreted as a function of $E(t, t_0)$ multiplied by a reduction factor (k (t, t_0)) as mentioned in JCI standards [8].

In view of the significance of this approach in design and analysis of long-term deformation of concrete structures, extensive research has been conducted to determine both ageing and creep coefficients. Different empirically-determined values have been proposed for the ageing coefficient [9–11]. In general, values reported for $\chi(t, t_0)$ varies from 0.4 to 1.0 [5]. A value of 0.80 is recommended for common creep and relaxation problems at early ages [12]. Similarly, different empirical models have been developed to determine time-dependent creep coefficients [13, 14]. Prediction models for creep are generally derived from fitting large number of creep in compression. In addition, these models are mainly concerned with long-term creep. Therefore, the applicability of such models for predicting short term creep especially for the case of early-age creep remains questionable [1]. The discrepancies between different creep models have been reported to cause erroneous estimations of ageing coefficients as well [15].

It is evident that the accuracy of determining the age-adjusted effective modulus method depends on many aspects, in particular at early ages. Firstly, the values for $\chi(t, t_0)$ obtained in literature vary widely. Secondly, determining early age tensile creep coefficients ($\phi(t, t_0)$) using the existing prediction models which are based on compressive creep of mature concrete is not appropriate. Furthermore, accurately quantifying the early age evolution of $E(t, t_0)$ is also a challenging task. Such unreliable data for $\chi(t, t_0)$, $\phi(t, t_0)$, and $E(t, t_0)$ may lead to inaccurate estimation of $E_{aaef}(t, t_0)$ at early ages.

In this paper, a simplified method is proposed to evaluate the concept of age-adjusted effective modulus ($E_{aaef}(t, t_0)$) of high-performance concrete. An experimental program is put forward to directly capture the age-adjusted effective modulus evolution using an advanced Temperature-Stress Testing Machine (TSTM) developed at the University of Queensland (UQ). The test results of $E_{aaef}(t, t_0)$ are compared with the that obtained for pure elastic modulus evolution. On that basis, values for $\chi(t, t_0)$ and $\phi(t, t_0)$ are back calculated. Then the determined values are compared with existing prediction models and an in-depth discussion on the suitability of the proposed method to quantify $E_{aaef}(t, t_0)$ in the context of early-age crack risk assessment is put forward. In addition, a convenient approach to determine $k(t, t_0)$ in high performance concrete is proposed on the basis of the developed experimental method in this paper.

2 Experimental Program

2.1 Materials and Mix Proportions

A high-performance concrete mix with w/b of 0.25 was used for the experiments in this paper as shown in Table 1. The oxide composition of all binders used are mentioned in Table 2. The mean cylinder 28-day compressive strength of this mix was 80 MPa. The maximum size of the coarse aggregates used was 10 mm. The water content was adjusted in both fine and coarse aggregates to maintain Saturated Surface Dry (SSD) condition. The whole concrete casting process was performed following the relevant Australian standard [16].

Table 1 Mix proportions of the concrete mix used in this study

Constituents	kg/m^3
Cement	560
Fly ash	120
Silica fume	40
Water	180
Superplasticizer	4.6
10 mm aggregate	930
Coarse sand	250
Fine sand	240

Table 2 Chemical composition (%, by mass) of the binders

	Cement	Fly ash	Silica fume
CaO	63.9	3.80	–
SiO_2	19.5	59.4	89.9
Al_2O_3	5.40	20.4	–
SO_3	2.64	0.20	0.84
Fe_2O_3	2.83	10.9	–
MgO	1.40	1.50	–
Na_2O	0.32	0.67	0.11
K_2O	0.42	1.42	0.23
Loss on ignition	3.11	0.70	3.80

2.2 *Test Setup and Methods*

An advanced Temperature Stress Testing Machine (TSTM) was utilized for all tests performed in this study. This test setup is kept in an environmentally controlled room with a temperature of 23 °C and a relative humidity of 55% which is the curing condition recommended in the Australia standard [16]. A schematic representation of the TSTM setup is shown in Fig. 1. This setup consists of a restrained deformation test specimen and a free deformation test specimen, allowing simultaneous measurements under restrained and free conditions. The free and restrained test specimens are of the same dimensions, with a 800 mm-long central part having a cross section of 80 × 80 mm^2. While the free specimen is unrestrained, the restrained specimen is fixed at one end and movable at the other end. The movable end is connected to a stepped motor which enables testing either under deformation controlled or stress-controlled conditions. The stress recordings are continually monitored through a load cell with a capacity of 100 kN. This setup consists of an advanced deformation capturing system via laser sensors. More importantly, the curing histories of the concrete specimens can be controlled using an advanced thermal regulation system. The whole test setup and its functions are controlled through an integrated program to facilitate the test process. Further information can be found in [17, 18, 19].

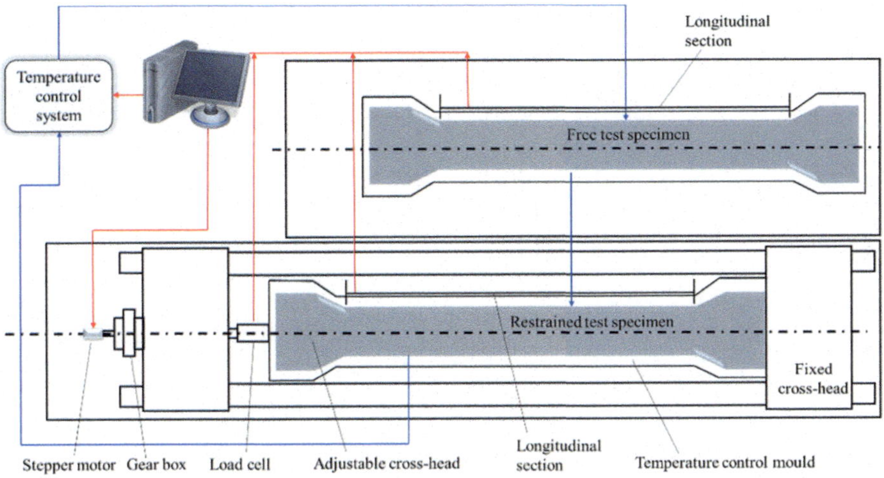

Fig. 1 Schematic view of the TSTM test setup consisting of restrained and free test specimens in a closed-loop system

Two main tests were performed to investigate the age-adjusted effective elastic modulus ($E_{aaef}(t, t_0)$) in tension at early age. All tests were performed maintaining a constant curing temperature at 23 °C using the thermal regulation system of TSTM setup. Both tests were initiated at an age of 3 days after casting to avoid the risk of premature failure of specimens and also due to the difficulty of deformation measurements at very early ages. In Test 1, evolution of the pure tensile elastic modulus ($E(t_0)$) of the concrete mix mentioned in Table 1 was measured. For this test, no sustained loading was applied. In Test 2, a sustained tensile stress of about 0.6 MPa was applied in the restrained deformation test specimen at 3 days after casting. The sustained stress for Test 2 was about 30% of the 3-day tensile strength. In both tests, elastic modulus evolutions were measured by performing instantaneous loading and unloading cycles up to 0.6 MPa. Each loading and unloading cycle only lasted for a period less than 5 min. The schematic diagrams of stress cycles imposed in Test 1 and Test 2 are shown in Fig. 2a and b, respectively.

3 Results and Discussion

The measured values of elastic modulus in Test 1 and Test 2 are plotted in Fig. 3. For Test 1 (in which test specimen was not subjected to any sustained load), measured values increased from 24.5 to 34 GPa between 3 to 10 days. The elastic modulus measured at 3 days in Test 2 (25.3 GPa) was comparable to that obtained in Test 1. However, the increasing rate of the measured values in Test 2 was lower when compared to Test 1. In Test 2, the measured value at 10 days was 15% lower than that in Test 1. This means that the sustained stress in Test 2 reduced the corresponding

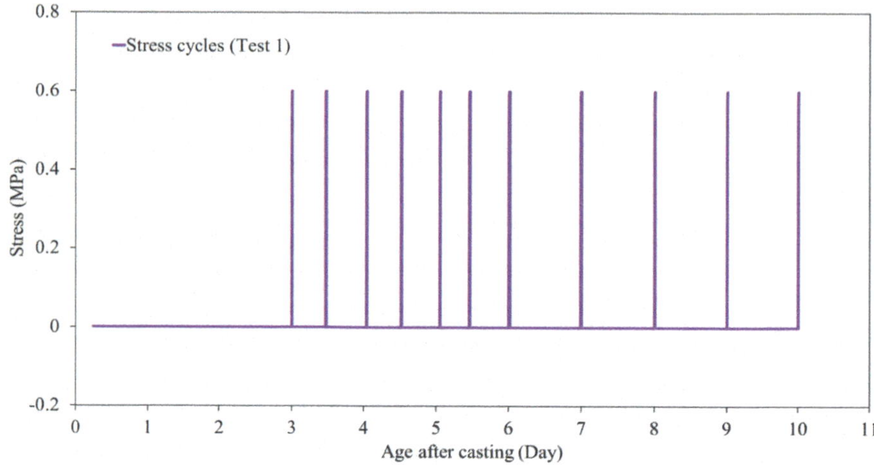

(a) Test 1

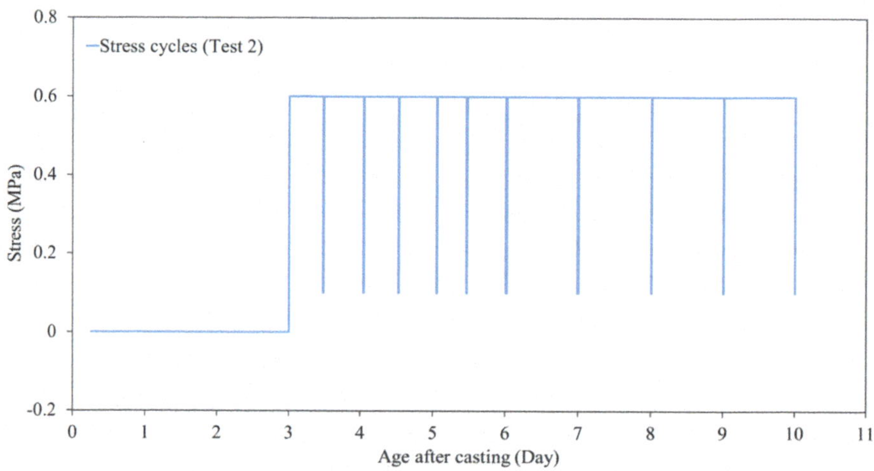

(b) Test 2

Fig. 2 Schematic overview of the stress cycles imposed to measure. **a** Pure tensile elastic modulus and **b** age-adjusted effective elastic modulus over time

$E_{aaef}(t, t_0)$ values when compared to $E(t, t_0)$ values in Test 1. However, the reduction factor for the considered concrete mix in this study was about 15% at the end of the test duration. This is significantly lower than that in normal concrete which can be typically up to 35–40% as mentioned in [3, 4]. Therefore, such profiles of reduction factors ($k(t, t_0)$) should be accurately determined for high-performance concrete to improve early-age crack risk assessment models.

On the basis of measured $E(t_0)$ and $E_{aaef}(t, t_0)$ test data, the product of the ageing and creep coefficients ($\chi(t, t_0)\ \phi(t, t_0)$) was computed using Eq. (1). The evolution mentioned in the Introduction, a value of 0.80 which has been commonly used as the ageing coefficient especially for concrete loaded less than 20 days, was used in the analysis. Such values for creep coefficient determined based on Test 2 are shown in Fig. 5. It is seen that the creep coefficients obtained from Test 2 gradually increased up to about 0.2 at 10 days. For comparison purposes, creep coefficients predicted by fib model code 2010 [14] are also plotted. It should be noted that the test data is measured at 23 °C following relevant Australia standard, whereas, predicted values in fib Model Code 2010 corresponds to that obtained at 20 °C. To consider the influence of temperature on maturity of concrete, the age was adjusted according to the simplified temperature-adjusted approach in fib Model Code 2010 (see Fig. 5). In contrast to the computed values, the predicted values tend to immediately increase within the first day after initial loading. Then onwards the increase rate gradually lowered over time. However, it is noteworthy that using such predicted tensile creep coefficients may underestimate the $E_{aaef}(t, t_0)$ evolution. For example, the predicted creep coefficient (at 10 days) resulted in reducing the age-adjusted effective elastic modulus by 30% when compared to the actual reduction factor of 15%. Therefore, using such creep prediction models for determining $E_{aaef}(t, t_0)$ may not be appropriate and result in underestimating the crack risk of high-performance concrete.

This study proposes a new approach for estimating the age-adjusted effective elastic modulus evolution which is closely related to that mentioned in JCI guidelines [8]. The JCI guidelines recommends a method to compute $E_{aaef}(t, t_0)$ of which the elastic modulus ($E(t_0)$) is multiplied by a reduction factor to take into account the effect of creep. According to this guideline, this reduction factor is primarily a function of time and curing history. For concrete subjected to semi adiabatic temperature histories, a reduction factor of 0.42 is assumed until maximum temperature is reached which gradually increases to 0.65 at 1 day after the peak temperature is

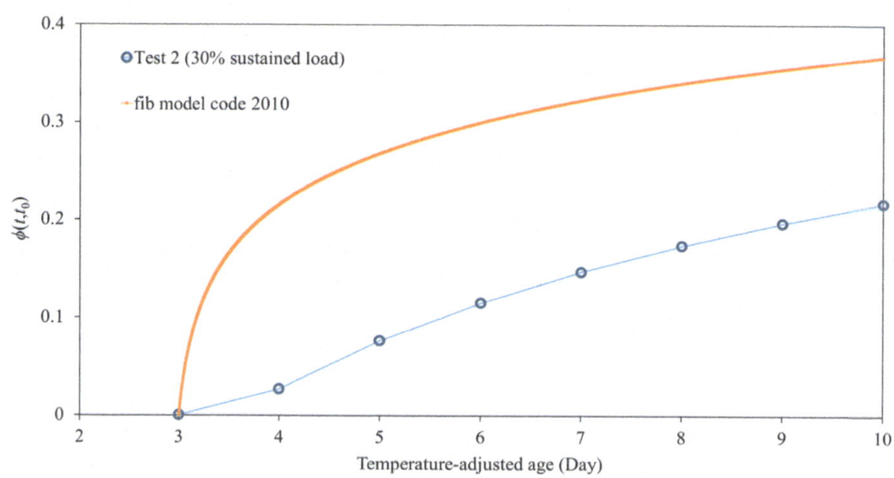

Fig. 5 Evolution of creep coefficients using the test data and fib model predictions

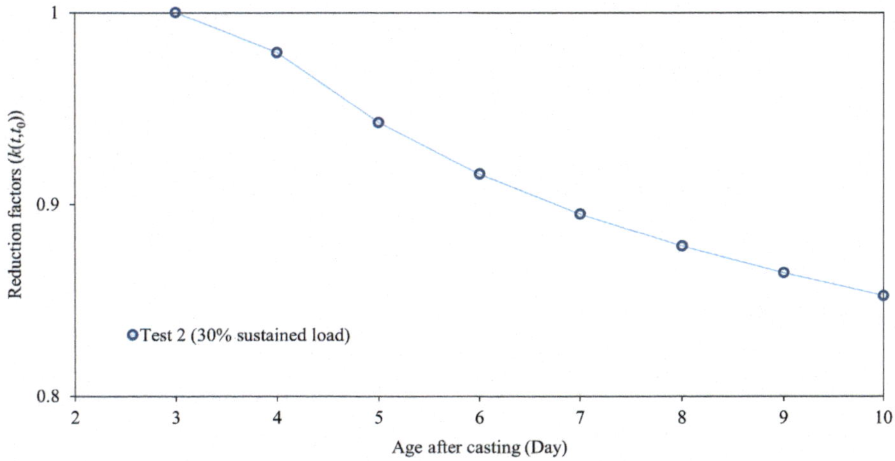

Fig. 6 Evolution of the reduction factors $k(t, t_0)$ computed based on the test data

reached. The values in between these two periods are computed using linear interpolation. On the other hand, the reduction factors computed based on the test data in this study showed a non-linear trend over the tested period (Fig. 6). This profile better reflects the early-age relaxation in the tested concrete due to creep effects. Thus, obtaining $E_{\text{aaef}}(t, t_0)$ using the time-dependent profile of the reduction factors $k(t, t_0)$ seems to be rather a rational and a convenient approach compared to that using ageing and creep coefficients. Such $k(t, t_0)$ profiles can be conveniently obtained using the experimental method proposed in this paper. Based on the initial outcomes of this study, it is recommended to generate a comprehensive database for $k(t, t_0)$ using different stress levels, initial loading ages, and curing histories to accurately model the early age evolution of age-adjusted effective elastic modulus of high-performance concrete. The ongoing experimental program in this study also aims to generate similar test data for normal strength concrete types as well.

4 Conclusion

In this paper, a simplified experimental method is proposed to evaluate the concept of age-adjusted effective modulus for high-performance concrete. Based on the experimental and analytical investigations, following main conclusions can be made:

- The proposed experimental method can be used to conveniently determine the early age evolution of age-adjusted effective elastic modulus ($E_{\text{aaef}}(t, t_0)$).
- Common assumptions on the reduction of stress due to creep (ranging between 35 and 40%) are not suitable for assessing ($E_{\text{aaef}}(t, t_0)$) in high-performance concrete at early age.

- Using existing tensile creep coefficient prediction models for determining the $E_{aaef}(t, t_0)$ may not be appropriate and result in underestimating the crack risk of high-performance concrete at early ages.
- Directly obtaining the time-dependent profiles of the reduction factors using the proposed experimental method may be a rather rational approach for accurately determining the age-adjusted effective elastic modulus at early age.

References

1. Maruyama, I., Lura, P.: Properties of early-age concrete relevant to cracking in massive concrete. Cem. Concr. Res. **123**, 105770 (2019)
2. Boumiz, A., Vernet, C., Tenoudji, F.C.: Mechanical properties of cement pastes and mortars at early ages: evolution with time and degree of hydration. Adv. Cem. Based Mater. **3**(3), 94–106 (1996)
3. Gilbert, R.I.: Cracking caused by early-age deformation of concrete—prediction and control. Procedia Eng. **172**, 13–22 (2017)
4. Bamforth, P.B.: Early-Age Thermal Crack Control in Concrete (CIRIA C660). CIRIA, London, UK, p. 112 (2007)
5. Cheng, Z., et al.: Ageing coefficient for early age tensile creep of blended slag and low calcium fly ash geopolymer concrete. Constr. Build. Mater. **262**, 119855 (2020)
6. Trost, H.: Implications of the principle of superposition on creep and relaxation problems in concrete and pre-stressed concrete, vol. 10, pp. 230–238, 261–269. Beton-und Stahlbetonbau, Berlin-Wilmersdorf (1967)
7. Bazant, Z.P.: Prediction of concrete creep effects using age-adjusted effective modulus method. ACI J. Proc. **69**(4) (1972)
8. Japan Concrete Institute: Guidelines for control of cracking of mass concrete 2016. Tokyo (2016)
9. Gilbert, R.I.: Time Effects in Concrete Structures. Elsevier, Amsterdam (1988)
10. Ghali, A., Favre, R., Elbadry, M.: Concrete Structures: Stresses and Deformations: Analysis and Design for Serviceability. CRC Press (2018)
11. Bazant, Z.P.: Mathematical Modeling of Creep and Shrinkage of Concrete. Wiley (1988)
12. fib, fib Bulletin No. 70: Code-type models for structural behaviour of concrete: background of the constitutive relations and material models in the fib model code for concrete structures 2010, p. 196 (2013)
13. James, E., et al.: ACI 209 R-92 (Reapproved 1997) Prediction of creep, shrinkage, and temperature effects in concrete structures reported by ACI Committee 209 (1998)
14. fib: fib Model Code for Concrete Structures 2010. Wiley-VCH Verlag GmbH & Co. KGaA, Weinheim, Germany (2013)
15. Wang, W., Gong, J.: New relaxation function and age-adjusted effective modulus expressions for creep analysis of concrete structures. Eng. Struct. **188**, 1–10 (2019)
16. Australian Standard, AS 1012.2:2014: Methods of testing concrete—preparing concrete mixes in the laboratory. Standards Australia (2014)
17. Nguyen, D.H., et al.: Temperature-stress testing machine—a state-of-the-art design and its unique applications in concrete research. Cement Concr. Compos. **102**, 28–38 (2019)
18. Li, L., Dabarera, A.G.P., Dao, V.: Time-zero and deformational characteristics of high performance concrete with and without superabsorbent polymers at early ages. Construct. Build. Mater. **264** (2020)
19. Li, L., Dabarera, A.G.P., Dao, V.: Basic tensile creep of concrete with and without superabsorbent polymers at early ages (submitted). Construct. Build. Mater. (2020)

Plain Geopolymer Concrete Cross-Section Surface Analysis After Creep and Shrinkage Tests in Compression and Tension

Rihards Gailitis, Andina Sprince, Leonids Pakrastins, Kinga Korniejenko, and Tomass Kozlovskis

Abstract Low calcium alkali-activated cement composite known as geopolymer has been around for more than 40 years. The main benefit of geopolymer based composites is the environmental aspect—it is partially made by utilizing waste products, such as fly-ash, slags, and others. It has been estimated that geopolymer binder production makes up to 6 times less CO_2 than the production of Portland cement. Due to the polymerization or in other words nature of the geopolymer binding process, there are some differences in creep and shrinkage development. Because of this microstructure of the specimen could be dissimilar to ordinary Portland cement. There has been an absence of investigations regarding the geopolymer composite long-term properties and micro-analysis. Also, the conditions affecting the long-term properties of the geopolymer composites have been little studied. The subject of the research is geopolymer concrete that has been tested for creep and shrinkage in compression and tension. The specimens for microstructure analysis were acquired from the cylindrical shape (compression) and compact tension (tension) specimens. Polished sections were used for SEM microanalysis. Acquired polished section image cross-sections were analyzed by determining the amount of geopolymer binder, filler, and air void in the analyzed cross-section. The results were cross-referenced with creep and shrinkage test results. After creep and shrinkage tests in compression and tension specimen cross-section zones that have been subjected to the highest stresses were chosen and analyzed. The article's main aim is to determine the geopolymer composite microstructure and applied load influence on long-term properties.

Keywords Geopolymer concrete · Polished section microanalysis · Long-term properties · Compression · Tension

R. Gailitis (✉) · A. Sprince · L. Pakrastins · T. Kozlovskis
Riga Technical University, Riga, Latvia
e-mail: rihards.gailitis@edu.rtu.lv

K. Korniejenko
Cracow University of Technology, Cracow, Poland

F. Kanavaris et al. (eds.), *International RILEM Conference on Early-Age and Long-Term Cracking in RC Structures*, RILEM Bookseries 31,
https://doi.org/10.1007/978-3-030-72921-9_2

1 Introduction

Alkali activated cement composites based on industrial waste products such as fly ash, blast furnace slag, etc., have been considered a cement for the future [1, 2]. As the cement consumption, year by year, go up and now is responsible for 1.5 billion tonnes of CO_2 emissions annually. It becomes a significant issue around 36% of global energy consumption to research viable alternatives for less polluting binder usage with comparable properties regarding workability [3]. The use of alkali-activated materials is beneficial to CO_2 reduction. It is positive from a sustainable environment standpoint as it incorporates such industrial by-products as fly ash and slag [4]. The issue regarding wide usage of geopolymer is mainly due to the binder hardening or the polymerization process. This process requires heat; the temperature can vary from 40 to 100 °C (depending on fly ash or slag type and alkali activator) and the polymerization time from 12 to 48 h and more, therefore, excluding on-site construction works due to difficulties in achieving satisfactory structural performance [5].

Geopolymer is a low calcium alkali-activated cement composite. It is formed due to a silicon and aluminium reaction activated by hydroxide silicates from sodium and potassium alkali activating solution [6, 7].

Geopolymer concrete has similar compressive strength to regular Portland cement (PC) based composites. Unlike regular PC, geopolymer composites 85% of their final compressive strength can reach in 48 h [8]. Long-term property wise geopolymer composites have 78% less shrinkage and 50% less creep strains than foamed regular concrete and a bit worse than regular PC composites [9].

Creep and shrinkage are well-known phenomena for cement and cement-like based composite materials, and it may influence the lifetime of structures. Most of the creep and shrinkage effects develop in the first ten years after construction. It is expected that the creep and shrinkage development after the first ten years are not significant and have a small impact on the performance of the structure [10–12]. Concrete and similar materials are considered to insufficient strain capacity and low tensile strength. And, consequently, they are brittle and susceptible to cracking. For cementitious composites under compression damages first happen in the paste-aggregate interface. The tensile stresses are necessary to determine long-term tensile properties for these materials [13]. Furthermore, because of the difficulties of performing tensile creep tests and differences in creep mechanisms in tension and compression, it is equally important to determine the factors that influence creep properties in compression and tension [14].

The paper focuses on the microstructure differences in specimens that have been used in creep tests in compression, tension, and shrinkage tests. Therefore, microstructure images were acquired and analysed. Results of image analysis were cross-referenced with the creep and shrinkage curves to determine whether there are notable correlations.

2 Materials and Methods

2.1 Geopolymer Mix Preparation

Geopolymer specimen matrixes were based on fly ash sourced from the power plant based in Skawina city (Poland). The fly ash contains spherical aluminosilicate particles and contains oxides such as SiO_2 (47.81%), Al_2O_3 (22.80%). The high value of SiO_2 and Al_2O_3 allows polymerization [15].

Geopolymer specimens were prepared using sodium promoter, fly ash, and sand (sand and fly ash ratio—1:1). The geopolymer activation process has been made by 10M NaOH solution and the sodium silicate solution (at a rate of 1:2.5). The technical NaOH in flake form and tap water with sodium silicate R–145 solution is used to make the composite solution. The alkaline solution was prepared by pouring sodium silicate and water over solid sodium hydroxide into sodium silicate and water aqueous solution. The solution was mixed, and the temperature was stabilized. The fly ash, sand, and alkaline solution were mixed for about 15 min using a low-speed mixing machine (to achieve homogenous paste). Then the geopolymers were poured into the plastic moulds, as is shown in Fig. 1. The specimens were hand-formed, and then the air bubbles were removed by vibrating the mass. Moulds were heated in the laboratory dryer for 24h at 75 °C. Then, the specimens were unmolded. All the geopolymer specimen preparation was done at Cracow University of Technology (CUT), Poland.

(a) (b) (c)

Fig. 1 Geopolymer composite preparation (**a**) and moulding process (**b** and **c**), CUT lab

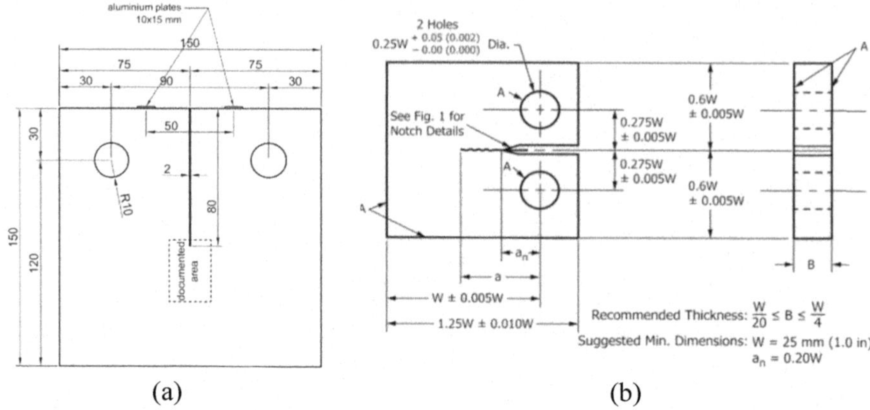

Fig. 2 Geometry of the Compact Tension (CT) specimen [17, 18]

2.2 Test Specimen Preparation

For creep testing in compression, specimens were prepared according to RILEM TC 107-CSP recommendations [16]. All of the specimens were Ø 46 × 190 mm or approximately 1:4 diameter to height ratio. For the dial gauge attaching six aluminium plates were glued on specimens intended for creep testing in compression. Afterward, dial gauges were attached to those plates. For shrinkage specimens one aluminium plate was glued at the bottom and top part of each specimen. After that, shrinkage specimens were placed in a stand for shrinkage measurements.

For creep measuring in tension, compact tension (CT) shaped specimens were used [17]. Specimens were cut out from a cube that was 150 × 150 × 150mm. Each cut CT specimen was 15 mm thick. Afterward, the notch was cut as well as two bore holes were made (for attaching within a loading rig), as shown in Fig. 2.

The 2 mm wide notch in the CT specimen was sawn using a Proxxon MICRO MBS 240/E bandsaw. According to Fig. 2a, the aluminium plates were glued to specimens intended for creep and shrinkage tests. Plates were glued 25 mm to each side from the notch center. There were prepared 12 cylinders and 12 CT specimens.

2.3 Experimental Setup

When the specimens' preparation was done, compressive strength and tensile strength ultimate values were determined. The procedure is shown in Fig. 3.

The ultimate compressive load was determined using Controls Mod. Nr C56G2 press with a speed of 0.8 MPa/sec. The ultimate tensile load was determined using INSTRON 3000 All-Electric Dynamic Test Instrument with speed 0.15mm/sec.

(a) (b)

Fig. 3 Compressive (**a**) and tensile (**b**) strength determination, RTU lab

Determined strength values were compiled in Table 2. Following strengths determination, creep specimens were placed into lever test stands and were loaded with a constant static load equal to 20% of the ultimate load values (see Fig. 4). With these stands, it is possible to apply constant loading to the specimens and to keep it uniform over a long period. Strains were measured using mechanical dial gauges "ИЧ" with a scale interval of 1/100 mm and maximum measuring range of 10 mm.

To determine basic creep behavior, similarly shaped shrinkage specimens were placed in equivalent environmental conditions, and their strain changes were monitored (no load applied to the shrinkage specimens). Conclusions were made based on subtracting shrinkage strain values from the creep values. Figure 5 shows the test setups for shrinkage tests. All specimens were kept in a dry atmosphere of controlled relative humidity in standard conditions: temperature 20 ± 1 °C and relative humidity $48 \pm 3\%$. The geopolymer specimen preparation and strength, long-term tests were done at Riga Technical University (RTU), Latvia.

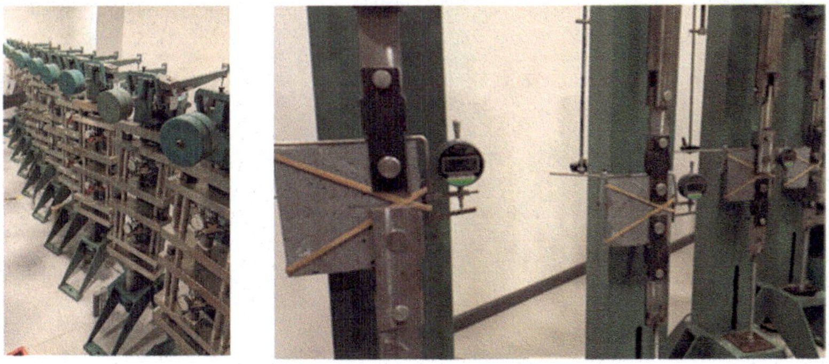

(a) (b)

Fig. 4 Creep specimen placement into compression (**a**) and tension (**b**) test stands, RTU lab

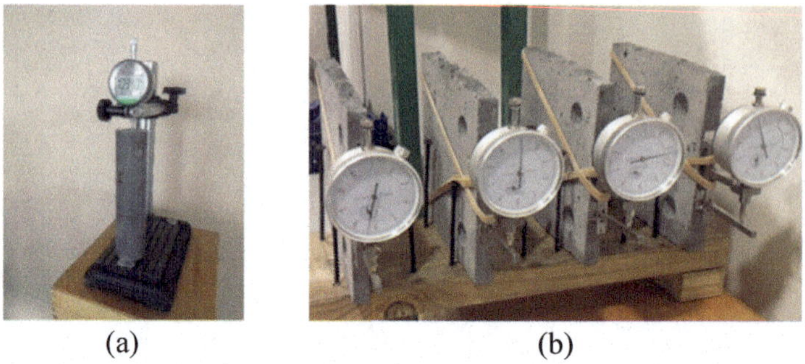

Fig. 5 Shrinkage test setup for compression (a) and tension (b) specimens, RTU lab

2.4 Microstructure Composition Specimen Preparation

After all long-term tests, specimen cross-section parts for the scanning electron microscope (SEM) were prepared for microstructure composition determination. Figure 6 shows prepared compression and tension specimen samples for microstructure analysis.

For the specimens that have been subjected to compression tests, cylinders middle parts were cut into disc shape samples with a thickness of 15mm. Compact tension specimens (CT) middle part where notch ends were drilled. The drilled samples were with Ø 48 and 15 mm thickness. Afterward, all samples were polished according to the sequence described in Table 1.

When the microanalysis samples were prepared, they were delivered to Cracow University of Technology (CUT) and covered with gold. For each sample, the characteristic cross-section areas were chosen and marked. The characteristic cross-section areas were analyzed. These sample areas were shown in Fig. 7.

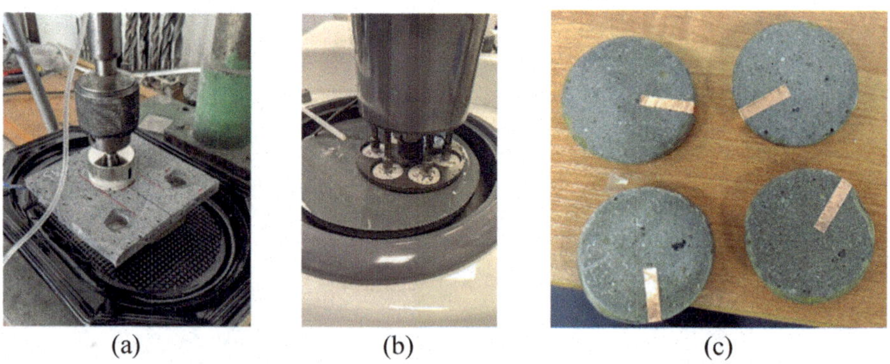

Fig. 6 SEM samples preparation (a) SEM samples polishing (b) and prepared samples before the gold plating (c), RTU lab

Table 1 Specimen surface polishing sequence

Polishing stage number	Polishing compound (sandpaper or paste grade) type	Polishing cycle time, minutes	Compression force to specimen polishing surface, daN
1	P180	2	2.5
2	P320	2	2.5
3	P600	2	2.5
4	P1000	2	2.5
5	P1200	2	2.5
6	3 μm	4	2.5

(a) (b)

Fig. 7 Compression (**a**) and tension (**b**) SEM samples, CTU lab

For specimens that have been subjected to compression testing, the cross-section was divided into five squared (10x10mm) section parts distributed into the central and peripheral part of the specimen cross-section. Still, for the specimens subjected to tensile loads (CT specimens), the microanalysis is done to the cross-section part near the notch and deeper into the specimen.

The SEM microanalysis was done in JEOL JSM-820. The achieved SEM images afterward were compiled together and divided into layers using Adobe Photoshop CC. The division into layers was based on partition type within cross-section (matrix, filler, air voids). For each of these partitions, the RGB tone was allocated. The process is shown in Fig. 8. The layer dividing process begins with filler layer separation that was continued with the void layer.

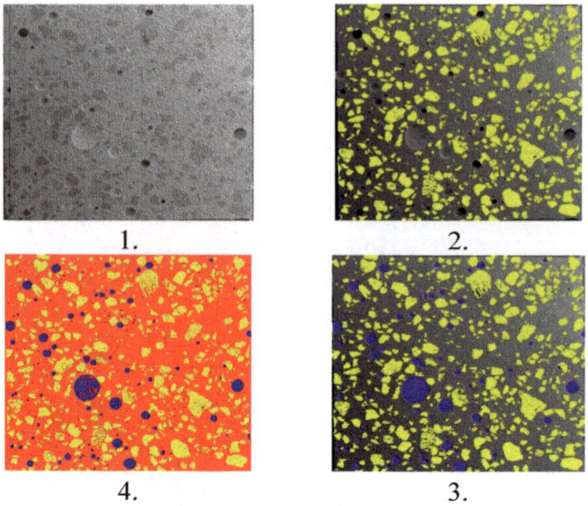

Fig. 8 Image dividing sequence in layers and tone allocation

When the image dividing and RGB tone allocation was done, the specific tone image pixels were counted and registered. By doing so, the composition amount of the studied cross-section was acquired.

3 Results and Discussion

The compressive and tensile strength of the tested specimens is compiled in Table 2. Specimens intended for creep tests were subjected to a load equal to 20% of the load values shown in Table 2.

After the compression and tension ultimate load tests, the creep and shrinkage tests were carried out for 91 days (more than three months). Tests were started on the 7th day since the preparation of the specimens. The creep and shrinkage curves for compression and tension specimens are shown in Fig. 9.

The curves in Fig. 9a and b show that throughout testing time creep strains for CT specimens are almost half of the creep strains in compression. The difference on average is 46%. It is also apparent that the amount of elastic strain at the beginning

Table 2 Compressive and tensile ultimate load values

Specimen type	Ultimate load value, average (kN)	Average compressive and tensile strength, MPa
Cylinders, plain geopolymer	60.35	61.44
CT, plain geopolymer	0.28	5.13

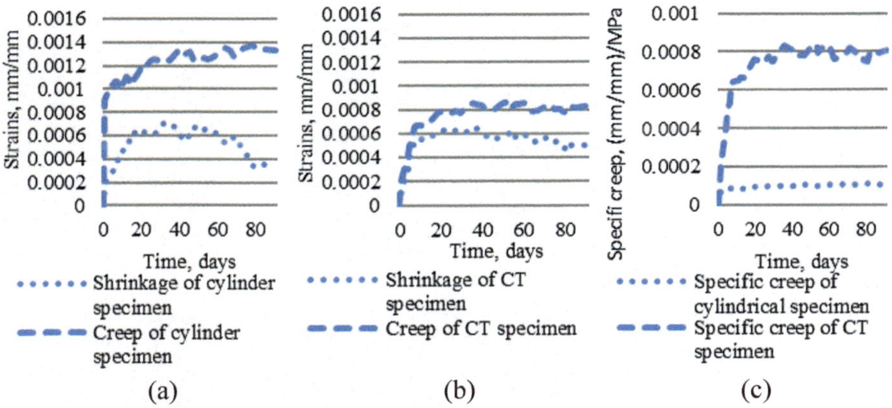

Fig. 9 Creep and shrinkage curves of compression (**a**) and tension (**b**) specimens and specific creep (**c**) of compression and tension specimens

of the tests and further strain development characteristics are different for specimens in compression and tension. Furthermore, specific creep in tension (Fig. 9c) is more than 7 times greater that in compression. It leads to a conclusion that there are significant microstructural differences to the microstructural development of the specimens tested in compression and tension, and also, plain geopolymer composite has similar creep properties as the regular Portland cement-based composites. Also, the shrinkage curve for CT specimens in Fig. 9b clearly shows that specimens have properly polymerized and achieved their modulus of elasticity close to what could be considered as the final modulus of elasticity value. Therefore, the notch is opening, unlike Portland cement composites that due to hydration would close the notch.

The obtained cross-section composition results are shown in Table 3.

First of all, it becomes apparent that the cylinder's air void wise was much better shape than the cube used to make CT specimens. The all in all cross-section composition analysis show that porosity for CT shaped specimens on average is from 24% to 32% higher than cylinder-shaped specimens. It means that due to the cube's dimension, it is much harder for air to escape from the middle parts of the cube while it was vibrated than it is for the air in the cylinder-shaped specimens.

Table 3 Compressive and tensile ultimate load values

Test type	Specimen type	Matrix amount in cross-section, %	Filler amount in cross-section, %	Air void amount in cross-section, %
Shrinkage	Cylinder	73.48	20.08	6.44
	CT	73.61	16.94	9.45
Creep	Cylinder	73.76	19.62	6.62
	CT	75.64	15.60	8.76

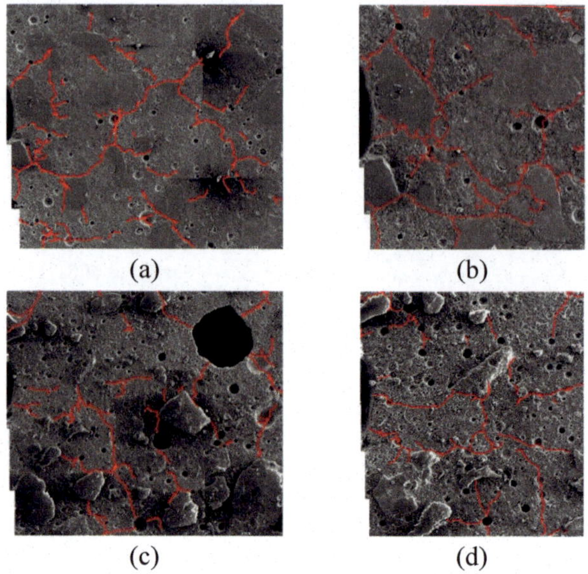

Fig. 10 Crack assessment of shrinkage specimen notch base at 100 times (**a**) and 200 times (**b**) magnification and creep specimen notch base at 100 times (**c**) and 200 times (**d**) magnification

The notch cross-section part's analysis was done to further determine the low amount of elastic strains for CT specimens.

In Fig. 10, the notch part (tip) is in the middle of the left-hand side of each image. Here it is apparent that there are a significant number of cacks in the notch area. Furthermore, the crack amount for creep specimens is close to shrinkage specimens with a slight increase to creep specimens. All that leads to thinking that due to early age testing, shrinkage plays a considerable role in the crack development, making specimens undergo larger plastic strains.

To further analyze the load impact to CT specimen cross-section notch zone (3mm from the beginning of the notch) was measured. The notch's overall general area and the notch's length and width from six equally spaced measurements along the notch length. The results are shown in Table 4.

It is clear to see that creep specimen notch basis was more deformed than shrinkage specimens. In the length of 3mm from the notch base, the notch area is average 5.7% bigger; thus, it was deformed than the notch part of the shrinkage specimens. Furthermore, while the width of the analyzed shrinkage specimens' notch part stays the same for the creep specimens, width increases on average by 0.168mm or 28.11%.

Table 4 Notch base part analysis

Test type	Average length of the notch, mm	Difference, %	Average width of the notch, mm	Difference, %	Average area of the notch, mm2	Difference, %
Shrinkage	2.979	1.2	0.684	5.4	1.974	5.7
Creep	3.016		0.723		2.094	

4 Conclusions

1. Compact tension (CT) specimens on average have a 5.15% higher amount of air voids than cylinder type specimens. The filler amount in the analyzed CT specimen cross-sections is 7.16% less than cylindrical specimen cross-sections, while the matrix amount stays the same. Therefore, while the cube specimens as a base of the CT specimen preparation for long-term tests are not bad, the CT specimen making directly in the right shape moulds would be considered a better practice for air void filler distribution wise.
2. The creep strain amount for the compression specimens is 35.8% higher than the creep strains for tension specimens. In contrast to ultimate load values, the difference is 99.54% in favor of the compression intended specimens.
3. Specific creep for specimens in compression is on average 85.92% less than for CT specimens. Therefore, geopolymer composites have 7.5 times larger creep strains in tension than in compression.
4. From the creep and shrinkage strain curves and notch base part cracks analysis, it is apparent that tension specimens, in this case, CT specimens, have lower elastic strain part and, in early stages, develop cracks in the base of the notch. Tension specimen elastic strains at the beginning of tests are on average 90.9% less than compression specimens.
5. Due to early age testing and lack of fiber reinforcement, the shrinkage strains play a considerable role in the crack development into the CT specimens and, therefore, the increased amount of plastic strains of the tension specimens. Tension specimens have 0.000379 mm/mm or 47.6% higher plastic strains than was determined by creep compression specimens.

Acknowledgements

1. This work has been supported by the European Regional Development Fund within the Activity 1.1.1.2 "Post-doctoral Research Aid" of the Specific Aid Objective 1.1.1 "To increase the research and innovative capacity of scientific institutions of Latvia and the ability to attract external financing, investing in human resources and infrastructure" of the Operational Programme "Growth and Employment" (No.1.1.1.2/VIAA/3/19/401). The authors acknowledge the support of the PROM programme no. PPI/PRO/2019/1/00013/U/001 which is co-financed by the European Social Fund under the Knowledge Education Development Operational Programme. This publication was supported by Riga Technical University's Doctoral Grant programme.
2. The authors acknowledge the support of the PROM programme no. PPI/PRO/2019/1/00013/U/001 which is co-financed by the European Social Fund under the Knowledge Education Development Operational Programme.
3. This publication was supported by Riga Technical University's Doctoral Grant programme.

References

1. Shi, C., Jiménez, A.F., Palomo, A.: New cements for the 21st century: the pursuit of an alternative to Portland cement. Cem. Concr. Res. **41**(7), 750–763 (2011)
2. Tu, W., Zhu, Y., Fang, G., Wang, X., Zhang, M.: Internal curing of alkali-activated fly ash-slag pastes using superabsorbent polymer. Cem. Concr. Res. **2019**(116), 179–190 (2018)
3. Kermeli, K., Edelenbosch, O.Y., Crijns-Graus, W., van Ruijven, B.J., Mima, S., van Vuuren, D.P., Worrell, E.: The scope for better industry representation in long-term energy models: modeling the cement industry. Appl. Energy **2019**(240), 964–985 (2018)
4. Rashad, A.M., Essa, G.M.F.: Effect of ceramic waste powder on alkali-activated slag pastes cured in hot weather after exposure to elevated temperature. Cement Concr. Compos. **2020**(111), 103617 (2019)
5. Kang, S.H., Jeong, Y., Kim, M.O., Moon, J.: Pozzolanic reaction on alkali-activated Class F fly ash for ambient condition curable structural materials. Construct. Build. Mater. **218**, 235–244 (2019)
6. Yan, S., He, P., Jia, D., Wang, J., Duan, X., Yang, Z., Wang, S., Zhou, Y.: Effects of high-temperature heat treatment on the microstructure and mechanical performance of hybrid C f -SiC f -(Al 2 O 3p) reinforced geopolymer composites. Compos. B: Eng. **114**, 289–298 (2017)
7. Linul, E., Korniejenko, K., ŞSerban, D.A., Negru, R., Marşavina, L., Łach, M., Mikuła, J.: Quasi-static mechanical characterization of lightweight fly ash-based geopolymer foams. In: IOP Conference Series: Materials Science and Engineering, vol. 416 (2018)
8. Assi, L.N., Carter, K., Deaver, E., Ziehl, P.: Review of availability of source materials for geopolymer/sustainable concrete. J. Cleaner Prod. **263**, 121–477 (2020)
9. Amran, M., Fediuk, R., Vatin, N., Lee, Y.H., Murali, G., Ozbakkaloglu, T., Klyuev, S., Alabduljabber, H.: Fibre-reinforced foamed concretes: a review. Materials **13**(19), 1–36 (2020)
10. Nastic, M., Bentz, E.C., Kwon, O., Papanikolaou, V., Tcherner, J.: Shrinkage and creep strains of concrete exposed to low relative humidity and high temperature environments. Nucl. Eng. Design **352**, 110–154 (2019)
11. Boumakis, I., Di Luzio, G., Marcon, M., Vorel, J., Wan-Wendner, R.: Discrete element framework for modeling tertiary creep of concrete in tension and compression. Eng. Fract. Mech. **200**, 263–282 (2018)
12. Rossi, P., Tailhan, J.L., Le Maou, F.: Comparison of concrete creep in tension and in compression: influence of concrete age at loading and drying conditions. Cement Concrete Res. **51**, 78–84 (2013)
13. Ranaivomanana, N., Multon, S., Turatsinze, A.: Basic creep of concrete under compression, tension and bending. Construct. Build. Mater. **38**, 173–180 (2013)
14. Cheng, Z.Q., Zhao, R., Yuan, Y., Li, F., Castel, A., Xu, T.: Ageing coefficient for early age tensile creep of blended slag and low calcium fly ash geopolymer concrete. Construct. Build. Mater. **262**, 119–855 (2020)
15. Korniejenko, K., Łach, M., Hebdowska-Krupa, M., Mikuła, J.: The mechanical properties of flax and hemp fibres reinforced geopolymer composites. IOP Conf. Series: Mater. Sci. Eng. **379**(1) (2018)
16. Acker, P., Agullo, L., Auperin, M., Carol, I., Carreira, D., Catarino, J., Chem, J.-C., Chiorino, M., Dougill, J., Huet, C., Kanstad, T., Kim, J.-K., Křístek, V., Republic, C., S Muller, H., Byung, G., Oh, H., Ožbolt, J., Reid, S., Wittmann, F.: RILEM TC 107-CSP: creep and shrinkage prediction models: principles of their formation. Recommendation Measurement of time-dependent strains of concrete. Mater. Struct. **31**, 507–512 (1998)
17. ASTM. E647—standard test method for measurement of fatigue crack growth rates. ASTM Book of Standards. 03(July), pp. 1–49 (2016)
18. Sprince, A., Pakrastinsh, L., Baskers, B., Gaile, L.: Crack development research in extra fine aggregate cement composites. Vide. Tehnologija. Resursi - Environment, Technology, Resources. **1**, 205–208 (2015)

Crack Width Estimation in Steel Fibre Reinforced Concrete (SFRC): Design and Limitations

Ned Gould, Fragkoulis Kanavaris, and Nuno Ferreira

Abstract The estimation of crack widths in steel fibre reinforced concrete (SFRC) with and without conventional reinforcement is still relatively uncharted territory. Engineers are faced with many challenges and uncertainties when designing SFRC members, such as precast SFRC tunnel segments, steel fibre sprayed concrete tunnel linings and other underground structures, owing to the peculiarities associated with SFRC behaviour, particularly when it is used without conventional reinforcement. At present, Eurocode 2 focuses on the design conventional reinforced concrete, as such, its recommended methodology for crack control is not applicable to SFRC. However, other guidance documents, such as fib MC2010, Concrete Society TR63, RILEM TC 162-TDF σε method, DafStb Guideline and research works by Johnson et al., are available and provide some recommendations with respect to crack width estimation in SFRC with and without conventional reinforcement. This paper reviews the aforementioned methods and evaluates their peculiarities, assumptions and differences with regards to the calculation of crack widths in SFRC.

Keywords Steel fibre reinforced concrete · Crack width · Crack control · Steel fibre sprayed concrete

1 Introduction

The estimation of crack widths in steel fibre reinforced concrete (SFRC) with or without conventional reinforcement is challenging, as there is no single clear guidance on the subject. Engineers are faced with many challenges and uncertainties when designing SFRC members, such as precast SFRC tunnel segments, steel fibre

N. Gould (✉)
Maritime Engineering Group, ARUP, London, UK
e-mail: Ned.Gould@arup.com

F. Kanavaris · N. Ferreira
Specialist Technology & Research, Advanced Digital Engineering, ARUP, London, UK

F. Kanavaris et al. (eds.), *International RILEM Conference on Early-Age and Long-Term Cracking in RC Structures*, RILEM Bookseries 31,
https://doi.org/10.1007/978-3-030-72921-9_3

sprayed concrete tunnel linings and other underground structures, owing to the peculiarities associated with SFRC behaviour, particularly in the absence of conventional reinforcement.

The use of SFRC, without conventional reinforcement, in underground structures is commercially advantageous and may even be safer than more traditional approaches, due to the higher production rates that can be achieved and reduced manual handling of reinforcing bars. Use of SFRC can also present a number of challenges including lack of uniformity of fibre content throughout the structure, possibly leading to weak spots where fibre content is lower and the lack of control over the orientation of fibres which are not always orientated in the direction of the tensile stress. In order to implement this technology, it is necessary to be able to reliably estimate the crack widths that might occur during the lifespan of the structure in order to ensure durability and prevent water ingress.

BS EN 1992-1-1 [1] does not provide guidance on the design of SFRC; therefore, the designer must turn to other guidance, such as RILEM TC 162-TDF σ-ε design method [2], fib MC2010 [3], Concrete Society TR63 [4] and the DafStb Guideline [5] or even research works by Johnson et al. [6], which provide some recommendations with respect to crack width estimation in SFRC with and without conventional reinforcement. Nevertheless, some of these methods are not applicable to sprayed concrete and most do not consider flexure in combination with axial compression which is the scenario encountered in the design of tunnel linings. It is also noted that many of the standards used to calculate crack widths due to imposed deformations, such as those reviewed in Refs. [7, 8], also do not provide crack control guidance for SFRC.

This paper reviews the aforementioned methods and evaluates their peculiarities, assumptions and differences with regards to the calculation of crack widths in SFRC with particular focus on whether each method accommodates the estimation of crack widths in SFRC without conventional reinforcement. Following the review of the available methodologies, the approach adopted for a particular project is then presented, along with the reasons for the choice.

2 Cracking Models for SFRC

The addition of steel fibres to the concrete matrix provides increased residual tensile strength post-cracking, increasing the ductility of the section. In order to evaluate crack widths in SFRC members, it is necessary to establish a suitable constitutive model that can be used in calculations and implemented in structural models.

2.1 RILEM TC 162-TDF

One of the first published pieces of guidance on calculation methods for crack widths in SFRC is that of RILEM TC 162-TDF. The proposed method is valid for SFRC with compressive strengths of up to 50 MPa. The stresses in SFRC in tension and compression are derived from the assumed stress-strain relationship shown in Fig. 1.

The post-cracking tensile capacity of the section is given in relation to the residual flexural tensile strength of the SFRC, which is a property of the SFRC that can be measured by testing. $f_{R,1}$ by definition is the minimum guaranteed residual flexural tensile strength of the SFRC at a crack mouth opening displacement of 0.5 mm; in this model, where crack widths are less than 0.5 mm, the residual tensile strength is always greater than 0.45 as can be seen in Fig. 2.

For SFRC sections without conventional reinforcement, that are part of statically indeterminate structures and where the neutral axis falls within the section depth, the designer is led to evaluate crack widths by assessing the compressive strain, then an

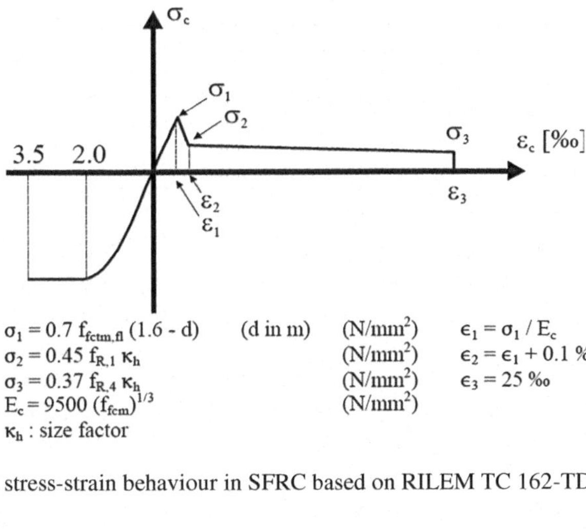

$\sigma_1 = 0.7\, f_{fctm,fl}\,(1.6 - d)$ (d in m) (N/mm²) $\epsilon_1 = \sigma_1 / E_c$
$\sigma_2 = 0.45\, f_{R,1}\, \kappa_h$ (N/mm²) $\epsilon_2 = \epsilon_1 + 0.1$ ‰
$\sigma_3 = 0.37\, f_{R,4}\, \kappa_h$ (N/mm²) $\epsilon_3 = 25$ ‰
$E_c = 9500\,(f_{fcm})^{1/3}$ (N/mm²)
κ_h : size factor

Fig. 1 Assumed stress-strain behaviour in SFRC based on RILEM TC 162-TDF

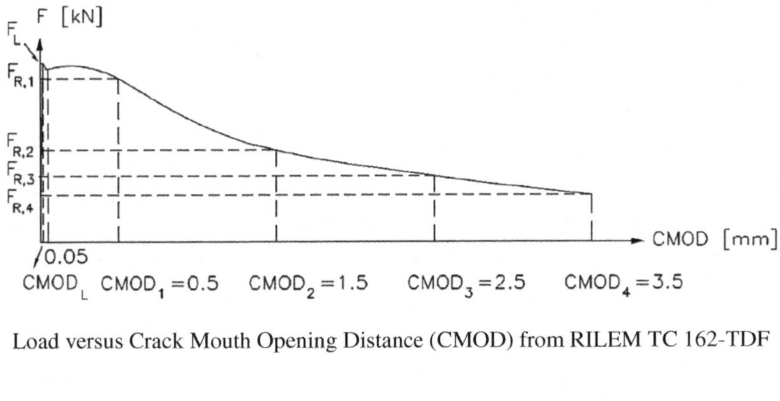

Fig. 2 Load versus Crack Mouth Opening Distance (CMOD) from RILEM TC 162-TDF

"idealised" tensile strain, assuming a linear variation of strain through the element, which is then factored by the crack spacing to give a crack width. Formulae and a methodology for evaluating crack widths in SFRC with conventional reinforcement are provided, but not for SFRC without conventional reinforcement.

A modified version of the Eq. 7.1 from BS EN 1992-1-1 clause 7.3.2 for the minimum area of reinforcement is provided in the guidance document (see Eq. 1).

$$A_s = \left(k_c \times k \times k_p \times f_{ct,ef} - 0.45 \times f_{Rm,1}\right) \times (A_{ct}/\sigma_s) \quad (1)$$

where A_s and A_{ct} is the area of reinforcement and concrete in the tensile zone, respectively, k_c, k and k_p, coefficients accounting for stress distribution, self-equilibrated stresses and prestressing effect, respectively, $f_{ct,ef}$ the effective concrete tensile strength, $f_{Rm,1}$ the average residual flexural tensile strength of SFRC at the time of cracking and σ_s the maximum stress permitted in the reinforcement immediately after formation of the crack.

If the minimum area of steel, A_s, is calculated to be less than zero, the guidance states that no conventional reinforcement is required. As such, by equating the expression in parenthesis to zero, it is possible for the designer to determine whether SFRC without conventional reinforcement is likely to be acceptable for the stress distribution in the section which represented by the coefficient k_c.

The process of calculating a design crack width value is similar to that in conventional concrete with rebar, i.e. the product of calculated crack-inducing strain and crack spacing, accounting for the fact that the tensile stress in SFRC is not zero but equal to $0.45 \times f_{Rm,1}$.

With regards to limiting crack widths values, for exposure class 1, equivalent to a dry indoor environment, the crack width is thought to have no influence on durability. For exposure class 2, equivalent to a humid environment, where crack widths are greater than 0.2 mm, the height of the cracked zone has to be reduced by 10 mm. Finally, for exposure class 3 and above, equivalent to humid environment with frost and de-icing salts, seawater environment and aggressive chemical environment, it is recommended that special measures should be taken. However, as previously mentioned, this method does not encompass recommendations with respect to crack width estimation in SFRC without conventional reinforcement.

2.2 fib *Model Code 2010*

The *fib* Model Code 2010, presents detailed stress-strain distributions for SFRC that are similar to that used in RILEM TC 162-TDF, but with a distinction drawn between SFRC that exhibits softening or hardening behaviour post-cracking. Sections that undergo strain hardening are more akin to conventional reinforced sections which exhibit strain hardening after cracking, provided the codified minimum amounts of reinforcement are provided.

Two simplified constitutive models for SFRC are given. The linear model equates the residual tensile strength of the concrete in the serviceability limit state (f_{Fts}) to the value given by the following equation, in which f_{R1} represents the flexural residual tensile strength for a crack mouth opening displacement of 0.5 mm.

$$f_{Fts} = 0.45 \times f_{R1} \quad (2)$$

This relationship is similar to the formula for σ_2 in RILEM (shown in Fig. 1) but without the size factor κ_h.

In the serviceability limit state, the fib Model Code limits compressive and tensile stresses in the concrete to 60% of their characteristic values. The following formula (Eq. 3) is then given for calculating the design crack widths in SFRC in Sect. 7.7.4.2 of the *fib* MC2010; however, only in sections with conventional reinforcement. Guidance is given on the minimum amount of reinforcement required for crack control which is similar to the RILEM approach.

$$\begin{aligned} w_k = {} & 2 \times [k \times c + 1/4 \times \varphi_s/\rho_{s,ef} \times (f_{ctm} - f)/\tau_{bm}] \\ & \times 1/E_s \times (\sigma_s - \beta \times \sigma_{sr} + \eta_r \times \varepsilon_{sh} \times E_s) \end{aligned} \quad (3)$$

The definition of each parameter is given in Sect. 7 of the *fib* MC2010. As mentioned earlier, no guidance is given to the designer on how to estimate of the crack widths in SFRC without conventional reinforcement. Also, no limiting crack widths values specifically for SFRC are given.

2.3 German Committee for Reinforced Concrete: DafStb Guideline

The DafStb Guideline for SFRC is presented as additions and changes to Eurocode 2. The document explicitly states that it is not applicable to steel fibre sprayed concrete, without further reasoning, something which restricts its usefulness in the design of underground structures, as many tunnel linings are made of sprayed concrete. The guideline is limited to SFRC where the fibres have a positive mechanical anchorage; i.e. where the material exhibits strain hardening, as addressed by the stress-strain distributions in the fib Model Code. Similar to the RILEM guidance, for SFRC sections without conventional reinforcement, the guidance is limited to statically indeterminate systems.

When calculating the axial residual tensile strength, a factor to account for the size of the section is taken into account, which is similar to the RILEM approach. In addition, another factor is included which accounts for the orientation of the fibres. The crack widths in SFRC with conventional reinforcement are calculated through the following equation:

$$w_k = S_{r,max} \times (\varepsilon_{sm}^f - \varepsilon_{cm}) \tag{4}$$

For which the difference between the mean strains of concrete and reinforcement in SFRC $(\varepsilon_{sm}^f - \varepsilon_{cm})$ may be calculated as follows:

$$\begin{aligned}(\varepsilon_{sm}^f - \varepsilon_{cm}) &= 1/E_s \times \left(1 - a_f\right) \times (\sigma_s - 0.4 \times f_{ct,eff} \times 1/\rho_{p,eff}) \\ &\geq 0.6 \times \left(1 - a_f\right) \times \sigma_s/E_s\end{aligned} \tag{5}$$

where a_f, σ_s, $f_{ct,eff}$, $\rho_{p,eff}$ and E_s are a parameter to account for steel fibre reinforcement, reinforcement stress in the crack without taking into account the fibre effect, effective axial tensile strength, effective reinforcement ratio and elastic modulus of steel, respectively. A formula is also provided for crack spacing. Most notably, this guidance offers a crack width calculation method for SFRC without conventional reinforcement (rebar). This is as follows:

$$w_k = S_w^f \times \varepsilon_{ct}^f \tag{6}$$

where ε_{ct}^f is the strain in the SFRC and S_w^f is 140 mm, for sections under bending stress only. It is, however, noted that no direct guidance is provided to the designer as to how to evaluate the strain in the SFRC nor what crack spacing, S_w^f, should be used where the section is subject to flexure and axial load, as is frequently encountered in the design of tunnel linings. Durability driven limited crack widths are nonetheless provided, more specifically 0.4 mm for X0, XC1; 0.3 mm for XC2, XC3 and 0.2 for XC4, XD1, XS1. If the above limiting crack width conditions are not satisfied in SFRC, as per Eq. 7, then it is mandated that conventional reinforcement is also used.

2.4 Concrete Society Technical Report No. 63

This technical report acknowledges that, at the time of publication, there are no agreed design approaches for many of the current applications of SFRC and that there is limited confidence in calculating crack widths in SFRC without conventional reinforcement. It states that, in most structural applications, conventional reinforcement is required to provide sufficient load bearing capacity and that the amount of steel fibre dosing required to replace the conventional reinforcement is high and outside scope of the report.

The report states that it is "*only possible to realistically calculate crack widths in statically indeterminate structure without conventional reinforcement by carrying out a non-linear analysis*" and that "*crack widths are not related to stresses given by elastic analysis once the section capacity is exceeded and softening develops*".

Furthermore, it notes that crack widths cannot be controlled in statically determinate members with only steel fibres unless a dosage of fibres greater than 80 kg/m$_3$ is provided to give a strain hardening response.

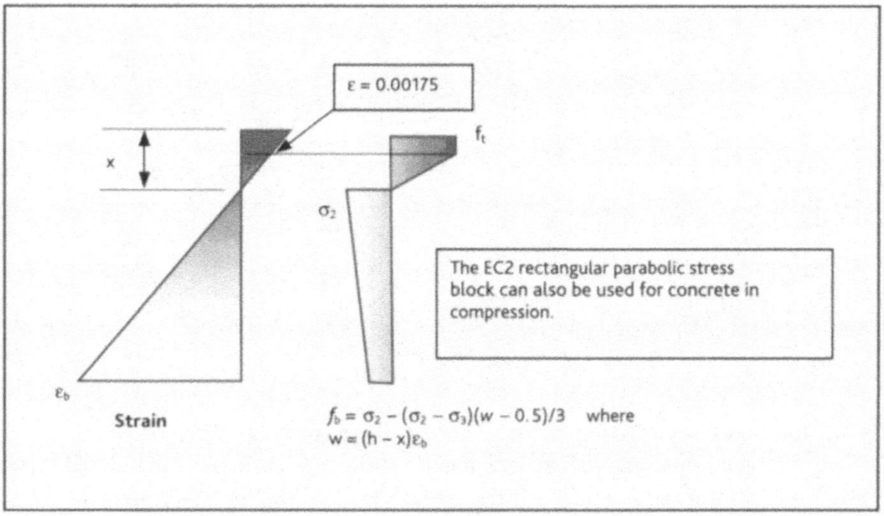

Fig. 3 Simplified stress block for deriving M-w response for SFRC according to Concrete Society TR63

This guidance document adopts the same equation for minimum area of reinforcement presented with the RILEM guidance (Eq. 1). In Sect. 6.2.3 of the report, equilibrium equations for a SFRC section without conventional reinforcement, subject to flexure but not axial load, are defined using a simplified stress block for the tensile zone based on RILEM guidance. These are shown in Fig. 3. The derived formulae are useful for the designer in some situations, but as an axial load is not included, the usefulness for underground structures is limited.

2.5 *Method by Johnson, Psomas and Eddie, 2017*

This paper presents one of the only methodologies for calculating crack widths in SFRC which takes into account axial force, i.e. the circumferential hoop force that is encountered in the design of tunnel linings and other underground structures, in combination with moment actions. The design methods presented in the paper are described as "potential" rather than being fully validated, the authors pointing to more test-based evidence being required.

The strain and stress distributions shown in Fig. 4 form the basis of the methodology. For crack widths up to 0.5 mm the tensile resistance of the section is assumed to be constant ($f_t = f_b = 0.45 \times f_{R1}$). This is the same value as used in the linear constitutive model in the fib Model Code (2010) and is similar to the formula for $\sigma 2$, the peak tensile resistance in the RILEM stress-strain relationship (Fig. 1), but without the size factor κ_h.

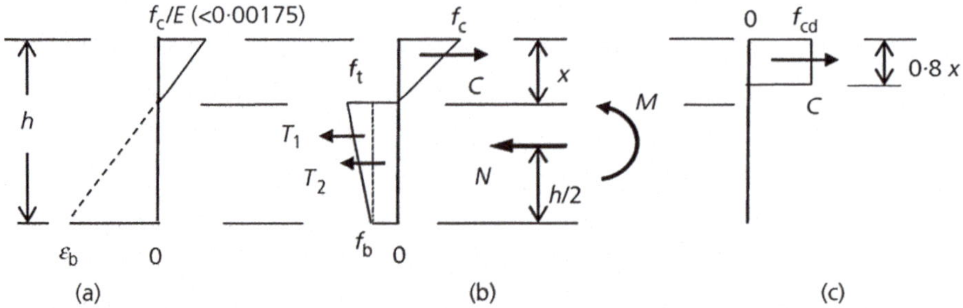

Fig. 4 Assumed distributions at a cracked cross-section of a beam, for SLS verification: **a** strain **b** elastic compressive stress **c** plastic compressive stress, after [6]

The methodology is based on determining the maximum compressive strain from equilibrium relationships; the surface tensile strain, εb, is then estimated by extrapolating the compressive strain profile. To associate the tensile strain with a crack width, the assumed spacing between cracks is taken as the distance between the neutral axis and tensile face. This is the same value as used in the Concrete Society TR63. As such, with this method, one can calculate the relationship between bending and axial loads for attaining a crack width of 0.3 mm (which arguably is the rather frequently adopted limiting design crack width in concrete structures not relying on limiting crack widths for water-tightness), based on different section thicknesses, as shown in Figs. 5 and 6. Figure 6 shows that the impact of implementing the size factor κ_h according to RILEM TC 162-TDF, is limited in the serviceability limit state. According to [6] the size factor can have a "profound negative effect" in the ultimate limit state.

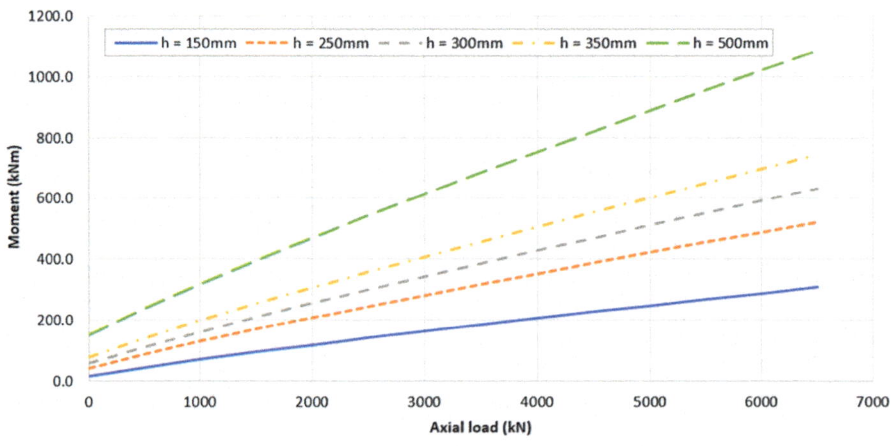

Fig. 5 Relationships between bending and axial load for attaining a 0.3 mm crack width in SFRC members with different thicknesses based on method in [6] without inclusion of a size factor

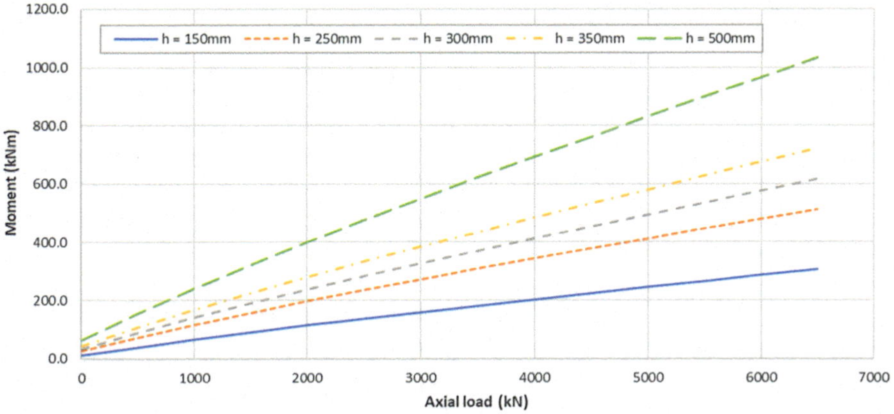

Fig. 6 Relationships between bending and axial load for attaining a 0.3 mm crack width in SFRC members with different thicknesses based on method in [6] including the size factor from [2]

3 Experiences from Design and Practice

This review into the different approaches to assessing crack widths in SFRC without conventional reinforcement was prompted by the need to assess durability and water-tightness for a major tunneling project using primarily single-pass sprayed concrete linings. The tunnel sections were assessed using 2D PLAXIS models with a simple linear constitutive model used for the concrete lining.

The inner surface of the tunnel lining was considered to be a dry environment and, therefore, it was thought that no specific limit on crack widths was necessary; however, to minimise crack widths and to comply with project requirements, it was deemed necessary to assess crack widths in the single-pass tunnel linings.

The crack widths were estimated using the approach in Ref. [6] assuming the stress profile shown in Fig. 4b. To ensure the sections remained watertight, the neutral axis depth was checked to ensure that it was compliant with clause 7.3.1 (112) of BS EN 1992-3:2006 [9] which states:

> To provide adequate assurance for structures of classes 2 or 3 that cracks do not pass through the full width of a section, the design value of the depth of the compression zone should be at least x_{min} calculated for the quasi-permanent combination of actions… The values of x_{min} for use in a country may be found in its National Annex. The recommended value for x_{min} is the lesser of 50 mm or 0,2 h where h is the element thickness.

The results of the PLAXIS modelling showed that, in the hoop direction, the tunnel sections were predominantly under compression with limited flexure. Through cracking of the sections was not predicted nor were crack widths found to be above a nominal limit of 0.2 mm.

4 Summary and Outlook

The design and assessment of SFRC sections is not covered by Eurocode 2; however, there are other published guidelines that cover the topic but the information is limited and there no single universal approach. Much of the guidance does not provide simplified practical approaches that can be used directly by a designer; rather the designer is required to develop their own approach choosing between various approaches that are subtly different.

There are two guidelines which present alternative constitutive models for SFRC, namely the RILEM TC 162-TDF recommendations and the *fib* Model Code 2010, the latter distinguishing between SFRC that exhibits softening or hardening post cracking. In evaluating the performance of SFRC in the serviceability limit state, a key parameter is the post-cracking tensile strength of the material which is dependent of the minimum guaranteed residual flexural tensile strength for a crack mouth opening distance of 0.5 mm. The value of this property of the material can be obtained from a standard beam test. Both RILEM TC 162-TDF recommendation and the *fib* Model Code 2010 apply the same factor to the value obtained through testing in order to obtain the design value; however, the approach by RILEM TC 162-TDF includes an additional size factor.

The DafStb Guideline and Concrete Society Technical Report 63 are based on the RILEM TC 162-TDF constitutive model. The DafStb Guideline explicitly states that it is not applicable to sprayed concrete, for unclarified reasons, limiting its usefulness in the design of underground structures. The Concrete Society Technical Report 63 includes formulae for evaluating a SFRC section subject to flexure but without axial load, again limiting its usefulness in the design of SFRC tunnel linings and other such structures. These guidelines differ on the suggested minimum crack spacing. DafStb recommend a crack spacing of 140 mm whereas Concrete Society Technical Report 63, RILEM TC 162-TDF and the fib Model Code recommend the depth of the section in tension is used.

Johnson, Psomas and Eddie, 2017 includes a methodology, based on the *fib* Model Code 2010, that allows for the evaluation of SFRC sections without conventional reinforcement subject to both flexure and axial load. This approach is not fully validated but is directly applicable to the design of underground structures.

The use of SFRC, without conventional reinforcement, in underground structures has commercial benefits and can potentially enhance safety. To facilitate wide adoption of this technology this paper identifies the need for clearer guidance regarding the design of SFRC, particularly in the serviceability limit state and for sections under both axial load and flexure. Proposed further work, to follow on from this study, will consist of a comparison between different methods for calculating the amount of conventional reinforcement required for a given crack width and the impact of adding steel fibres to the concrete matrix.

References

1. BS EN 1992-1-1:2004+A1:2014: Eurocode 2: Design of concrete structures—Part 1-1: General rules and rules for buildings (2004)
2. RILEM TC 162-TDF: Test and design methods for steel fibre reinforced concrete, σ-ε-design method. In: Final Recommendation, Materials and Structures, vol. 36, pp. 560–567 (2003)
3. Fib Model Code for Concrete Structures 2010. Wiley (2013)
4. Concrete Society TR 63: Guidance for the design of steel-fibre-reinforced concrete (2007)
5. German Committee for Reinforced Concrete, DafStb Guideline: Steel fibre reinforced concrete, draft, November 2012
6. Johnson, R.P., Psomas, S., Eddie, C.: Design of steel fibre reinforced concrete tunnel linings. In: Proceedings of the Institution of Civil Engineers—Structures and Buildings, vol. 170(SB2), pp. 115–130 (2017)
7. Kanavaris, F., Gardner, A., Ito, S., Chen, F., Schlicke, D., Zych, M., Jędrzejewska, A., Azenha, M.: A comparison between crack control reinforcement estimates from different codified approaches. In: Barros, J., Pereira, E., Figueiredo, F. (eds.) 3rd RILEM Spring Convention and International RILEM Conference, March 2020, Guimaraes, Portugal (2020)
8. Kanavaris, F., Kaethner, S.: Ciria guide C766: an overview of the updated Ciria C660 guidance on control of cracking in reinforced concrete structures. In: Gabrijel, I., Grosse, C., Skazlic, M. (eds.) International RILEM Conference on Sustainable Materials, Systems and Structures—SMSS19', March 2019, Rovinj, Croatia (2019)
9. BS EN 1992-3: Eurocode 2: Design of concrete structures—Part 3: Liquid retaining and containment structures (2006)

Cracking Analysis of Partially Prestressed Concrete Tie Under the Effect of Primary and Secondary Cracks

Maurizio Taliano

Abstract A mathematical model for the cracking analysis of a concrete tie reinforced by ordinary and prestressing steels, subjected to a statically applied axial force, is proposed. The two types of steels do not have only different diameters, but also different bond properties, since in general the prestressing steel presents a lower bond quality with respect to the ordinary steel. The model is based on the assumptions of linear elastic σ-ε laws for both types of steels, as well as for the concrete, and on the bond law proposed in *fib* Model Code 2010. Both the crack formation stage and the stabilized cracking stage are analyzed. In particular, in the stabilized cracking stage, under the assumption that the crack spacing is maximum and equal to twice the transmission length, the effect of primary and secondary cracks is taken into account. This leads to distinguish, in the stabilized cracking stage, an initial transitional phase where the transfer length of ordinary steel is smaller than the transfer length of prestressing steel, followed by a fully developed phase, where the two transfer lengths are equal to each other. Finally, the theoretical results of the crack spacing and the crack width, obtained with this refined model are compared to the experimental data available in the literature.

Keywords Partially prestressed concrete · Cracking · Maximum crack spacing · Maximum crack width

1 Introduction

Partially prestressed concrete, in which ordinary steel is present with the prestressing steel, is a suitable alternative to reinforced or fully prestressed concrete. Among the various advantages of simultaneously adopting ordinary and prestressing steels, partially p.c. beams present an improved structural serviceability with reduced crack widths in comparison to fully prestressed concrete beams, thanks to the better bond

M. Taliano (✉)
Department of Structural, Geotechnical and Building Engineering, Politecnico di Torino, Turin, Italy
e-mail: maurizio.taliano@polito.it

F. Kanavaris et al. (eds.), *International RILEM Conference on Early-Age and Long-Term Cracking in RC Structures*, RILEM Bookseries 31,
https://doi.org/10.1007/978-3-030-72921-9_4

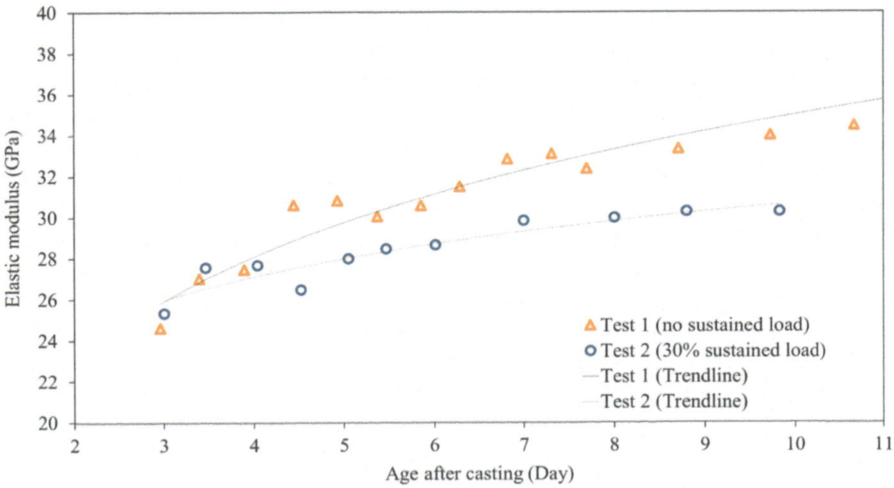

Fig. 3 Measured elastic modulus of concrete with and without sustained tensile load

of this product is plotted in Fig. 4. The computed values increase over time and are well below the constant value recommended as 0.65 which corresponds to a 40% stress reduction as mentioned in [3]. Thereby, this study highlights that such common assumptions not be applicable for determining the early age evolution of $E_{aaef}(t, t_0)$ in high-performance concrete.

The creep coefficients can be back calculated from Eq. (1) incorporating measured $E(t_0)$ and $E_{aaef}(t, t_0)$ test data and assuming a specific ageing coefficient. As

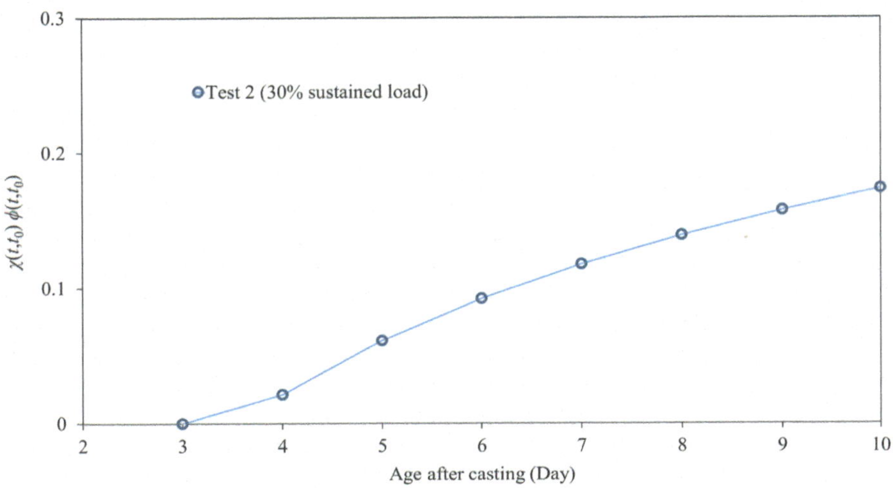

Fig. 4 Evolution of the product of the ageing and creep coefficients ($\chi(t, t_0)\,\phi(t, t_0)$)

of ordinary steel to the concrete. In effects, for partially prestressed beams international standards (fib Model Code 2010 [1], Eurocode 2 [2]) permit the formation of cracks under the frequent load combination, with crack width limits dependent on the environmental exposure class.

The study of the cracking behavior of a concrete tie reinforced with ordinary and prestressing steels is here developed with a mathematical model, named general model. This model is based on a previous work performed by Taliano [3] where the cracking analysis of a concrete tie reinforced with ordinary bars of different diameters is performed taking into account the effect of secondary cracks or Goto cracks. But, here, a further extension of this refined crack prediction model is developed, where the two types of steel do not have only different diameters, but also different bond properties. According to *fib* Model Code 2010 both the crack formation phase and the stabilized cracking phase are considered. In particular, in the stabilized cracking stage it is assumed that the crack spacing is maximum and equal to twice the transmission length of the prestressing steel.

2 Refined Model for the Cracking Analysis

A partially prestressed concrete tie subjected to a statically applied axial force is considered. Both ordinary and prestressing steels and concrete have linear elastic behaviour. For the sake of simplicity, the concrete section is reinforced by one prestressing tendon ($n_p = 1$) located at the centre of the section and four reinforcing bars ($n_s = 4$) located at the four corners (Fig. 1a). As the static conditions of the concrete vary, passing from compression to tension before and after decompression, the considerations made hereafter concern the behavior that occurs beyond the decompression of the concrete section. In other words, forces, deformations, stresses and strains are intended, in the following, as forces, deformations, stresses and strains beyond the decompression.

The formation of a crack determines the redistributions of stresses, strains and deformations around the crack that can be studied under two main situations: crack formation stage (Fig. 1b–d) and stabilized cracking stage (Fig. 2b–d). In both stages, around the crack the assumption of plane section is lost and, therefore, the elongations of ordinary and prestressing steels differ from the deformation of the concrete. It means that slips, s_s and s_p, occur between concrete and ordinary and prestressing steels:

$$s_s = u_s - u_{c,s} \tag{1}$$

$$s_p = u_p - u_{c,p} \tag{2}$$

where u_s and u_p, and $u_{c,s}$ and $u_{c,p}$ are the elongations of the ordinary bars, prestressing steel and concrete, respectively, from the relevant zero slip sections of the ordinary

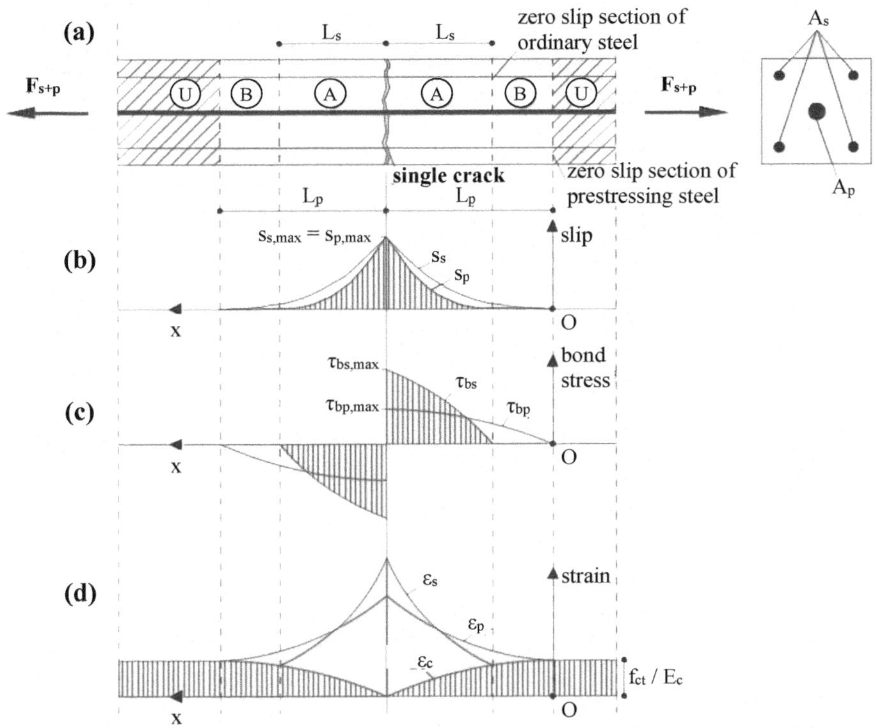

Fig. 1 Crack formation stage: **a** concrete tie reinforced with ordinary and prestressing steels; **b** distribution of the slips; **c** bond stresses; **d** longitudinal strains of ordinary and prestressing steels and concrete

and prestressing steels till the considered section. In particular, $u_{c,s}$ represents the elongation of the concrete determined from the zero slip section of the ordinary bars, while $u_{c,p}$ represents an analogous concrete elongation but referred to the prestressing steel. As a consequence, in general on the same section these two deformations differ because the starting points for their calculation, that means the zero slip sections for ordinary and prestressing steels, do not coincide.

The equations used to solve the mathematical problem are formally identical to that ones introduced by Taliano [3] in case of concrete tie reinforced with ordinary bars of different diameters. But, when the prestressing steel assumes the form of single strands or strand bundles, some modifications are needed in these equations. In effects, while the area of the transversal section, A_p, is associated to the nominal diameter ϕ_p, the external surface U_p, in contact to the concrete, is determined from the equivalent diameter, $\phi_{p,eq}$. And according to *fib* Model Code 2010 the equivalent diameter can be determined as follows: $\phi_{p,eq} = 1.75 \cdot \phi_{wire}$ for single 7-wire strands, $\phi_{p,eq} = 1.20 \cdot \phi_{wire}$ for single 3-wire strands and $\phi_{p,eq} = 1.6 \cdot \sqrt{A_p}$ for strand bundles, being ϕ_{wire} the wire diameter. This is a theoretical aspect that characterizes the

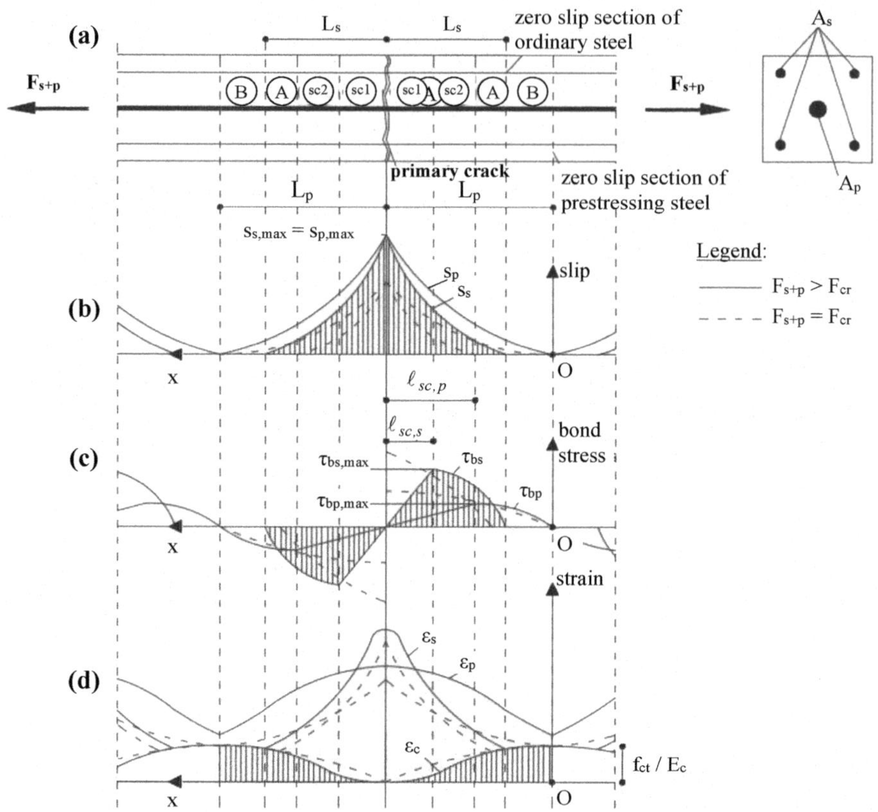

Fig. 2 Stabilized cracking stage when $L_s < L_p$: **a** concrete tie reinforced with ordinary and prestressing steels; **b** distribution of slips; **c** bond stresses; **d** longitudinal strains of ordinary and prestressing steels and concrete

cracking analysis of partially prestressed concrete tie that brings to do not simplify the nominal and equivalent diameters to each other in the differential equations that describe the mathematical model.

2.1 Bond Laws for Ordinary and Prestressing Steels

As far as the ordinary steel, the bond stress, τ_{bs}, can be expressed, in accordance with *fib* Model Code 2010, as a function of the slip, s_s, that occurs between steel and concrete, setting that the conditions for pull-out failure are met:

$$\tau_{bs} = \tau_{\max} \cdot \left(\frac{s_s}{s_1}\right)^{\alpha} = K_s \cdot s_s^{\alpha} \quad 0 \le s_s \le s_1 \tag{3}$$

being τ_{max} the bond resistance ($\tau_{max} = 2.5 \cdot \sqrt{f_{cm}}$ for good bond conditions); s_1 the value of slip corresponding to the attainment of τ_{max} ($s_1 = 1$ mm); α a numerical coefficient that can vary from 0 to 1. It has been shown in Debernardi et al. [4, 5] that a good value of the exponent α is equal to 0.25 in case of RC ties.

Actually, *fib* bond law for ordinary reinforcement is formed by various branches. But, as the maximum slip does not overcome 0.2 mm, a value that is obtained by dividing by two the maximum allowable crack width given by standards, for cracking analyses *fib* bond law can be limited to the first branch expressed by Eq. (3).

Prestressing steel presents reduced bond properties when compared to ordinary steel [6]. The effect of the different bond behavior of prestressing and ordinary steels was studied experimentally and theoretically by Rudlof and Hegger [7, 8] in case of post-tensioned ties. These Authors show that the bond strength of prestressing steel can be assumed as directly proportional to that one of the ordinary steel, through a bond factor, ξ, dependent on the type of prestressing steel (Table 1). Eurocode 2 as well as *fib* Model Code 2010 also provide values of the ξ factor, as shown in Tables 2 and 3. The values of ξ-factor proposed by Eurocode 2 refer to many cases, both for pre-tensioned and post-tensioned bonded steels with normal and high strength concretes: they are in agreement with the experimental values obtained in [7] in case of post-tensioned steels with normal strength concretes under short-term loading as well as the values proposed by *fib* Model Code 2010 in case of pre tensioned steels. In case of post-tensioned steels with high strength concretes Eurocode 2 also gives values that are in agreement with those obtained experimentally by Rudlof [7] under

Table 1 Bond factors ξ provided by Rudlof [7]

Type of prestressing steel	Pre-tensioned	Post-tensioned (bonded)	
		NSC	HSC
Smooth bars and wires	n.a.	0.30 (0.23)	0.18 (0.17)
Bundle of strands	n.a.	0.55 (0.41)	0.56 (0.29)
Indented wires	n.a.	n.a.	n.a.
Ribbed bars	n.a.	0.70 (0.56)	0.19 (0.13)

Table 2 Bond factors ξ provided by Eurocode 2

Type of prestressing steel	Pre-tensioned	Post-tensioned (bonded)	
		NSC	HSC
Smooth bars and wires	n.a.	0.30	0.15
Bundle of strands	0.60	0.50	0.25
Indented wires	0.70	0.60	0.30
Ribbed bars	0.80	0.70	0.35

Table 3 Bond factors ξ provided by *fib* Model Code 2010

Type of prestressing steel	Pre-tensioned	Post-tensioned (bonded)
Smooth bars and wires	n.a.	0.2
Bundle of strands	0.6	0.4
Indented wires	n.a.	n.a.
Ribbed bars	0.8	1.0

long-term loading, unless for ribbed bars where the bond factor is overestimated. As far as *fib* Model Code 2010, it does not distinguish between normal and high strength concretes, however it provides values that are, on average, in agreement with the values of Eurocode 2, except for ribbed bars for which the bond factor is assumed equal to 1.0 and so largely overestimated. In the following the bond factors proposed by *fib* Model Code 2010 (Table 3) are taken as reference in the calculations. Therefore, the adopted bond law for prestressing steel has a form that is analogous to that of ordinary steel, as a function of the slip, s_p, with a bond strength reduced by the ξ factors given in Table 3 (NB: $\xi < 1$):

$$\tau_{bp} = \xi \cdot \left[\tau_{\max} \cdot \left(\frac{s_p}{s_1}\right)^{\alpha}\right] = K_p \cdot s_p^{\alpha} \quad 0 \le s_p \le s_1 \tag{4}$$

2.2 Crack Formation Stage

A crack occurs in a section of the concrete tie when the concrete stress due to the applied axial load overcomes its tensile strength. The cracking force after decompression is:

$$F_{cr} = f_{ct} \cdot A_c \cdot \left(1 + \alpha_e \cdot \rho_s + \alpha_p \cdot \rho_p\right) \tag{5}$$

where in case of rectangular section $A_c = B \cdot H - A_s - A_p$.

The compatibility condition of the displacements requires that, at the crack, the distance between the two faces of the crack remains constant over the section, with the two sides of the crack parallel to each other and normal to the member axis. Therefore, at the crack the relative deformations (slips) of ordinary or prestressing steels and the concrete are the same (Fig. 1b), assuming the maximum value equal to the half of the crack width:

$$s_{s,max} = s_{p,max} = w/2 \tag{6}$$

As the assumption of plane section is lost, at the cracked section strains and stresses of ordinary and prestressing steels are unknown as well as the distances, L_p

and L_s, from the crack, where the condition of perfect bond (null slips) is restored for prestressing and ordinary steels (Fig. 1d).

Three different zones can be observed around a single crack, each with different behaviour (Fig. 1a):

- zone A, adjacent to the crack and extending for a distance equal to Ls from the crack, where all the steels are in slipping contact with the concrete ($s_p \geq s_s > 0$);
- zone B, extending from a distance L_s to a distance L_p from the crack, where the ordinary bars are perfectly bonded to the concrete ($s_s = 0$ and $\varepsilon_s = \varepsilon_c$), although the perfect bond condition is not observed for the prestressing steel ($s_p > 0$ and $\varepsilon_p > \varepsilon_c$);
- zone U (undisturbed), where both bars are perfectly bonded to the concrete ($s_s = s_p = 0$ and $\varepsilon_s = \varepsilon_p = \varepsilon_c$).

The two unknowns of the problem, that is, the transmission lengths of ordinary and prestressing steels, L_s and L_p, are determined through an iterative procedure. Convergence is reached when the concrete stress at the crack is equal to zero, and the ordinary and prestressing steels have the same maximum slip.

The following equations are used, with reference to an x-axis with origin at the zero slip section of the prestressing steel:

- in zone U, where perfect bond conditions can be applied, strains, stresses and deformations of materials are known;
- in zone B, only the ordinary bars are in perfect bond conditions with the concrete, while for prestressing steel it is possible to determine the distribution of slip of the prestressing steel, sp, in a closed form:

$$s_p(x) = \left[\frac{(1-\alpha)^2}{2 \cdot (1+\alpha)} \cdot \frac{\tau_{\max}}{s_1^\alpha} \cdot \frac{U_p}{E_p \cdot A_p} \cdot \frac{1 + \alpha_e \cdot \rho_s + \alpha_p \cdot \rho_p}{1 + \alpha_e \cdot \rho_s} \cdot x^2 \right]^{\frac{1}{1-\alpha}} \tag{7}$$

  from which the strains, stresses and deformations of the concrete and steels can be obtained.
- in zone A, the following two second-order differential equations can be derived to study the slipping contact between the ordinary and prestressing steels and the concrete

$$\begin{cases} \ddot{s}_s(x) = \frac{4 \cdot K_s \cdot (1+\alpha_e \cdot \rho_s)}{E_s \cdot \varphi_s} \cdot s_s^\alpha + \frac{(\xi \cdot K_s) \cdot U_p \cdot \alpha_p \cdot \rho_p}{E_p \cdot A_p} \cdot s_p^\alpha \\ \ddot{s}_p(x) = \frac{4 \cdot K_s \cdot \alpha_e \cdot \rho_s}{E_s \cdot \varphi_s} \cdot s_s^\alpha + \frac{(\xi \cdot K_s) \cdot U_p \cdot (1+\alpha_p \cdot \rho_p)}{E_p \cdot A_p} \cdot s_p^\alpha \end{cases} \tag{8}$$

2.3 *Stabilized Cracking Stage with Maximum Crack Spacing*

The crack spacing is here assumed as maximum and equal to twice the transmission length of the prestressing steel, L_p, determined at the crack formation stage. For compatibility of the deformations, ordinary and prestressing steels present the same maximum slip at the crack (see Eq. (6)).

At the beginning of this stage, the transmission length of the ordinary bars, L_s, is smaller than that of the prestressing steel, L_p, so that continuity with the previous stage is maintained. However, the increase in the length L_s goes on until the situation in which the transmission length of the ordinary bars, L_s, equals the transmission length of the prestressing steel, L_p. Therefore, when the axial load increases, two different phases form the stabilized cracking stage. An initial transitional phase, during which the transfer length of ordinary steel is smaller than the transfer length of prestressing steel, but it increases as the axial load increases. This phase is followed by a fully developed phase, in which the transfer length of ordinary steel remains unchanged and equal to the transfer length of prestressing steel.

In both these two phases, as in case of an RC tie in the stabilized cracking stage under the assumption that the crack spacing is maximum [4, 5], the equilibrium condition requires to modify the distributions of bond stresses close to the primary crack. Linear distributions of bond stresses on ordinary and prestressing steels are adopted which, on the physical point of view, represent the effect of secondary cracks that occur around the primary crack. The lengths, over which bond stresses are reduced, are named as lengths of reduced bond of ordinary and prestressing steels, $\ell_{sc,p}$ and $\ell_{sc,s}$. These length tend to increase when the axial load increases.

Initial transitional phase ($L_s < L_p$). Two new zones occur around each primary crack, because of the presence of secondary cracks. Starting from a primary crack, four different zones can be observed (Fig. 2):

- zone SC1, adjacent to the crack, where all the steels are in slipping contact with the concrete, and secondary cracks are observed on both ordinary and prestressing steels ($s_p \geq s_s > 0$). The extent of this zone is equal to the length, $\ell_{sc,s}$, of the internal cracking of ordinary steels;
- zone SC2, which extends over a part of the length $\ell_{sc,p}$, beyond the length $\ell_{sc,s}$, and is equal to $\ell_{sc,p} - \ell_{sc,s}$. Ordinary and prestressing steels are in slipping contact with the concrete ($s_p \geq s_s > 0$), but secondary cracks are only observed on the prestressing steel;
- zone A, over the length $\ell_{sc,p}$, delimited on the other side by the zero slip section of the ordinary bars, whose extent is $L_s - \ell_{sc,p}$. Ordinary and prestressing steels are in slipping contact with the concrete ($s_p \geq s_s > 0$), but no secondary cracks form;
- zone B, which falls between the two zero slip sections of the ordinary and prestressing steels, whose extent is $L_p - L_s$. The ordinary bars are perfectly bonded to the concrete ($s_s = 0$), while the prestressing steel is in slipping contact with the concrete ($s_p > 0$).

The unknowns of the problem are the transmission length of the ordinary bars, L_s, and the lengths of the internal cracking, $\ell_{sc,s}$ and $\ell_{sc,p}$. They can be determined through an iterative procedure, verifying that, at the cracked section, the concrete stress is equal to zero and the ordinary and prestressing steels have the same maximum slip, and that the ratio between the maximum bond stresses of the prestressing and ordinary steels that occur in two different sections, at distances $\ell_{sc,s}$ and $\ell_{sc,p}$ from the primary crack, respectively, is constant and equal to the ξ-factor, that means:

$$t_{bp,max} = \xi \cdot \tau_{bs,max} \tag{9}$$

The calculation method is based on the following differential equations:

- for zone SC1 ($L_s - \ell_{sc,s} \leq x \leq L_p$), because of the presence of internal secondary cracks it is assumed that bond stresses τ_{bs} and τ_{bp} reduce linearly close to a primary crack, according to the model proposed by Debernardi et al. [4, 5]:

$$\begin{cases} \ddot{s}_s(x) = \frac{4\cdot(1+\alpha_e\cdot\rho_s)}{E_s\cdot\varphi_s}\cdot\frac{\tau_{bs,\max}}{\ell_{sc,s}}\cdot\left(L_p - x\right) + \frac{U_p\cdot\alpha_p\cdot\rho_p}{E_p\cdot A_p}\cdot\frac{\tau_{bp,\max}}{\ell_{sc,p}}\cdot\left(L_p - x\right) \\ \ddot{s}_p(x) = \frac{4\cdot\alpha_e\cdot\rho_s}{E_s\cdot\varphi_s}\cdot\frac{\tau_{bs,\max}}{\ell_{sc,s}}\cdot\left(L_p - x\right) + \frac{U_p\cdot(1+\alpha_p\cdot\rho_p)}{E_p\cdot A_p}\cdot\frac{\tau_{bp,\max}}{\ell_{sc,p}}\cdot\left(L_p - x\right) \end{cases} \tag{10}$$

  where $\tau_{bs,max}$ and $\tau_{bp,max}$ are the maximum bond stresses of the ordinary and prestressing steels that occur at the abscissas $x = L_p - \ell_{sc,s}$ and $x = L_p - \ell_{sc,p}$, respectively;
- for zone SC2 ($L_p - \ell_{sc,p} \leq x \leq L_p - \ell_{sc,s}$):

$$\begin{cases} \ddot{s}_s(x) = \frac{4\cdot K_s\cdot(1+\alpha_e\cdot\rho_s)}{E_s\cdot\varphi_s}\cdot s_s^\alpha + \frac{U_p\cdot\alpha_p\cdot\rho_p}{E_p\cdot A_p}\cdot\frac{\tau_{bp,\max}}{\ell_{sc,p}}\cdot\left(L_p - x\right) \\ \ddot{s}_p(x) = \frac{4\cdot K_s\cdot\alpha_e\cdot\rho_s}{E_s\cdot\varphi_s}\cdot s_s^\alpha + \frac{U_p\cdot(1+\alpha_p\cdot\rho_p)}{E_p\cdot A_p}\cdot\frac{\tau_{bp,\max}}{\ell_{sc,p}}\cdot\left(L_p - x\right) \end{cases} \tag{11}$$

- for zone A ($L_{s2} - L_{s1} \leq x \leq L_{s2} - \ell_{sc,p}$):

$$\begin{cases} \ddot{s}_s(x) = \frac{4\cdot K_s\cdot(1+\alpha_e\cdot\rho_s)}{E_s\cdot\varphi_s}\cdot s_s^\alpha + \frac{(\xi\cdot K_s)\cdot U_p\cdot\alpha_p\cdot\rho_p}{E_p\cdot A_p}\cdot s_p^\alpha \\ \ddot{s}_p(x) = \frac{4\cdot K_s\cdot\alpha_e\cdot\rho_s}{E_s\cdot\varphi_s}\cdot s_s^\alpha + \frac{(\xi\cdot K_s)\cdot U_p\cdot(1+\alpha_p\cdot\rho_p)}{E_p\cdot A_p}\cdot s_p^\alpha \end{cases} \tag{12}$$

- for zone B ($0 \leq x \leq L_p - L_s$), no equation is needed for ordinary bars that are perfectly bonded to the concrete, while the differential equation for the prestressing steel is:

$$\ddot{s}_p(x) = \frac{(\xi \cdot K_s)\cdot U_p}{E_p \cdot A_p}\cdot\left(\frac{1+\alpha_e\cdot\rho_s+\alpha_p\cdot\rho_p}{1+\alpha_e\cdot\rho_s}\right)\cdot s_p^\alpha \tag{13}$$

Fully developed phase ($L_s = L_p$). Both steels are in slipping contact with the concrete from the zero slip section to the cracked section. Three different zones can be distinguished around a primary crack (Fig. 3):

- zones SC1 and SC2, which are under the same conditions as the SC1- and SC2-zones described above for the initial phase;
- zones A ($0 \leq x \leq L_p - \ell_{sc,p}$), where non-linear distributions of bond stresses are observed on both ordinary and prestressing steels.

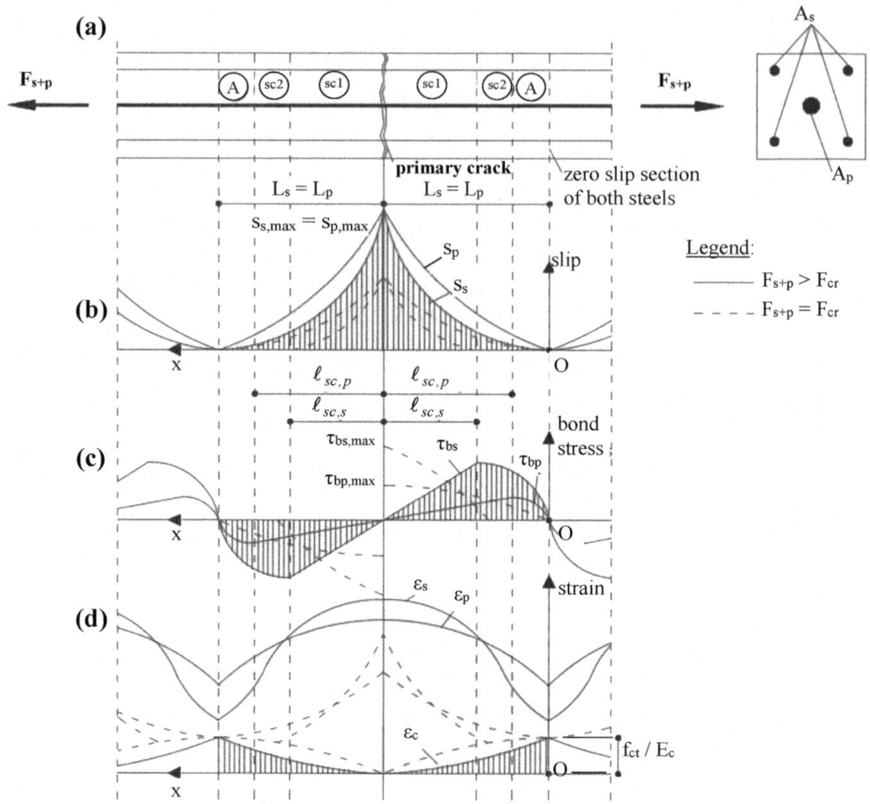

Fig. 3 Stabilized cracking stage when $L_s = L_p$: **a** concrete tie reinforced with ordinary and prestressing steels; **b** distribution of slips; **c** bond stresses; **d** longitudinal strains of ordinary and prestressing steels and concrete

The mathematical problem presents three unknowns, that means the lengths of the internal cracking, $\ell_{sc,s}$ and $\ell_{sc,p}$, and the stress, σ_{pE}, of the prestressing steel at the zero slip section. They are determined through iterative calculations considering, as boundary conditions, a null concrete stress at the crack, a maximum slips for ordinary and prestressing steels at the crack and a constant ratio between the maximum values of the bond stresses of ordinary and prestressing steels equal to the ξ-factor (see Eq. (9)). The last boundary condition is consistent with the initial transitional phase, where the maximum bond stresses of ordinary and prestressing steels are proportional to each other through the ξ-factor. On the other hand, a previous analysis concerning the cracking behaviour of a concrete tie reinforced with small and large ordinary bars has shown that this assumption is acceptable [3].

3 Comparison Between Experimental and Theoretical Data

The theoretical results obtained with the refined model are compared to the experimental data obtained by Rudlof [7]. These data refer to tests on concrete ties reinforced with four ordinary bars and post-tensioned with different types of prestressing steels all located in the centre of the section, without stirrups, with lengths of 1800 mm and square sections of 220 mm width. A groove at the mid-length of the ties marked the location of the first crack. In this section, in order to determine the forces that act on the ordinary and prestressing steels, strain gauges were glued on each ordinary bar, being applied after removing locally the ribs for their application. The measurements of crack width were performed by a mechanical extensometer (Staeger) with gauge length of 100 mm whose measuring points (3 × 18) form a grid on the two faces of the specimens. From these measurements it was possible to determine the axial mean deformations of ties on a measuring base of 1700 mm.

Here, the sample named K7 is considered, reinforced by four bars of 10 mm diameter and one central tendon composed by 3 strands, each of 0.6″ nominal diameter ($A_p = 420$ mm^2). Concrete has a medium compressive strength ($f_c = 51.6$ MPa) as well as the grout ($f_{c,mortar} = 47.3$ MPa) and a splitting tensile strength of 2.46 MPa. The theoretical calculations are performed assuming $\xi = 0.5$. The comparison is made as a function of the applied axial force, in terms of stresses of ordinary and prestressing steels at the cracked section (Fig. 4a), mean strain of ordinary steel (Fig. 4b), twice of the transmission lengths of ordinary and prestressing steels (Fig. 5a) and crack opening (Fig. 5b). From the comparison of the experimental and theoretical values of the mean strain of ordinary steel (Fig. 4b) it results a quite good agreement between experimental and theoretical data. From Fig. 5a it appears that the mean value of the experimental crack spacing, which is equal to 180 mm, is included between the

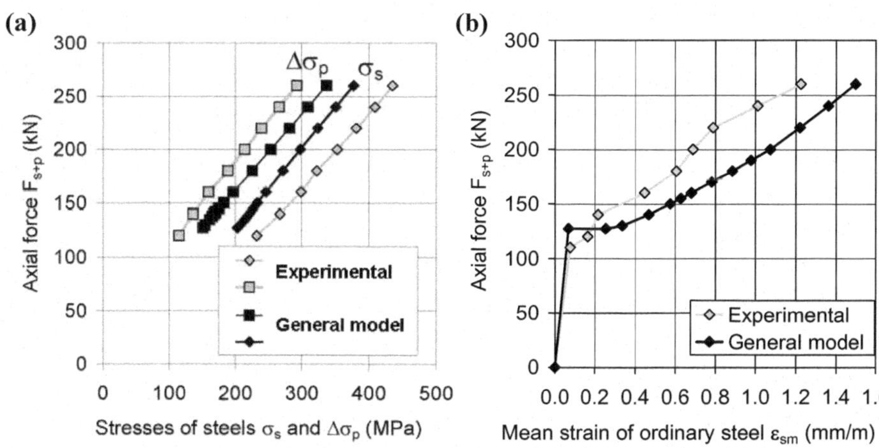

Fig. 4 Comparison between experimental and theoretical results (specimen K7): **a** stresses of ordinary and prestressing steels; **b** mean strain of ordinary bars

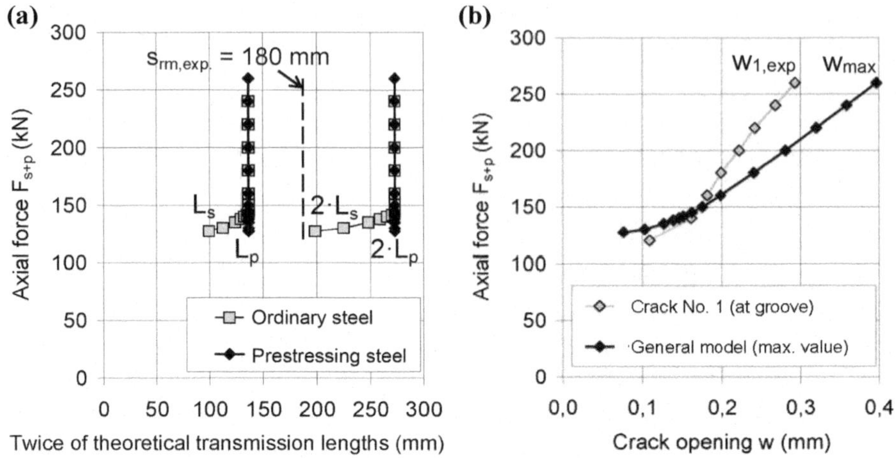

Fig. 5 Specimen K7: **a** twice of the transmission lengths of ordinary and prestressing steels; **b** local values (experimental) and maximum values (theoretical) of crack opening

maximum and minimum values of the theoretical crack spacing. This leads to a safe overestimation of the maximum theoretical crack opening (Fig. 5b).

4 Conclusions

A refined model for the cracking analysis of a concrete tie reinforced with ordinary and prestressing steel, subjected to a statically applied axial force, is proposed. The theoretical results obtained with the general model are compared to the experimental data available in the literature allowing the stress redistributions between prestressing and reinforcing steels and the strains and the deformations to be determined. This leads to a safe overestimation of the maximum theoretical crack opening.

References

1. Fédération internationale du béton (*fib*): *fib* Model Code for Concrete Structures 2010. Ernst & Sohn, Berlin (2013)
2. CEN: EN 1992-1-1. Eurocode 2: Design of concrete structures—Part 1-1: General rules and rules for buildings (2004)
3. Taliano, M.: Cracking analysis of concrete tie reinforced with two diameter bars accounting for the effect of secondary cracks. Eng. Struct. **144**, 107–119 (2017)
4. Debernardi, P.G., Guiglia, M., Taliano, M.: Effect of secondary cracks for cracking analysis of reinforced concrete tie. ACI Mater. J. **110**(2), 209–216 (2013)
5. Debernardi, P.G., Taliano, M.: An improvement of the Eurocode 2 and fib Model Code 2010 methods for the calculation of crack width in r.c. structures. Struct. Concr. **17**(3), 365–376 (2016)

6. Eligehausen, R., Bigaj-van Vliet, A.: Structural Concrete. Textbook on Behavior, Design and Performance. fib Bulletin 51, 2nd edn. Fédération internationale du béton (fib), Lausanne (2009)
7. Rudlof, M.: Zur Spannungsumlagerung zwischen Spannstahl und Betonstahl bei Spannbeton mit nachträglichem Verbund. Dissertation 82, RWTH Aachen (1998)
8. Hegger, J., Will, N.: Spannbetonbauteile mit nachträglichem Verbund aus hochfestem Beton Untersuchungen zu den Auswirkungen des unterschiedlichen Verbundverhaltens von gemischter Bewehrung aus Betonstahl und Spannglied. Bericht 72, RWTH Aachen (2002)

Design, Evaluation and Specification for Concrete Mixtures with Low Potential for Cracking

Jose Pacheco

Abstract The development of concrete mixtures with low potential for cracking is necessary for durable, sustainable, and efficient concrete construction. Due to the wide range of crack-inducing parameters, efforts on the development, evaluation, and specification of concrete mixtures with low potential for cracking are paramount. On the design phase, criteria related to the selection and performance of concrete mixture constituents, mixture proportions, and laboratory testing are necessary for crack mitigation. During concrete production, efforts to maintain control of batch water and water-to-cementitious ratio are paramount. Results presented in this paper will compare the performance of concrete mixtures with low and high potential for cracking in both the laboratory and field.

Keywords Cracking · Concrete mixture · Specification · Design

1 Introduction

Modern concrete construction necessitates the use of advanced concrete technology to develop concrete mixtures capable of meeting multiple and challenging specification requirements. Still, project specifications that overlook the effect of restraint, mixture materials and proportions, and curing typically result in detrimental performance of cast-in-place concrete. In such cases, cracking of newly constructed reinforced concrete liquid containment walls, bridge decks and overlays, or structural repairs is often present and, in most cases, exceed the acceptable level of cracking and thus necessitating repair.

In the design phase, the required performance of cast-in-place concrete is determined from the structural calculations and detailing. From a materials perspective, the Designer identifies the materials—and in some less frequent cases, the proportions of such materials—that are allowed for the selection and use of concrete mixtures during construction. The specification requirements that are developed from such

J. Pacheco (✉)
CTLGroup, Skokie, IL 60077, USA
e-mail: jpacheco@ctlgroup.com

F. Kanavaris et al. (eds.), *International RILEM Conference on Early-Age and Long-Term Cracking in RC Structures*, RILEM Bookseries 31,
https://doi.org/10.1007/978-3-030-72921-9_5

task include other performance requirements such as compressive strength, drying shrinkage, restrained shrinkage, and others as applicable.

With respect to volumetric changes, the evaluation of candidate concrete mixtures for drying shrinkage and restrained shrinkage are the most common. In the US, drying shrinkage is evaluated in accordance with ASTM C157 [1] and restrained shrinkage per ASTM C1581 [2]. For repair projects, a modification to ASTM C157 has been proposed by ACI 364.3R [3] which reduces the duration of moist curing of concrete specimens to simulate field application conditions. This modification is often specified for projects in which moist curing will be significantly shortened. The outcome of the evaluation program should be the identification of concrete mixtures with a low potential for cracking for construction.

During construction, batching tolerances specified in ASTM C94 [4] must be complied with so that the production of ready-mix concrete be reliable to provide accurate batching of concrete materials and to maintain consistent total cementitious content and water-to-cementitious ratio (w/cm), respectively. Variations in the cementitious content or w/cm will inherently result in a performance that can result in unintended volumetric changes leading to cracking.

This paper provides a review of the commonly specified performance requirements for controlling drying and restrained shrinkage of concrete. Other causes for cracking such as thermal gradients, imposed deformations or others are not in the scope of this manuscript. Results presented herein show the performance of different concrete mixtures subject to laboratory testing. Finally, recommendations for the specification of concrete mixtures with a low potential for cracking are provided.

2 Typical Performance Specifications for Shrinkage

The typical performance requirement for maximum allowable shrinkage of cast-in-place concrete are as follows:

- Drying Shrinkage: Maximum of 500 microstrain (0.05%) after 28 days of drying when determined in accordance with ASTM C157. In some cases, a maximum of 300 microstrain (0.03%) may be specified.
- Restrained Shrinkage: No cracking observed after 28 days of drying when evaluated in accordance with ASTM C1581. For repair or a highly restrained condition, no cracking observed after 14 days may be specified. Guidance for the classification of the potential for cracking of concrete mixtures is provided in Appendix X1 of ASTM C1581 as shown in Table 1.

Although other requirements may be specified for special projects, the above listed represent typical requirements for cast-in-place concrete construction in the US. These requirements are in addition to compressive strength, modulus of elasticity, tensile strength or others.

Table 1 Classification of potential for cracking of concrete mixtures per ASTM C1581

Net time-to-cracking, t_{cr} days	Average stress rate, S (MPa/day)	Potential for cracking
$0 < t_{cr} \leq 7$	$S \geq 0.34$	High
$7 < t_{cr} \leq 14$	$0.17 \leq S < 0.34$	Moderate-high
$14 < t_{cr} \leq 28$	$0.10 \leq S < 0.17$	Moderate-low
$t_{cr} > 28$	$S < 0.10$	Low

3 Evaluation

3.1 Drying Shrinkage

The evaluation of linear length change due to drying is determined in accordance with ASTM C157. The test method consists on the fabrication of concrete prisms with a cross section that depends on the maximum coarse aggregate size. For a maximum coarse aggregate size of 25 mm, the cross section shall be 75 mm. For larger coarse aggregates up to 50 mm in size, the cross section will be 100 mm. The length for each prism regardless of the cross section is 285 mm.

Concrete batching needs to be conducted in accordance with ASTM C192 [5] and specimen fabrication and curing should follow the requirements of ASTM C157. Upon fabrication, the specimens are removed from the molds after 23–24 h after water-addition to the concrete mixture and immersed in limewater for 15 or 30 min for prisms with a cross section of 75 mm or 100 mm, respectively. Subsequently, the initial length of teach prism is taken and then the specimens are immersed in limewater at 23 ± 2 °C for the duration of moist curing. Since the tolerances for ambient temperature, time of demolding, and curing are strict, laboratory batching and specimen fabrication is always preferred.

The effect of the duration of moist curing is critical for evaluating the drying shrinkage performance of concrete mixtures, especially if these contain supplementary cementitious materials such as fly ash, slag cement or silica fume. To illustrate this, concrete prisms were fabricated from two different concrete mixture proportions shown in Table 2.

Although the mixtures contain two different SCM replacement levels, the total cementitious content of 356 kg/m^3, batch water content of 130 kg/m^3, w/cm ratio of 0.36, and paste content of 31.6% by volume remained constant for both. Four sets of three prisms (total of 12 prisms) with a cross section of 75 mm were fabricated. Each set was subject to curing for 1, 3, 7, and 28 days, respectively. Environmental conditions during drying were maintained at 23 ± 2 °C and 50 ± 2% RH for the entire experimental program.

Drying shrinkage measurements for each set of concrete prisms for each concrete mixture are shown in Fig. 1. The results clearly demonstrate the effect of concrete

Table 2 Concrete mixture materials and proportions for drying shrinkage

Material	SG	M1	M2
Cement—ASTM C150 type I/II, kg/m^3	3.15	196	237
Slag—ASTM C989 grade 100, kg/m^3	2.82	142	–
Fly ash—ASTM C618 Class F, kg/m^3	2.58	–	107
Silica fume—ASTM C1240, kg/m^3	2.23	18	12
Coarse aggregate—ASTM C33 12.5mm, kg/m^3	2.61	445	445
Coarse aggregate—ASTM C33 25mm, kg/m^3	2.62	653	653
Fine aggregate—ASTM C33 5 mm, kg/m^3	2.62	655	651
Water, kg/m^3	1.00	130	130
SCM replacement, %		45.0	33.3
Total cementitious content, kg/m^3		356	356
w/cm ratio, (–)		0.36	0.36
Air content, %		7.5	7.5
Paste content, % by volume		31.6	31.6

materials, SCM replacement and duration of moist curing on the measured length change after 28 days of drying.

Length change measurements shown in Fig. 1 show that the difference in 28-day microstrain for M1 between 1 and 28 days of curing was nearly four times lower. Length change microstrain values for M1 ranged between 200 and 780 for 1-day and 28-day curing, respectively. In fact, the effect of curing duration on length change in M1 specimens was significantly greater than that observed in M2. This is caused by the higher SCM replacement in M1 (45%) which necessitates extended moist curing to allow for the SCM materials to react [6]. When the duration of moist curing is reduced, mixtures with higher SCM replacement levels may not perform as intended and can potentially develop length change measurements that exceed the typical requirements for cast-in-place concrete construction.

Although, it is generally accepted that a reduction in drying shrinkage of concrete is expected when increasing the SCM replacement [7], Fig. 1 shows that the selection of the type of SCM plays a crucial role. Even at a lower SCM replacement in M2 (33%), microstrain values after 28 days of drying ranged between 400 and 550 regardless of the duration of moist curing.

3.2 Restrained Shrinkage

Determination of the susceptibility to cracking of concrete mixtures is crucial for placements where restraint can lead to the formation of significant early-age tensile stresses and, consequently, cracking. For the evaluation of restrained shrinkage, laboratory testing conducted in accordance with ASTM C1581 can help in identifying

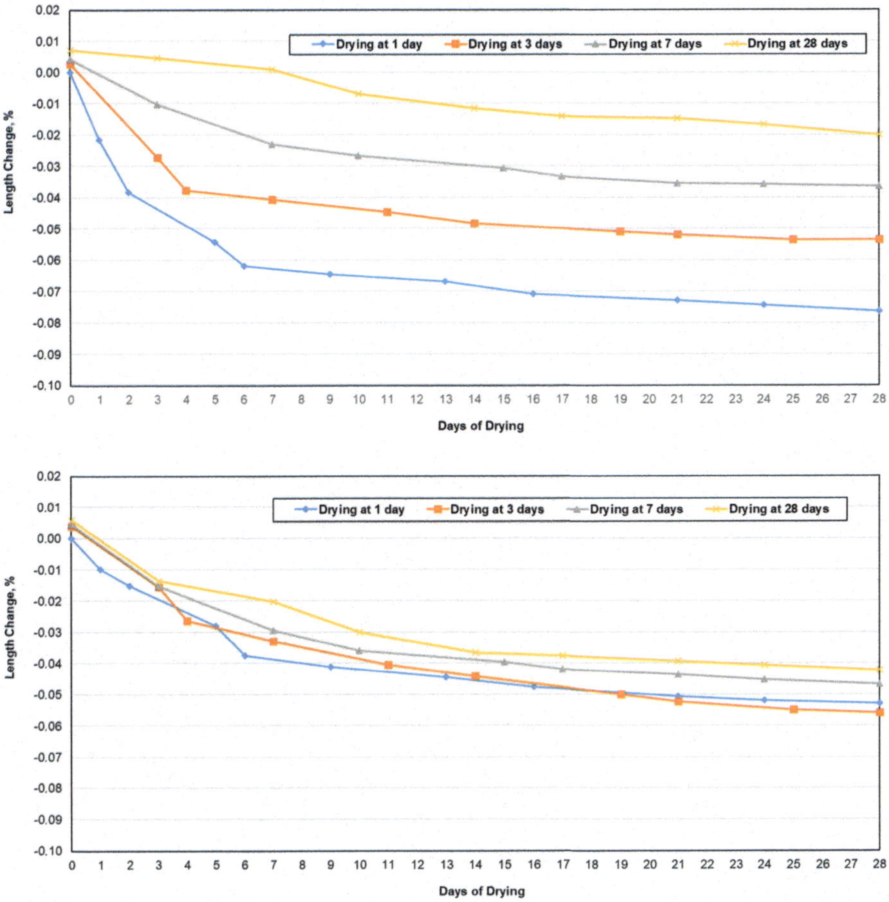

Fig. 1 Drying shrinkage measurements of two different concrete mixtures after different moist curing durations. Top: M1, bottom: M2

concrete mixture materials and proportions that result in a low potential for cracking performance. In this method, three rings with a thickness of 38 ± 3 mm and a height of 150 mm are fabricated after batching of concrete in accordance with ASTM C192 A steel inner ring with a thickness of 13 ± 1 mm is instrumented with strain gauges to measure and record strains on the inner surface of the concrete ring. The outer ring and base are fabricated of non-absorptive material (Fig. 2).

After fabrication, the specimens remain in the molds and covered with wet burlap for 24 h after water addition at 23 °C. Subsequently, the outer ring is removed and the top of the ring is sealed to prevent drying as shown in Fig. 3. The molded surface in contact with the outer ring is exposed to the environmental conditions of 23 ± 2 °C and 50 ± 2% RH for the duration of the test.

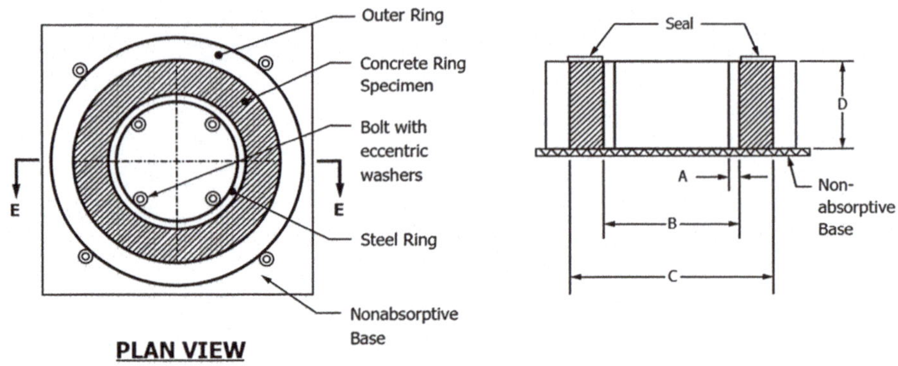

Fig. 2 Plan view and cross section of test specimens for ASTM C1581 testing

Fig. 3 Example of ASTM C1581 test specimen

Four overlay concrete mixtures using the materials and proportions shown in Table 3 were evaluated in accordance with ASTM C1581. The base mixture proportions identified as C1 are composed of a ternary blend of cementitious materials with a SCM replacement of 24% by mass. Mixtures C1-SRA and C1-L are composed of the same materials and proportions as C1 but include a shrinkage reducing admixture (dosage of 8.5 l/m^3), and a styrene-butadiene latex modifier (dosage of 12.1 l/m^3), respectively. Mixture C2 is composed of a concrete mixture without supplementary cementitious materials that also incorporates the styrene-butadiene latex modifier (dosage of 12.1 l/m^3). The w/cm ratio shown in Table 2 for mixtures C1-L and C2 includes water in the styrene-butadiene latex modifier which is not reported as batching water.

The results of ASTM C1581 testing are shown in Fig. 4 and Table 4. The results clearly indicate the effect of concrete materials and proportions have on the development of tensile strains and subsequent cracking in some of the concrete mixtures shown in Table 2. All test specimens of mixture C1 cracked before 28 days of drying

Table 3 Concrete mixture materials and proportions for restrained shrinkage

Material	S.G	C1	C1-SRA	C1-L	C2
Cement—ASTM C150 type I/II, kg/m^3	3.15	301	301	301	391
Fly ash—ASTM C618 class F, kg/m^3	2.46	81	81	81	–
Silica fume—ASTM C1240, kg/m^3	2.20	15	15	15	–
Coarse aggregate—ASTM C33 12 mm, kg/m^3	2.71	977	977	977	697
Fine aggregate—ASTM C33 5 mm, kg/m^3	2.63	764	764	764	1006
Water, kg/m^3	1.00	147	147	53[a]	67[a]
SCM replacement		24	24	24	0
Total cementitious content, kg/m^3		397	397	397	391
w/cm ratio		0.37	0.37	0.30	0.34
Air content, %		7.2	8.5	8.5	5.4
Paste content, %		35.2	35.2	35.2	36.3

[a]Amount of free batch water when accounting for water in styrene-butadiene latex modifier

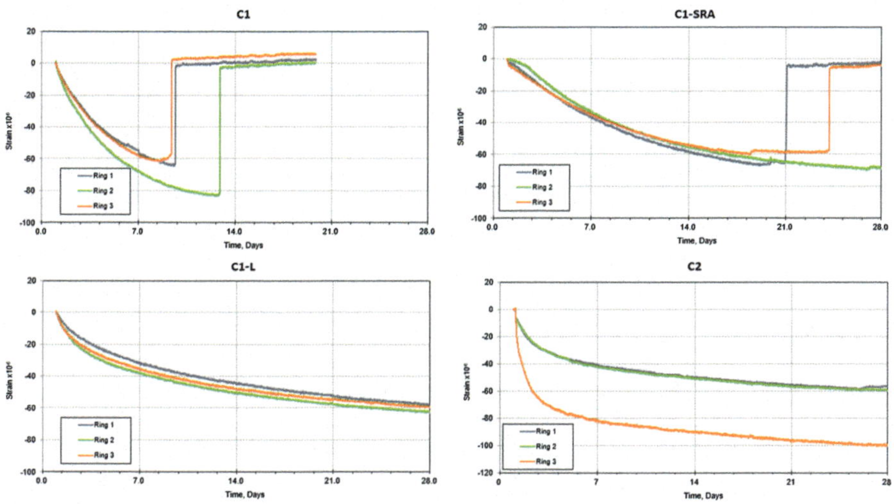

Fig. 4 ASTM C1581 test results of mixtures shown in Table 3

Table 4 Average time-to-cracking, initial strain, maximum strain, average strain rate, and stress rate of concrete in accordance with ASTM C1581

Mixture	Time to cracking, days	Initial strain, ε_o	Maximum strain, ε_{max}	Average strain rate factor, a_{avg}	Stress rate (q), MPa/day
C1	10.7	−2.6	−70	−25.2	0.27
C1-SRA	22	3.6	−67	−12.1	0.09
C1-L	>28	−2.0	−74	−6.7	0.03
C2	>28	−3.4	−73	−9	0.08

with an average net time-to-cracking, t_{cr}—average of the three test specimens—was 11 days. For C1-SRA, two test specimens cracked before 28 days (21 and 24 days, respectively) and the average net time-to-cracking was 26 days. With respect to C1-L and C2, all test specimens exceeded 28 days of drying without cracking.

The results presented in Fig. 4. Indicate that the restrained shrinkage performance can be significantly different even for mixtures with the same cementitious content, w/cm ratio, and paste content. The incorporation of SRA or latex modifiers contributes to a slower development of tensile strains in the concrete, except for one ring in mixture C2. The performance of this specific specimen (C2-Ring 3) was found to be influenced by a delay in the demolding process and unadvertent effects during handling of the specimen during demolding and the application of the surface sealer.

Using the criteria shown in Table 1, the classification of the potential for cracking of the four overlay concrete mixture is as follows: Moderate-High (C1), Moderate-Low (C1-SRA), Low (C1-L), and Low (C2). Similar results have been reported by others [9–12] on the effect of SRA or styrene-butadiene latex modifier which improved the restrained shrinkage performance of concretes.

4 Recommendations for the Specification of Shrinkage Performance

Specifying the shrinkage performance of concrete should be tailored to the specific project requirements. For elements in which volumetric changes are not critical, specification of length change due to drying can be used to limit the cementitious content, w/cm ratio, or paste content of concrete. In most cases, a maximum length change of 500 microstrain (0.05%) is typical for most placements in which restraint is not significant or where the presence of cracking does not impose a detriment on the performance of hardened concrete. Concrete elements subject to substantial restraint, or in which cracking would not be tolerated, i.e., water-containment structures, specification of restrained shrinkage performance—rather than length change due to drying—is paramount. In such cases, it is recommended that concrete mixtures are proportioned to resist at least 14 days without cracking when evaluated in accordance with ASTM C1581. Furthermore, it should be clarified that special projects may require different performance requirements than those presented above. The recommended performance criteria presented herein should be considered as typical and would not be applicable to all cases, and are intended for the use of individuals who are competent to evaluate the significance and limitations of them.

5 Summary and Conclusions

Specifying, selecting, and evaluating concrete for optimal shrinkage performance is challenging. In modern construction practices, the evaluation of length change due to drying can be insufficient when the potential for cracking of concrete can result in shrinkage cracked that cannot be tolerated. Experimental results presented in this paper exemplify the influence of material selection, concrete mixture proportioning on of linear shrinkage (ASTM C157) and restrained shrinkage (AST C1581) laboratory testing. The performance of two distinct concrete mixtures with the same cementitious content, w/cm ratio, paste content by volume showed significant differences in length change due to drying when different durations of moist curing were conducted. Mixtures with a substantial SCM replacement required longer durations of moist curing for achieving lower length change microstrain. With respect to restrained shrinkage, one control mixture was compared to three others incorporating SRA and a styrene-butadiene latex modifier. Results show that the incorporation of SRA or a latex modifier can significantly improve the restrained shrinkage performance of concretes, even when the cementitious content, w/cm ratio and paste content by volume remain comparable. Finally, recommendations for length change and restrained shrinkage performance are provided for typical performance requirements for construction.

References

1. ASTM C157/C157M-17: Standard Test Method for Length Change of Hardened Hydraulic-Cement Mortar and Concrete. ASTM International, West Conshohocken, PA (2017)
2. ASTM C1581/C1581M-18a: Standard Test Method for Determining Age at Cracking and Induced Tensile Stress Characteristics of Mortar and Concrete Under Restrained Shrinkage. ASTM International, West Conshohocken, PA (2018)
3. ACI 364.3R-09: Guide for Cementitious Repair Material Data Sheet. American Concrete Institute, Farmington Hills, MI (2009)
4. ASTM C94/C94M-20: Standard Specification for Ready-Mixed Concrete. ASTM International, West Conshohocken, PA (2020)
5. ASTM C192/C192M-19: Standard Practice for Making and Curing Concrete Test Specimens in the Laboratory. ASTM International, West Conshohocken, PA (2019)
6. Kosmatka, S.H., Wilson, M.L.: Design and Control of Concrete Mixtures. Portland Cement Association, Skokie, IL (2016)
7. Baroghel-Bouny, V., Aïtcin, P.C.: International RILEM Workshop on Shrinkage of Concrete: International RILEM Workshop on Shrinkage of Concrete, Shrinkage 2000, Paris, France, 16–17 Oct 2000
8. Clear, K.C., Chollar, B.H.: Styrene-butadiene latex modifiers for bridge deck overlay concrete. FHWA-RD-78-35, Department of Transportation, Washington DC, p. 117 (1978)
9. Kovler, K., Bentur, A.: Cracking sensitivity of normal-and high-strength concretes. ACI Mater. J. **106**(6), 537 (2009)
10. Weiss, J., Lura, P., Rajabipour, F., Sant, G.: Performance of shrinkage-reducing admixtures at different humidities and at early ages. ACI Mater. J. **105**(5), 478 (2008)

11. Chaunsali, P., Lim, S., Mondal, P., Foutch, D.A., Richardson, D., Tung, Y., Hindi, R.: Bridge decks: mitigation of cracking and increased durability. Illinois Center for Transportation (ICT) (2013)
12. Raoufi, K.: Restrained shrinkage cracking of concrete: the influence of damage localization. Purdue University (2011)

Experimentation and On-Site Monitoring

Crack Initiation and Propagation in Fiber-Glass Reinforced Mortars

Pascale Saba, Tulio Honorio, Omar-Ateeq Mahmood, and Farid Benboudjema

Abstract Reinforced cement-based rendering mortars are used as the protection layer in External Thermal Insulation Composite Systems (ETICS). Thermal insulations, when used in renovation, have a big impact on the reduction of CO_2 emissions. The interactions with the environment changing temperature and relative humidity lead to thermal and hygral strains, which when restrained, may lead to stresses that can attain the tensile strength of the material causing then the mortar cracking. The eventual penetration of water inside the cracks may cause the insulator to lose partially its efficiency and durability. Here, we focus on the cracking development in the reinforced mortar layer using experimental techniques. To understand the crack initiation and propagation in the reinforced mortar layer, and the role of the fiber-glass mesh as reinforcement inside the mortar, a new mechanical setup is developed. This setup is designed to perform 3-point bending tests using in-situ X-ray tomography. The latter allows observing the cracks inside the mortar sample shedding lights on the reinforcement mechanisms of the fiber-glass mesh and its impact on the initiation and the propagation of the cracks. The role of the mortar heterogeneities is also analyzed and information about cracks characteristics such as openings and lengths may be extracted.

Keywords Mortar · Cracking · 3-points bending · X-ray tomography · Fiber-glass

P. Saba (✉) · T. Honorio · F. Benboudjema
LMT - Laboratoire de Mécanique et Technologie, Université Paris-Saclay, ENS Paris-Saclay, CNRS, 91190 Gif-sur-Yvette, France
e-mail: pascale.saba@ens-paris-saclay.fr

P. Saba · O.-A. Mahmood
Département Thermique, Mécanique et Modélisation, Saint-Gobain Recherche Paris, Quai Lucien Lefranc, 93300 Aubervilliers, France

F. Kanavaris et al. (eds.), *International RILEM Conference on Early-Age and Long-Term Cracking in RC Structures*, RILEM Bookseries 31,
https://doi.org/10.1007/978-3-030-72921-9_6

1 Introduction

Final energy use in buildings led to direct emissions of 10 Gigatons of CO_2 in 2019 worldwide [1]. Buildings were, therefore, responsible for 28% of global energy emissions. In France, home and building heating systems are responsible for 45% of total French energy consumption and about 25% of Greenhouse gas emissions [2]. The objective of the French government is to attain carbon neutrality by 2050 and to collectively set as an ambition the renovation of 500,000 housing units per year, half of which are occupied by low-income households [2]. Covid-19 Crisis pushed the government to further investigate in the field of energetic renovation with "France Relance" project [3] that insists on the ecological and energetic transition as the main part of the economic growth.

Two types of thermal insulation can be used in renovation projects: (i) External Thermal Insulation Composite Systems (ETICS) and (ii) Internal Thermal Insulation Systems. The popularity of the ETICS technology grew due to its advantages regarding the internal counterpart. ETICS guarantee the reduction of the thermal bridges and greater thermal comfort due to the higher interior thermal inertia, providing a finished appearance similar to the traditional rendering. From the construction point of view, ETICS allow thinner exterior walls and increase the facades' durability. Three very relevant aspects in the construction industry must be added to the advantages above: low cost, ease of application, and the possibility to be installed without disturbing the building's dwellers, which is particularly important in refurbishment [4]. As shown in Fig. 1, the protection layer of the insulator, expanded polystyrene, in ETICS is a base coat, which is a rendering mortar reinforced with fiberglass mesh.

Like all cementitious materials, rendering mortars are subject to cracking. The interactions with the environment changing temperature and relative humidity lead to thermal and hygral strains, which when restrained, may lead to stresses that can attain the tensile strength of the material causing then the mortar cracking as in Fig. 2. This is problematic for two main reasons. On one hand, the eventual penetration of water

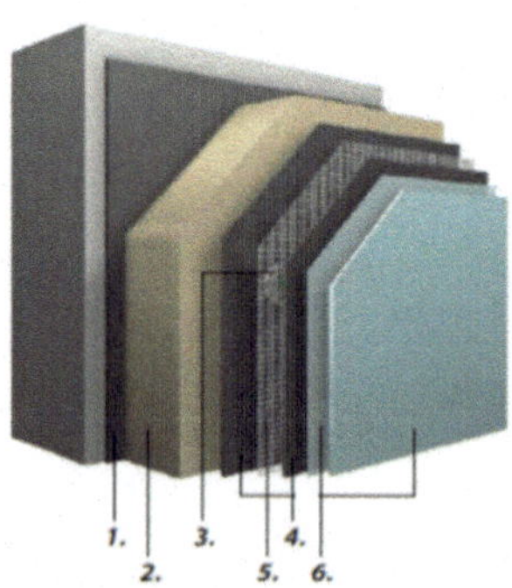

1. Adhesive
2. Thermal insulation material
3. Anchors (not mandatory)
4. Base-coat
5. Reinforcement, usually glass fiber mesh
6. Finishing layer: finishing coat with a key coat (optional) and/or decorative coat (optional)

Fig. 1 Typical system components in ETICS using anchors [5]

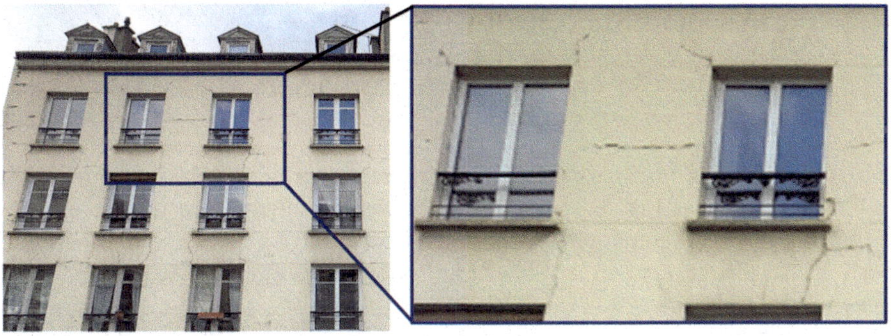

Fig. 2 Visible cracking on a facade

inside the cracks may cause the insulator to lose partially its efficiency and durability. On the other hand, the visible cracking on the newly renovated façades leads to unsatisfied clients which is an important marketing problem. Oriented and non-oriented cracking and visibility of joints are among the main anomalies encountered in ETICS [6–8]. The glass fiber mesh fabric is proposed as a solution for cracking in several types of applications including the base coat reinforcement for ETICS.

The reinforced cement composite used in ETICS is composed of (i) the mortar, which is a heterogeneous cement-based material containing three main phases at its mesoscale: cement paste matrix, sand particles, and porosity, and (ii) the fiber-glass mesh which is also a composite heterogeneous material containing three phases at its mesoscale: glass fibers, porosity, and coating. These heterogeneities present in the mortar, the porosity, the sand, or even the mesh can be the source of localization for cracks at the mesoscale. Nevertheless, cracking in the rendering mortar can also be initiated at the structure scale. Geometrical singularities, such as window corners and thermal joints, can be a source of localization of cracks.

Here, we focus on the mechanical characterization of the reinforced rendering mortar. The aim is to determine the main source of crack localization and crack propagation in the mortar in 3D, using X-ray tomography in an original setup. The tomography scans enable unveiling the role of heterogeneities with a characteristic size on the order of a hundred micrometers—namely, (macro)porosity, aggregates, and fiber-glass mesh—on crack localization and crack propagation in reinforced mortar. To the author's knowledge, it has never been done previously on this type of material.

2 Materials and Experimental Methods

2.1 Materials

In this study, 5 mm thickness prismatic reinforced mortar samples of 70 mm long and 12 mm large (see Fig. 3a) are tested in bending inside an X-ray tomograph. The mortar used in our samples is a commercial basecoat for ETICS. The fiber-glass mesh is also a commercial glass fiber mesh fabric (see Fig. 3b) that is specifically preconized for ETICS.

Three non-reinforced mortar samples ($4 \times 4 \times 16$ cm^3) were tested by 3-point bending at 28 days to determine its mechanical properties such as the Young modulus and the tensile strength. The experimental setup and procedure are described in [9]. The mortar has a Young modulus of 2.47 $\pm$ 0.34 GPa, and a tensile strength of 1.97 $\pm$ 0.04 MPa (mean value $\pm$ standard deviation for both parameters).

Mixing protocol for the mortar. The manufacturer of the basecoat recommends 6 L of water for 25 kg of mortar. He preconizes mixing the water and the premix using an electric hand mixer at low speed (500 rpm) for 3 min, then letting the mixture rests for 5–10 min before application.

In this study, some modifications have been made. The water/premix ratio is kept, but not the mixing protocol. This decision is justified by the non-repeatability of the procedure if using a hand mixer since the movement of the human hand or body can never be the same. To limit the variability related to human intervention, a stand electric mortar mixer (IBERMIX) is used.

Since the mixing speed of the stand mixer is limited, we decided to select the mixing protocol detailed in Table 1, after having studied the effects of mixing protocol on mortar's properties.

Curing conditions. After casting, the samples are protected from drying for 24 h until demolding using several layers of plastic film and a final layer of adhesive aluminum. After demolding, the samples are stored in a temperature and relative humidity-controlled room at 20 °C $\pm$ 1 °C and 60% $\pm$ 5% respectively. Weight loss of the samples is monitored.

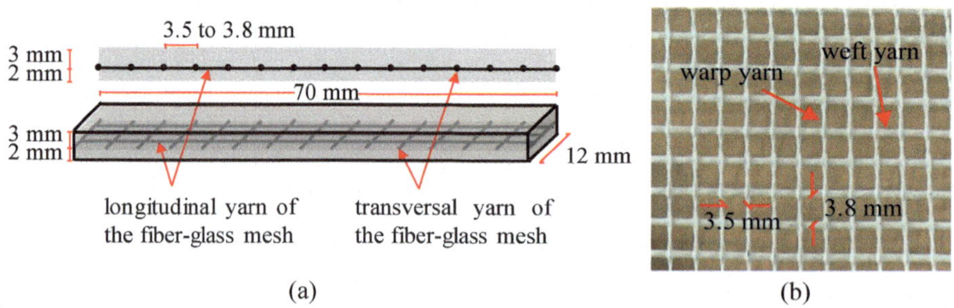

Fig. 3 **a** 2D and 3D representation of the fiber glass mesh reinforced mortar sample. **b** Glass-fiber mesh fabric structure

Table 1 Mixing protocol

Step	Speed [rpm]	Duration [s]
Mixing the premix before adding water	140	60
Mixing while progressively adding water	140	60
Mixing at high speed	300	90
Mixing with a trowel to scrape the sides	NA	60
Mixing at high speed	300	60
Leaving the mixture to "rest"	NA	5–10 min

2.2 *Experimental Methods*

To understand the crack initiation and propagation in the reinforced rendering mortar and the reinforcement mechanisms of the fiber-glass mesh, several specimens are tested using in-situ X-ray tomography. Specimens are loaded in 3-point and 4-point bending.

The rendering mortar is subject to drying shrinkage or hygral swelling in case of a decrease or increase of the relative humidity of the surrounding air, respectively. Only the external surface of the mortar will exchange humidity with the surrounding air. Therefore, only the external surface of the mortar will shrink or swell causing a "curling" effect [10] of the mortar as shown in Fig. 4. The "curling" effect puts the mortar under bending load. Similar resulting loads can be obtained with restrained thermal strains in case of an increase or decrease of the surrounding external temperatures caused by a difference in the thermal expansion coefficient between the mortar and the expanded polystyrene. For these reasons, we opted for bending tests as a primarily loading on the reinforced rendering mortar samples.

As mentioned in Sect. 1, cracks can be initiated on two different levels:

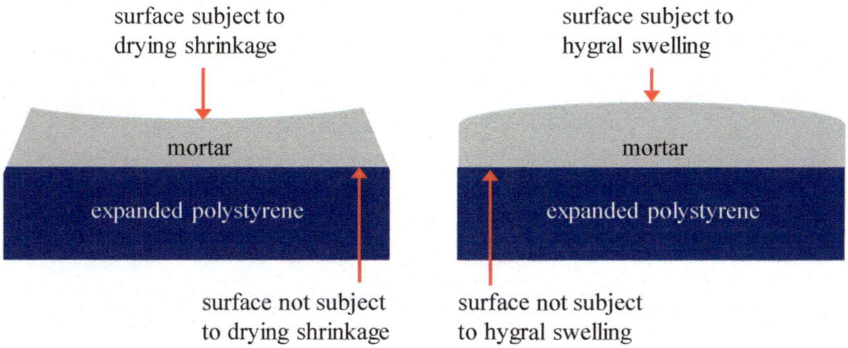

Fig. 4 "Curling" effect in the rendering mortar in ETICS

i. On the mesoscale, where the heterogeneities of the mortar (porosity, sand particles, and the mesh) are the only source of stress concentration.
ii. On the structural scale, where the geometrical singularities such as window corners and thermal joints take the lead of the stress concentration and crack initiation.

And so, 4-point bending tests, on one hand, are representative of the cracking that takes place on facades without geometrical singularities where the heterogeneities of the mortar are the main crack initiators. On the other hand, 3-point bending tests are representative of the cracking that occurs on the geometrical singularities where the stress is localized.

In-situ X-ray tomography, for instance, is the tool that will permit the visualization of the crack in the volume of the specimen. A 3D crack monitoring is possible at different times of the tests. It is also possible to extract information as crack openings, crack length, etc.

X-ray tomography. X-ray tomography is a non-destructive technique that allows obtaining a three-dimensional image. It is an imaging technique that generates a data set, called tomogram, which is a three-dimensional representation of the structure. Each three-dimensional point in the tomogram is called a voxel.

In a tomography experiment (see Fig. 5), 3D information is gathered by acquiring a series of 2D images while rotating the sample typically between 0 and 360°. Polychromatic X-rays from a micro-focus X-ray source are used to probe the specimen and an X-ray camera is used to record the X-ray transmission radiograph. To generate the tomogram, a series of radiographs are collected at different viewing angles by rotating the specimen. This set of radiographs, called projection data, are processed with a reconstruction algorithm to generate the tomogram of the specimen.

X-ray tomography is based on the variations in absorption coefficient along the path of the X-ray beam. The absorption coefficient is linked to the density and the atomic number of the different materials that the beam encounters as it passes through the sample. This is translated by Beer-Lambert law that links the transmitted beam intensity to material attenuation coefficient.

In case of monochromatic X-ray beams, the transmitted beam intensity is calculated as:

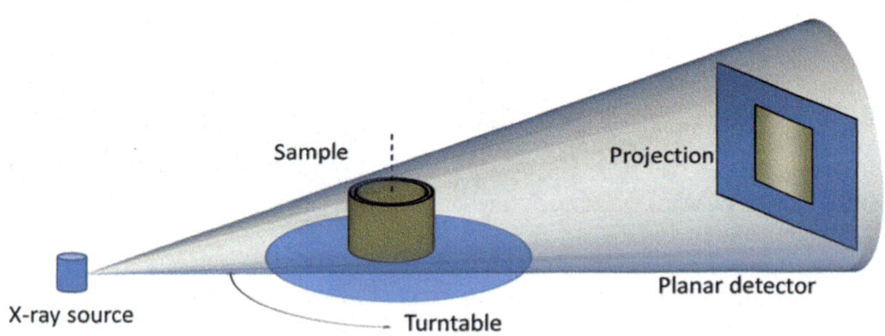

Fig. 5 Principal components of conventional X-ray tomography system (cone-beam example) [11]

$$I = I_0 \cdot exp\left[\sum_i (-\mu_i x_i)\right] \tag{1}$$

where I_0 and I are the X-ray intensities of the incident and transmitted beams, μ_i denotes the linear attenuation coefficient and x_i the linear extent of the *ith* material traversed by the beam. In case of polychromatic beams containing a large spectrum of energies, the beam flux is much higher and the exposure time is reduced. The transmitted beam intensity becomes:

$$I = \int I_0(E) \cdot exp\left[\sum_i (-\mu_i(E) x_i)\right] dE \tag{2}$$

where both I_0 and μ_i are related to the energy E of the incident beam.

Mechanical test assembly. In this contribution, we will be focusing on the 3-point bending setup. The latter is detailed in Fig. 6.

The specimen in this particular 3-point bending setup, unlike in traditional ones, is inclined. This change is what allowed us to obtain better quality images since the X-ray beam is traversing a smaller distance in the "scanned zone" shown in Fig. 6 compared to the total length of the specimen in case of a non-inclined bending setup. This can be related to the Eqs. (1) and (2) whereas the larger x_i, the smaller the transmitted beam intensity I.

Note that supports 2, 3, 4, and 5 of Fig. 6 are made of PMMA (*poly(methyl methacrylate)*) which has a density of about 1185 kg.m^{-3}. The density of the mortar being 1323 ± 5 kg.m^{-3} (mean value ± standard deviation) making it denser than the PMMA. Thus, the PMMA is more X-ray transparent.

Plate 7 in Fig. 6 is directly fixed on a 100 N load cell that will put the plate in an upward movement, hence the load is being transmitted to the specimen by the lower supports.

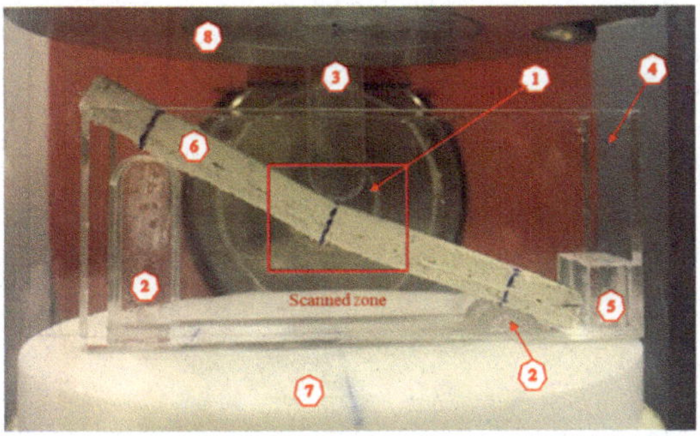

1. X-ray source
2. Lower supports
3. Upper support
4. Side supports
5. Longitudinal support
6. Specimen
7. Plate in an upward movement
8. Fixed plate

Fig. 6 3-point bending mechanical assembly

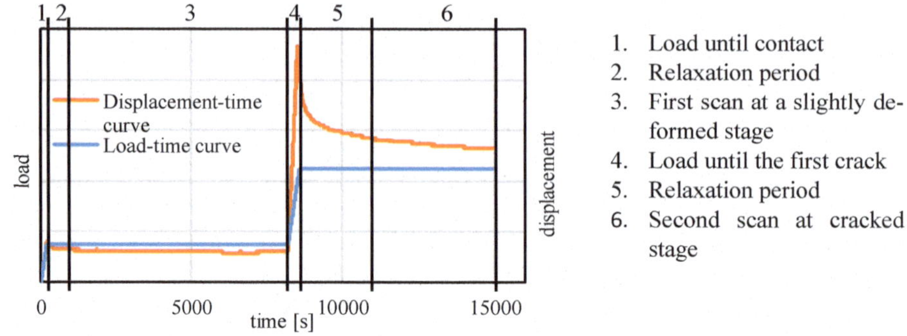

Fig. 7 The schematic course of the test described with the load vs. time and displacement versus time curves

The test is controlled by the displacement of the plate 7 at a rate of 1 $\mu m.s^{-1}$. The advantage of using a displacement controlled test compared to a load controlled one is that the former allows getting the post-peak curve and prevents creep strains during the scans that will complicate the analysis of displacement fields related to cracking. Two scans are done to compare the cracked stage of the sample to its slightly deformed elastic stage. Before each scan, a pause period is preferred to allow the material to relax. This relaxation period prevents differential strains (for instance between viscous cement paste and elastic aggregates) that will occur during the scan hence limiting the blurred image. The in-situ inclined 3-point bending test is divided into 6 phases that are detailed in Fig. 7.

3 Experimental Results and Discussions

3.1 Experimental Results

Thanks to the advanced technology of the X-ray in-situ tests, reconstruction algorithms, and image processing algorithms, we had access to the 3D volume of the cracked sample. Following the scheme in Fig. 8, sections from the 3D reconstructed volume of the crack samples are displayed in Fig. 9.

In a 3-point bending test, considering a homogeneous sample, the crack initiation is theoretically known to be exactly under the middle-upper support where the moment is maximum. The crack will propagate, from the bottom surface in tension

Fig. 8 Representation of a vertical section in the sample

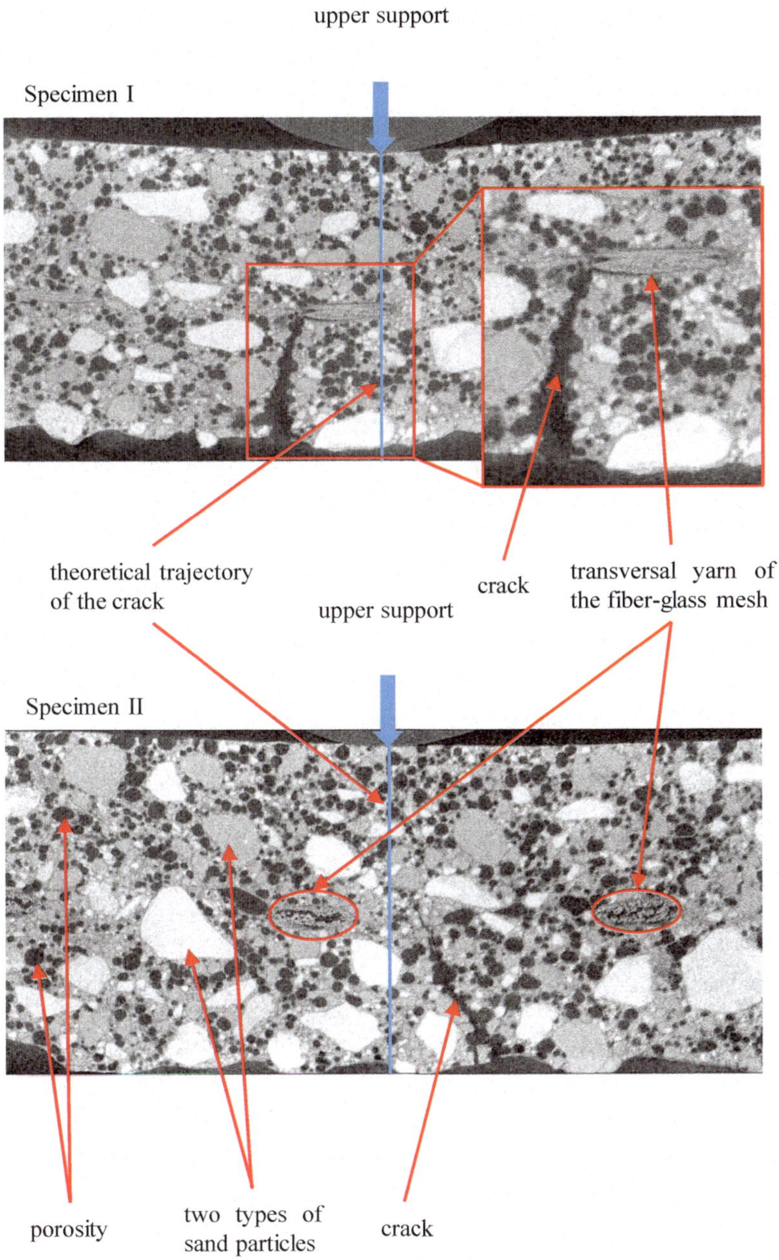

Fig. 9 Vertical sections inside the reconstructed 3D volumes of cracked specimens I and II

toward the upper surface, vertically due to the geometrical and mechanical symmetry of the classical 3-point bending set-up.

This theoretical cracking pattern is not obtained in our 3-point bending tests. The crack trajectory is neither located under the middle-upper support nor vertical. It is slightly shifted from the center and inclined as shown in Fig. 9.

3.2 Study of the Cracking

The theoretical cracking trajectory described above does not always apply to heterogeneous materials, since the heterogeneities present in the material may induce geometrical singularities, weak points, or, material incompatibilities that can relocate the crack initiation and redirect its propagation. In the reinforced fiber-glass mortar samples, three main categories of heterogeneities exist at the millimeter scale: (macro)porosity, sand particles (two types of sand can be differentiated), and, the fiber-glass mesh. Each of these heterogeneities can be a source of stress concentration.

Considering the heterogeneities as inclusions in the tensile zone, stress concentration, as well as crack trajectory, will be as shown in Fig. 10. Therefore, several phenomena are to be considered to determine the crack initiation. The crack in the mortar is initiated where the local stress attains the mortar's tensile strength, hence anywhere in the tensile zone and not only on the lower surface of the sample as in homogeneous materials.

Considering the quasi-brittle behavior of the mortar, the propagation of the crack once initiated is rapid. Since the X-ray 3D requires a steady-state of the scanned object hence cannot be carried out during the mechanical test but afterward, the exact location of the crack initiation cannot be determined. The vertical sections shown in Fig. 9 show the critical influence of the porosity on the trajectory of the crack: it is clear how the crack preferentially passes through several porosities on its way up. The fiber-glass fabric plays an important role as well.

The cracks in both samples of Fig. 9 did not cross the fiber-glass mesh. The latter stops the crack propagation and thus limits the crack opening. This first reinforcement mechanism directly influences the durability of the ETICS. As discussed in the introduction, the potential water leakage is limited by the smaller crack opening. In the reinforced mortar, emerging cracks should not easily attain the thermal insulator since they do not propagate through the entire thickness of the mortar.

Further, the analysis of specimen I of Fig. 9 suggests a potential influence of the transversal yarn on the localization of the crack. Nevertheless, it is difficult to presume, based on the 3-point bending test, that the transversal yarn will locate the cracking in the mortar because of the maximal bending moment in the middle of the sample. This potential mechanism that will be further investigated using 4-point bending and/or tensile tests might have a relevant impact on the cracking pattern. It is thus possible, using different dimensions of the fiber-glass mesh, to directly impact the number of the cracks as well as their openings to make them less visible

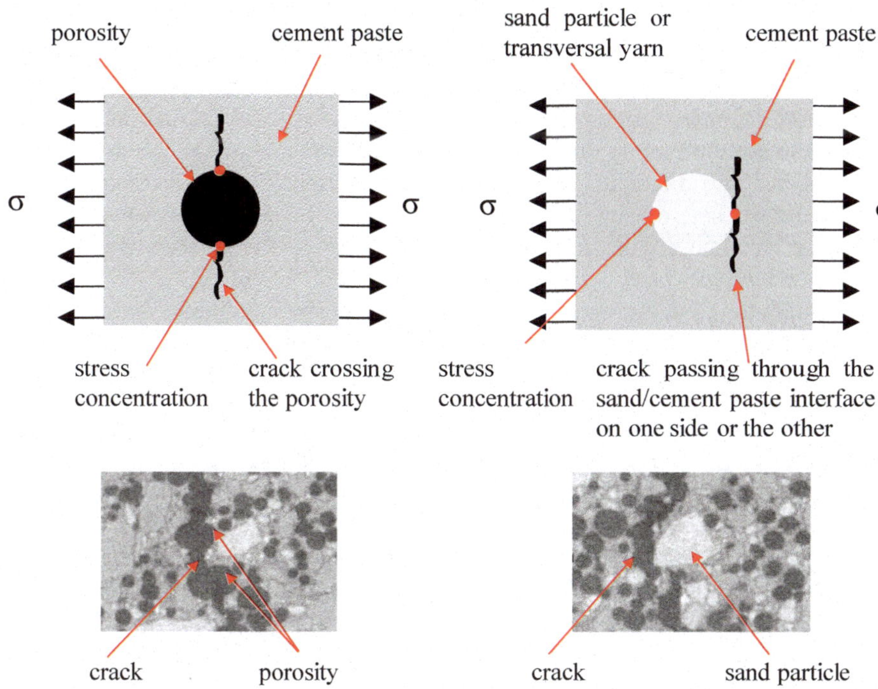

Fig. 10 Representation of the stress concentration caused by the heterogeneities in the mortar and thus the potential crack initiation and/or propagation

and more watertight. Other works investigated the influence of the fiber-glass mesh dimensions under different conditions on the mechanical behavior of the reinforced rendering mortar [12], only 2D surface observations were reported.

4 Conclusions

We studied cracking development in the reinforced mortar relevant for ETICS application using X-ray tomography. The classical mechanical testing methods even when combined with advanced technologies such as Digital Image Correlation (DIC) can only provide information on the chosen 2D surfaces. Therefore, to access the 3D information and dive into the depth of the 3D volume of the sample, a new testing setup is demanded. The newly developed X-ray in-situ inclined 3-point bending setup needed to ensure relatively small scale testing in order to obtain a sufficiently small image resolution (10 μm), X-ray transparency of the different parts of the set-up as well as geometrical adjustments in order to limit the noises and obtain good quality exploitable images.

The 3D images of the samples revealed a material crowded with different types of heterogeneities. Each of which can have an impact on the cracking pattern. Thus, the theoretical cracking pattern in a 3-point bending test is changed. The crack that should have appeared vertically in the middle of samples is shifted and inclined. This observation is mainly due to the competition between several phenomena; the maximal bending moment in the middle of the sample, the stress concentration induced by the heterogeneities in the mortar, the presence of the transversal yarn, and the real thickness of the mortar beneath the fiber-glass fabric caused by an uneven and coarse lower surface.

Although the initiation point of the crack could not be identified, two main observations have been made. First, the porosities and the sand part particles will dictate the trajectory of the crack. The crack will pass through the porosities and encounter the sand particles on its way up towards the fiber-glass mesh. Second, the fiber-glass mesh will stop the propagation of the crack. The crack will not propagate through the entire thickness of the mortar, thus limiting its opening. Therefore, the fiber-glass mesh plays an important role in the durability of the ETICS by reducing the potential water leakage within the thermal insulator.

Note that, using the 3D images, we can have access to the width of the crack opening, not only on the surface but through the length of the crack using Digital Volume Correlation (DVC).

As stated above, the transversal yarn of the grid, whether it is a warp or a weft yarn, might influence the localization of the crack. This assumption is to be further investigated along with the influence of other heterogeneities. X-ray in-situ 4-point and/or tensile tests will be carried out. The advantages of 4-point bending compared to 3-point, is the constant bending moment between the upper supports, thus limiting the competition between the transversal yarn and the maximal bending moment in the 3-point bending.

On the other hand, in order to identify the starting point of the crack and to ensure whether it initiates on the tensile surface or by one of the heterogeneities especially the transversal yarn, 2D X-ray radiography along with the 4-point bending set-up might be of great use.

In future studies, experiments will be carried out to understand the cracking behavior and the reinforcement mechanisms under the real environmental loading. Temperature and relative humidity fatigue cycles can be applied to a defined structure of ETICS. This environmental loading supported by an imaging technique can reveal the behavior of the fiber-glass reinforced mortar in "real-life" scenarios.

References

1. iea, Tracking Building. https://www.iea.org/reports/tracking-buildings-2020 (2020). Last accessed 2 Sept 2020
2. Ministry of ecological solidarity transition, Energy renovation of buildings: A plan to accelerate general mobilization. https://www.ecologique-solidaire.gouv.fr/renovation-energetique-

des-batiments-plan-accelerer-mobilisation-generale. Last accessed 2 Sept 2020 (in French)
3. Government, Press file: France relaunch. https://www.gouvernement.fr/sites/default/files/document/document/2020/09/dossier_de_presse_france_relance_-_03.09.2020.pdf. Last accessed 1 Sept 2020 (in French)
4. Barreira, E., de Freitas, V.P.: External Thermal Insulation Composite Systems (ETICS): An Evaluation of Hydrothermal Behavior. Springer (2015)
5. Eurorean Association for External Thermal Insulation Composite Systems, About ETICS. https://www.ea-etics.eu/etics/about-etics/. Last accessed 2 Sept 2020
6. Amaro, B., Saraiva, D., de Brito, J., Flores-Colen, I.: Inspection and diagnosis system of ETICS on walls. Constr. Build. Mater. **47**, 1257–1267 (2013)
7. Nilica, R., Harmuth, H.: Mechanical and fracture mechanical characterization of building materials used for external thermal insulation composite systems. Cem. Concr. Res. **35**(8), 1641–1645 (2005)
8. Pareira, C., de Brito, J., Silvestre, J.D.: Contribution of humidity to the degradation of façade claddings in current buildings. Eng. Fail. Anal. **90**, 103–115 (2018)
9. Chan, N.: Effects of initial stresses and strains in the prediction of cracking and failure of concrete structures by a discrete element approach, Ph.D., Paris-Saclay University, to be submitted (2021) (in French)
10. Chilwesa, M., Facconi, L., Minelli, F., Reggia, A., Plizzari, G.: Shrinkage induced edge curling and debonding in slab elements reinforced with bonded overlays: influence of fibers and SRA. Cement Concr. Compos. **102**, 105–115 (2019)
11. Chen, Y.: Damage mechanisms of SiC/SiC composite tubes: three-dimensional analysis coupled with tomographic imaging and numerical simulations, Ph.D., Paris-Est University (2017) (in French)
12. Colombo, I.G., Magri, A., Zani, G., Colombo, M., di Prisco, M.: Textile reinforced concrete: experimental investigation on design parameters. Mater. Struct. **46**, 1953–1971 (2013)

Distributed Optical Fibre Sensors for Strain and Temperature Monitoring of Early-Age Concrete: Laboratory and In-situ Examples

Rafał Sieńko, Łukasz Bednarski, and Tomasz Howiacki

Abstract Distributed fibre optic sensors (DFOS) provide new possibilities in structural technical condition assessment in comparison with traditional spot measurements. It is possible to analyze strains and temperature changes continuously over structural member length with spatial resolution starting from as fine as 5 mm. Thanks to the appropriate sensor construction and its installation before concreting it is possible to analyze material behaviour starting from its early stage, when thermal-shrinkage strains appear. This phenomenon depends on many factors, such as the type of concrete mix, dimensions of structural member, the way of concrete care, external conditions (temperature, humidity), formwork and constraints related to reinforcing bars or external friction and resulting with crack appearance. Stage of early-age concrete (hydration process) is thus very important for its final durability and performance. The article presents the very new measuring tools which allow for comprehensive analysis of concrete temperatures and strain state including all local nonlinearities (cracks). The attention was paid to hydration process, but sensors installed inside the structural members can be also effectively used during other phases such us: activation of prestressing tendons, construction stages or operation. Selected examples of laboratory tests as well as the unique in situ installations realized in Poland during last few years are presented and discussed hereafter. Except of traditional concrete, also other materials were examined, such as concrete made on lightweight sintered aggregate and fibre-reinforced concrete mixed with the ground.

Keywords Early-Age concrete · Optical fibres · Distributed measurements · DFOS · Strain · Temperature

R. Sieńko (✉)
Faculty of Civil Engineering, Cracow University of Technology, Krakow, Poland
e-mail: rsienko@pk.edu.pl

Ł. Bednarski
Department of Mechanics and Vibroacoustics, AGH University of Science and Technology, Krakow, Poland

T. Howiacki
SHM System, Jana Pawła, I82A Libertów, Poland

F. Kanavaris et al. (eds.), *International RILEM Conference on Early-Age and Long-Term Cracking in RC Structures*, RILEM Bookseries 31,
https://doi.org/10.1007/978-3-030-72921-9_7

1 Introduction

1.1 The Idea of Distributed Measurements

The main limitation of traditional extensometers and thermistors is the ability to carry out measurements only locally along defined measuring base (Fig. 1a). Sometimes some efforts are made to perform measurements along a given line by installing several sensors within this line (Fig. 1b), but this approach is expensive and thus occasionally used. Distributed optical fibre technology is based on light scattering and allows for strain or temperature measurements to be made with a spatial resolution starting from as fine as 5 mm along the length of the optical fibre [9]. From an engineering point of view such measurements can be considered as continuous in a geometric sense (Fig. 1c).

A number of studies concerning DFOS (distributed fibre optic sensing) have been conducted under laboratory and in situ conditions over the last few years. The attempts were made to find the best ways to embed fragile glass measuring fibres into a concrete [3, 4] detect and estimate crack width within concrete structural members [2, 16, 17] and analyse their strain and temperature distributions [6, 11] using different optical phenomena such us Brillouin [8] or Rayleigh scattering [1].

In all examples presented further optical backscatter reflectometer OBR 4600 manufactured by Luna Innovations (based on Rayleigh scattering) was applied for distributed measurements. The selected technical parameters of this device are summarised in Table 1.

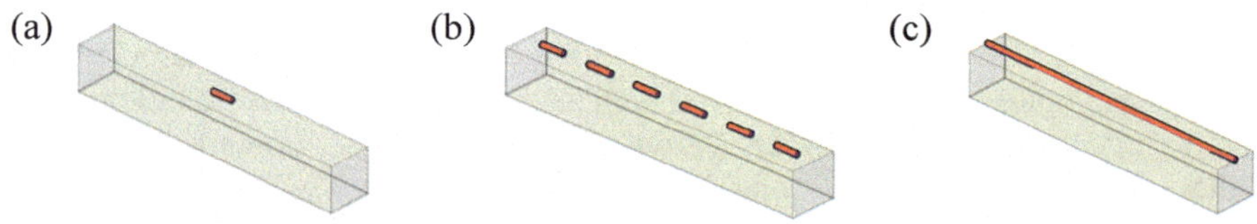

Fig. 1 The idea of measurements: **a** spot, **b** quasi-continuous, **c** distributed (geometrically continuous)

Table 1 Selected parameters of reflectometer used for distributed measurements

Parameter	Value	Unit	
Measuring range (standard mode)	70	m	
Optimal spatial resolution	10	mm	
Temperature resolution	±0,1	°C	
Strain resolution	±1,0	με	

1.2 Sensors Construction

Glass optical fibres used in the telecommunication field should be as pure as possible i.e. free from micro-imperfections that cause light wave energy losses over the length. The Rayleigh scattering phenomenon is one of the reasons and it occurs in every fibre cross section due to the particle structure of matter (heterogeneity of refractive index at micro-scale). However, this fact is favourable in terms of strain measurements. The light reflected from the imperfection of the glass structure moves backward relative to the original direction of motion. Scattering amplitude is a random but constant property for a given fibre and can by calibrated for mechanical or thermal strains by advanced reflectometers.

One of the most challenging tasks while constructing sensors for embedding into concrete is to provide appropriate strain transfer from surrounding medium (concrete) to the glass measuring core of the optical fibre. This phenomenon is influenced by the slip at the surface between the fibre external jacket and the concrete, as well as between the individual coatings and the optical glass core itself. Finite element (FE) analyses of this phenomenon are currently being conducted by a range of specialists in the field of structural mechanics and photonics together with the authors of this article.

One of the approaches is to use fibres with minimalized number of coatings (only with one primary coating applied directly in the production process of the fibre—see Fig. 2a), what reduce the risk of the slippage and thus the risk of disturbances. The limit strains for such standard SM telecommunication fibre could be more than $\pm 5\%$, so they are able to measure even very wide cracks. What is more, in this case the presence of the sensor inside the member do not affect its behaviour at all (the fibre with negligible diameter of 250 μm and thus with negligible stiffness). However due to the fragility and the lack of resistance to shearing forces, this solution can be used only in laboratory conditions, where installation process could be performed with high precaution and precision.

Another option is to apply specially designed fibres with different coatings, which improve the adhesion between the concrete. Example of such fibre with coarse-grained silica coating is shown in Fig. 2b. Disadvantage of this solution is that the high stiffness coating and its unregular geometry reduces the measuring range of the sensor (which is then less than $\pm 1\%$).

Fig. 2 Example of DFOS measuring tools: **a** standard SM 9/125 fibre with acrylic coating, **b** with experimental coarse-grained silica coating, **c** integrated inside composite core (EpsilonSensor—[5])

The most advanced solution dedicated for laboratory and in situ installations is to integrate the glass optical fibre with composite core characterized by appropriate parameters (elastic modulus E, limit strains ε_{lim}, strength f and others). These parameters could be determined in two ways: making from a sensor the reinforcement ($E_{sensor} \gg E_{concrete}$) or not influencing the medium behaviour at all ($E_{sensor} \leq E_{concrete}$). Example of such dedicated solution is presented in Fig. 2c (EpsilonSensor [5]) with nominal diameter of 3 mm and elasticity modulus of 3 GPa.

The optical fibres are sensitive both to mechanical and thermal strains. For measuring temperature changes only, it is necessary to isolate the fibre from mechanical strains, e.g. by placing it inside the tube or use DFOS technique insensitive to mechanical strains (like Raman scattering). However, there are many important details which should be taken into consideration. That is why it is recommended to consult the choice of strain sensors and possible ways of compensation with experienced specialists, as these issues are crucial for quality of measurements and data interpretation.

1.3 Early-Age Concrete

Structural health monitoring (SHM) of the structures made of concrete is especially difficult task, because this material is very heterogeneous and subjected to the rheological phenomena over time. What is more, concrete members usually work within cracked state [12] which has important impact on corrosion development and the final durability. This process starts from hydration when thermal-shrinkage strains appear [18]. This phenomenon depends on many factors [19, 20], such as the type of concrete mix, dimensions of structural member, the way of concrete care, external conditions (temperature, humidity), formwork and constraints related to reinforcing bars or external friction and resulting with crack appearance. Stage of early-age concrete is thus very complex and important for its final durability and performance so works are ongoing to develop effective measuring tools able to provide reliable information about this phenomenon which is not yet sufficiently investigated and described in the literature.

2 Laboratory Tests

2.1 Lightweight Concrete Specimens

The specimens under consideration were made from concrete on lightweight sintered aggregate [7] in the Laboratory of Building Material and Structures at Krakow University of Technology. Some of them were prestressed after achieving the appropriate concrete strength to observe the influence of intensified creep. Optical fibres with silica-grained coating were installed inside the formwork by specially designed mounting frames (Fig. 3a), which were removed after concreting leaving the measuring fibres inside the specimens. This allowed for appropriate positioning the initially pretensioned fibres longitudinally near the lower and upper surface of the specimen. The measurements of early-age concrete strains were performed according to the planned schedule and are currently being processed.

2.2 Fibre-Reinforced Concrete Mixed with the Ground

Similar mounting solution of the DFOS sensors (composite rods integrated with measuring fibre) was used to create the specimen made from the fibre-reinforced concrete mixed with the ground (Fig. 3b). Early-age material strains as well as mechanical strains induced in three-point bending tests (performed in the Laboratory of Building Material and Structures at Krakow University of Technology) were measured continuously in geometrical sense and then used for designing the real structural element (slurry wall, see also chapter 3.2). Some exemplary results of thermal-shrinkage strains are presented on spatial visualization (Fig. 4a) in the time and specimen length domain.

2.3 Concrete 'Tomograph'

One of the most challenging tasks was to create the concrete 'tomograph' (Fig. 3c) enable to measure strains and temperature distributions in all directions and in many sections of the concrete massive specimen. It was formed as a cube with 36 cm side insulated with extruded polystyrene. One of the challenges was to not break the pure optical fibre only in its primary coating during concreting, so self-consolidating concrete was applied. This process finally was succeed providing thousands of measuring data during first few weeks of concrete hardening without any influence of embedded sensors on the concrete behaviour.

Four optical fibres were used: three for measuring strains along three orthogonal directions XYZ and one for measuring the changes in temperature. Fibres were traced in such a way that 5 measuring cross sections, each with 5 measuring traces, were created in each direction, what means that 75 optical paths with total length of 27 000 mm was installed within the cube. Assuming the spatial resolution of measurements equal to 10 mm, this solution replaces 2700 traditional spot strain sensors, which installation would not be obviously possible in such conditions.

This type of experiments and results was not yet described in the literature. All gathered data are currently being processed in detailed. Exemplary plot showing the possibilities of geometrically distributed measurements is presented in Fig. 4b for the fibre located on X direction over its all 25 measuring traces.

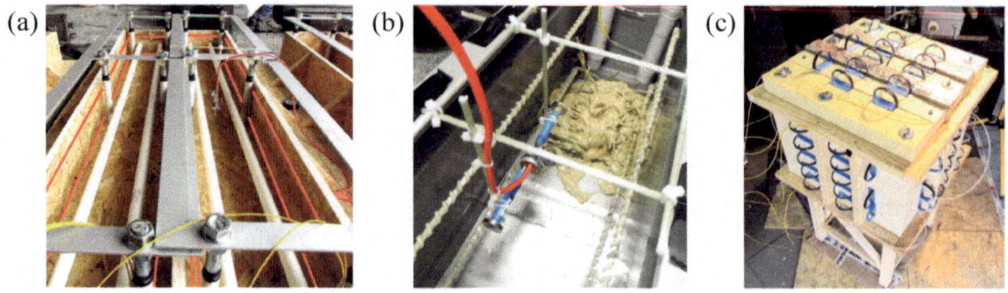

Fig. 3 Examples of laboratory specimens, where early-concrete strains were measured: **a** lightweight concrete, **b** fibre-reinforced concrete mixed with the ground, **c** tomograph for self-consolidating concrete (view during installation)

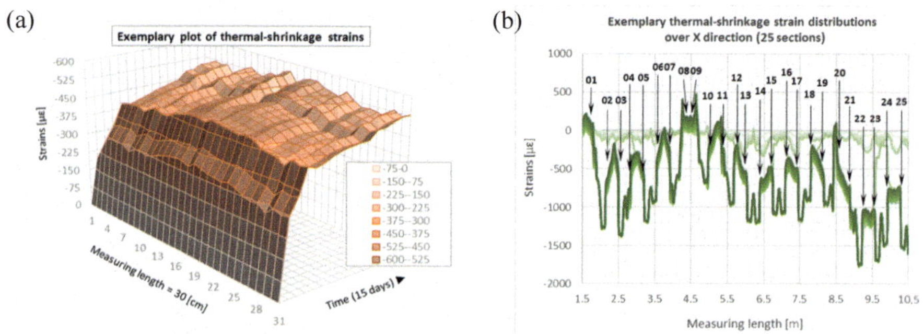

Fig. 4 **a** Exemplary plot of thermal-shrinkage strains of: **a** specimen made from fibre-reinforced concrete mixed with the ground, **b** concrete cube measured by the 'tomograph' fibre in X direction within 25 sections

3 In Situ Installations

3.1 Prestressed Truck Scale Platforms

Truck scales platforms are mainly used for weighting the car vehicles. Their main structural element is a slab which transfer the loads from vehicles into foundations through measuring devices. The crucial aspects during design phase are durability, bending stiffness and the dead weight. Thus, some platforms made from prestressed concrete were designed and implemented in Poland during last few years [15].

Platforms under consideration were prefabricated in the production hall in the technology of pre-tensioned concrete as fully prestressed, i.e. operating normally in a non-cracked condition. To reduce weight of the platforms, five oval Styrofoam inserts were placed along its length. Eight prestressing tendons were applied for the lower part of the cross section, and two for the upper. Moreover, reinforcing bars and stirrups were designed.

Several tests during different phases of the platform's lifecycle were performed, starting from hydration process (thermal-shrinkage strains), through tendons activation (strains regarding the transfer of compression forces from the tendons to the concrete) and finally during laboratory tests, when slabs were mechanically loaded in four-point bending test until destruction [14]. Distributed optical fibre measurement technology was used to record strain and temperature changes along selected measuring paths. For internal concrete strains, composite rods with integrated optical fibres were simply tied to the stirrups along the prestressing tendons—see Fig. 5a. The localization and numbering of all applied optical sensors are presented in Fig. 5b.

The measurements started immediately after concreting. Strains and temperatures were recorded every 30 min during first day of concrete hydration. During subsequent measuring sessions, the development of thermal-shrinkage strains was clearly observed. It is noteworthy that the distribution (shape) of strains along the member length is remarkably stable with respect to the shape from the beginning throughout all the following measurement sessions. Thus, the weakest points (those, where the local concentrations of strains occur) may be identified in the early stage, i.e. even then, when the concrete is not yet cracked or when only the micro-cracks invisible to the naked eye occur. The strain distribution plots with spatial resolution of 10 mm along the length of the platform under consideration during the first seven hour of hydration are presented in Fig. 6a for the lower (2) and Fig. 6b for upper (4) optical fibre rods.

During first hours of hydration the higher level of strains was observed for the lower optical fibre rod (2), because the concrete platforms were heated from below.

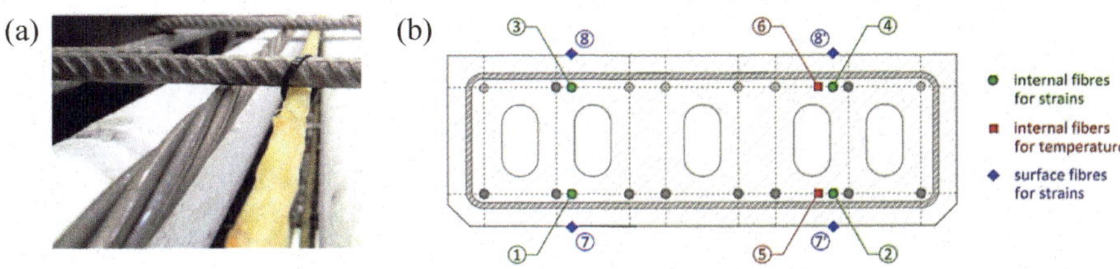

Fig. 5 **a** DFOS sensor tied to the stirrups before for concreting of the slab, **b** localization and numbering of all optical sensors

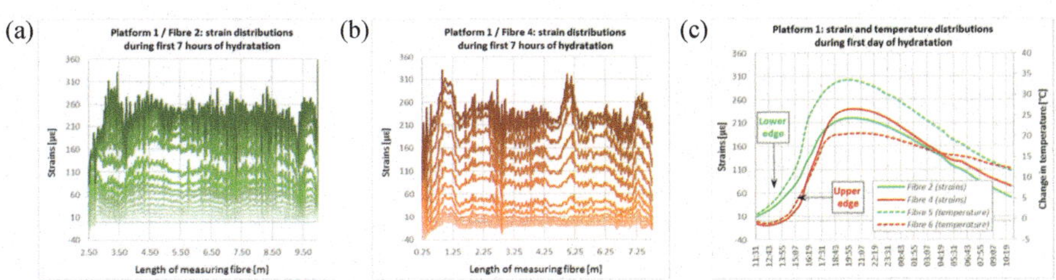

Fig. 6 Strains distributions: for fibre 2 (**a**) and 4 (**b**) over platform length during the first 7 h of hydration; averaged strains for fibre 2 & 4 over time during 23 h of hydration

Upper strains (4) at the very beginning were even slightly on the compressed side because of the external temperature domination before the start of concrete bonding end releasing hydration heat. The values of non-compensated strains (2, 4) and corresponding temperatures (5, 6) at lower and upper edges averaged over 1m base in the middle of the platform span are presented in Fig. 6c in time domain during first 23 h.

3.2 *Slurry Wall*

Other examples are a slurry walls made in technology of fibre-reinforced concrete mixed with the existing ground. These walls were executed within the research fields where their behaviour was monitored and analysed by composite optical fibre sensors (Fig. 7a) during the hydration process, but also during trenching the excavation (increasing the ground pressure—Fig. 7b), changing the thermal-moisture conditions, loading the surcharge with concrete slabs and loading the wall directly by the means of hydraulic jackets. Exemplary results of early-age concrete strains with simplified interpretation are presented below (Fig. 7c).

During the initial state, the shortening of the wall caused by the thermal contraction and shrinkage of the concrete-ground mixture was observed over entire length of the wall. However, due to the mechanical constraints (friction) between the wall and surrounding ground, the tensiled stress occurred causing the appearance of microcracks.

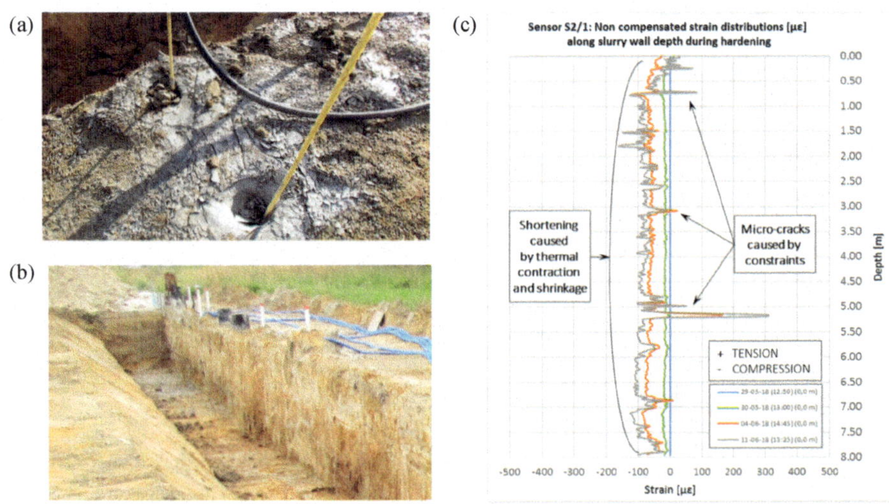

Fig. 7 **a** The view of optical fibre sensors installed in the slurry wall, **b** the view of slurry wall during trenching the excavation, **c** exemplary strain distributions of early-age concrete mixture

3.3 CFA Concrete Column

Continuous Flight Auger (CFA) columns are one of the most widely used ground improvement type because of their versatility, simplicity and efficiency. CFA concrete columns are formed by drilling the hollow to the required depth using special auger with inner tube, which is closed during driving and opened during extraction from the hole, allowing for down-up concreting in continuously way. This process is always monitored by an electronic system, but structural integrity of the columns is still affected by different factors such as concrete quality (low workability), cement consumptions, aggregate segregation or exudation. This is the reason why more and more effective and non-destructive ways of control the structural condition of the columns are developed.

One of the main advantages of installation internal optical fibre sensors is the possibility of measuring early-age concrete strains and temperature development. In research described further, distributed sensors were installed immediately after column execution. For this reason, special technology of installation (steel tubes with anchorages) was elaborated and implemented—see Fig. 8a.

In the first days of hardening, concrete reduced its volume (the length of the column was shorter and shorter) due to the phenomenon of thermal contraction as well as autogenous and drying shrinkage—see Fig. 8b. But, because the structural member under consideration is not totally free element (constraints through the friction within the shaft), tensile stress could be induced in the material. Depending on the level of constraints as well as the parameters of concrete mix used and conditions of hardening, this phenomenon can lead to the occurrence of microcracks and cracks.

(a) (b)

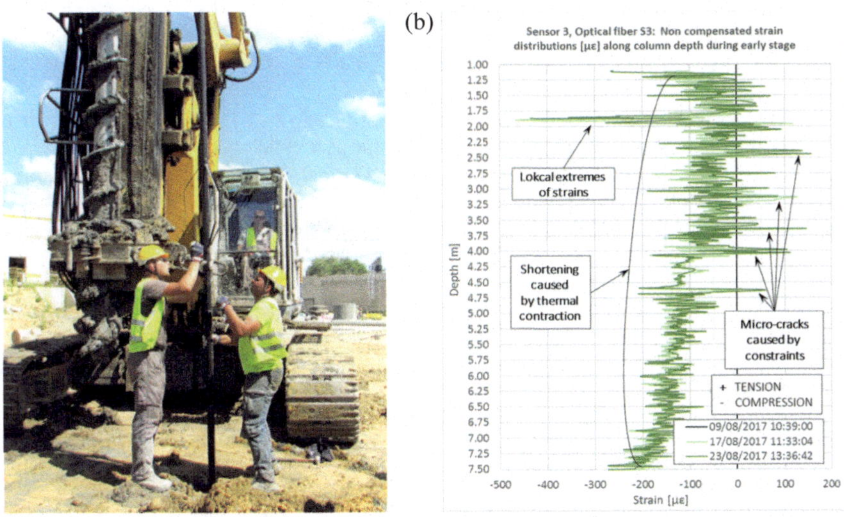

Fig. 8 **a** Installation of DFOS sensor inside CFA column, **b** exemplary strain distributions of early-age concrete

This is especially visible in the elements with reinforcement which is in fact additional internal constraint for the concrete. Thus, applying linear optical fibre sensor for early-concrete strain and temperature measurements along the whole column depth is very useful tool which allows for advanced structural condition assessment. With comparison with geological profiles, it is also possible to analyse cooperation of the column concrete with surrounding ground at selected depths. Optical fibres were also used for strain measurements during the load test of the CFA columns [13].

4 Conclusions

The pilot studies described in this article include comprehensive measurements of early-age concrete strains and temperatures on the examples of laboratory tests but also the pioneer in situ installations.

Strain and crack analysis of concrete structural members is crucial in the context of the assessment of their technical condition and safety [13] and this is especially challenging task for early-age concrete [10, 18], which behaviour depends on many factors and is not yet sufficiently investigated and described in the literature.

This is the reason why works and studies are ongoing to improve the methods used in this field. Based on the research carried out and presented in this article, it can be concluded that the distributed fibre optic sensor (DFOS) technology provides new opportunities in comparison with traditional spot measuring techniques (e.g. inductive, electrical resistance or vibrating wire spot sensors). It should be also emphasized that optical fibre sensors installed inside the structural members for early-age concrete strains measurements can be further used as a part of long-term structural health monitoring systems, significantly improving the possibility of structural condition assessment and thus safety of the structure and its performance.

Acknowledgements Authors would like to show gratitude to the SHM System company (Cracow, Poland, www.shmsystem.pl), which completed the project called "*Development of the new fibre optic sensor allowing for the determination of the vertical and horizontal displacements of the studied objects at the distances of up to 120 km*". This project was funded by the grant won at the National Centre for Research and Development within the framework of Intelligent Development Operational Program 2014–2020 (POIR.01.01.01–00-0550/15).

References

1. Barrias, A., Casas, J.R., Villalba, S.: A review of distributed optical fibre sensors for civil engineering applications. Sensors 16/2016 (2016)
2. Bersana, S., et al.: Distributed strain measurements in a CFA pile using high spatial resolution fibre optic sensors. Eng. Struct. **160**(2018), 554–565 (2018)
3. Composite-DFOS Sensors, Distributed fibre optic strain sensor EpsilonSensor, Data Sheet, SHM System, Krakow, Poland

4. Delepine-Lesoille, S., Merliot, E., Boulay, C.: Quasi-distributed optical fibre extensometers for continuous embedding into concrete: design and realization. Smart Mater. Struct. **15**(2006), 931–938 (2006)
5. Luna Innovations. User Guide, Optical Backscatter Reflectometer 4600 (2013)
6. Li, W., Bao, X.: High spatial resolution distributed fibre optic technique for strain and temperature measurements in concrete structures. In: International Workshop on Smart Materials & Structures, SHM and NDT for the Energy Industry, Calgary, Alberta, Canada (2013)
7. Mieszczak, M., Domagała, L.: Lightweight aggregate concrete as an alternative for dense concrete in post-tensioned concrete slab. Mater. Sci. Forum. **926**, 140–145 (2018). ISSN: 1662-9752
8. Minardo, A., et al.: Long term structural health monitoring by Brillouin fibre-optic sensing: a real case. Smart Mater. Struct. **9**(2012), 64–69 (2012)
9. Samiec, D.: Distributed fibre-optic temperature and strain measurement with extremely high spatial resolution. Photonik International (2012)
10. Seruga, A., Zych, M.: Thermal Cracking of the Cylindrical Tank under Construction. I: Case Study. Am. Soc. Civ. Eng. ASCE. J. Perform. Constr. Facil. **29**(4), 04014100-1^04014100-9 (2015)
11. Sieńko, R., et al.: Application of distributed optical fibre sensor technology for strain measurements in concrete structures. COST TU1402, Quantifying the Value of Structural Health Monitoring, Copenhagen, Denmark (2016)
12. Sieńko, R., et al.: Application of distributed optical fibre sensor for strain and temperature monitoring within continuous flight auger columns. In: 4th World Multidisciplinary Earth Sciences Symposium WMESS, Prague, Czech Republic (2018)
13. Sieńko, R., et al.: Strain and crack analysis within concrete members using distributed fibre optic sensors. Struct. Health Monit. https://doi.org/10.1177/1475921718804466 (2018). Last accessed 8 Oct 2018
14. Sieńko, R., Bednarski, Ł., Howiacki, T.: About distributed internal and surface strain measurements within prestressed concrete truck scale platforms. In: 3rd World Multidisciplinary Civil Engineering—Architecture—Urban Planning Symposium WMCAUS, Prague, Czech Republic (2018)
15. Szydłowski, R., Ślaga, Ł, Łabuzek, B.: 30 tons truck scale platform made from concrete prestressed with unbonded tendons. Przegląd Budowlany **6**(2017), 33–39 (2017)
16. Yang, D., et al.: Fatigue crack monitoring using plastic optical fibre sensor. Procedia Struct. Integr. **5**(2017), 1168–1175 (2017)
17. Zhou, Z., Wang, B., Ou, J.: Local damage detection of RC structures with distributive FRP-OFBG sensors. In: Second International Workshop on Structural Health Monitoring of Innovative Civil Engineering Structures, Winnipeg, Canada (2004)
18. Zych, M.: Research on thermal cracking of a rectangular RC tank wall under construction. II: comparison with numerical model. Am. Soc. Civ. Eng. ASCE. J. Perform. Constr. Facil. **30**(1), 04014199-1^04014199-10 (2016)
19. Zych, M.: Degree of external restraint of wall segments in semi-massive reinforced concrete tanks: part i rectangular segments. Struct. Concret (2018). https://doi.org/10.1002/suco.201700036
20. Zych, M.: Degree of external restraint of wall segments in semi-massive reinforced concrete tanks: part ii rectangular and cylindrical segments. Struct. Concret (2018). https://doi.org/10.1002/suco.201700037

Experimental Investigations of Cracking in Reinforced Concrete Beams of Different Depth

Aleksandr Sokolov, Deividas Rumsys, Karolis Sakalauskas, Darius Bacinskas, and Gintaris Kaklauskas

Abstract The study carried out at Vilnius Gediminas Technical University (Vilnius Tech) reports test results on cracking of seven RC beams of the rectangular section. The beams were subjected to a short-term loading under a four-point bending configuration. The experimental program was aiming at evaluating the effect of section height, reinforcement ratio, and bar diameter on cracking of RC beams. The beam cross-sections were designed to have two different cases of section heights (200 and 600 mm), reinforcement ratios (0.36% and around 1.9%), and bar diameters (Ø10 and Ø20 mm). The paper reports test results on mean and maximum crack spacings, as well as mean and maximum crack widths recorded using digital cameras and an electronic microscope. These characteristics were compared to the predictions by Eurocode 2 and Model Code 2010. For the beams with high reinforcement ratio, the mean and maximum crack width predictions by the Eurocode 2 and Model Code 2010 were close to the test results. However, the calculated values for the beams with low reinforcement ratio at most load levels notably exceeded the test results.

Keywords RC beams · Crack spacing · Mean crack width · Maximum crack width · Design code predictions

1 Introduction

Cracking is one of the most complex phenomena of the behaviour of reinforced concrete. Primary and secondary cracks of various orientations, shapes and lengths form due to the low tensile strength of concrete and interaction between reinforcement and concrete. These cracks have an adverse effect on stiffness, durability and

A. Sokolov · D. Rumsys · K. Sakalauskas · D. Bacinskas · G. Kaklauskas (✉)
Department of Reinforced Concrete Structures and Geotechnical Engineering, Vilnius Gediminas Technical University, Vilnius, Lithuania
e-mail: gintaris.kaklauskas@vgtu.lt

A. Sokolov
Research Institute of Innovative Building and Bridge Structures, Vilnius Gediminas Technical University, Vilnius, Lithuania

F. Kanavaris et al. (eds.), *International RILEM Conference on Early-Age and Long-Term Cracking in RC Structures*, RILEM Bookseries 31,
https://doi.org/10.1007/978-3-030-72921-9_8

appearance of the structure and make the stress-strain state in the tensile zone highly unpredictable. Much has been learned already about crack width and spacing, but a completely rational and general method is still unavailable.

Relations between reinforcement strain, $\varepsilon_s(x)$, bond stress, $\tau(x)$, slip and crack width, w, can be established using this differential equation [1, 2]:

$$\frac{d\varepsilon_s(x)}{dx} = \frac{4}{E_s \O_s}\tau(x) \tag{1}$$

where E_s is the modulus of elasticity of reinforcement; $\O_s$ is the bar diameter.

To simplify sophisticated calculation procedures, Eurocode 2 and Model Code 2010 assumed a constant bond stress. This results in a linear strain function of the reinforcement making the transfer length and crack spacing proportional to the ratio $\O_s/\rho_{ef}$, where ρ_{ef} is the reinforcement ratio assuming the *effective area* of concrete in tension, the characteristic established empirically. The current versions of the Eurocode 2 [3] and Model Code 2010 [4] suggests to combine the concrete cover and the classical bond concepts using the following generalized equation of crack spacing:

$$s_r = A_1 c + A_2 \frac{\O_s}{\rho_{ef}} \tag{2}$$

where c is the clear cover.

Crack width is assessed as

$$w = \int_{s_r} (\varepsilon_s - \varepsilon_c) dx \tag{3}$$

where ε_s and ε_c are the strains in reinforcement and concrete, respectively.

A limited number of accurate tests on cracking behaviour of RC beams were reported in the literature [5–7]. The current study presents new results of accurately performed short-term tests on cracking of RC beams. The experimental program was aiming at evaluating the effect of section height, reinforcement ratio, and bar diameter on cracking of RC beams. The beam cross-sections were designed to have two different cases of section heights (200 and 600 mm), reinforcement ratios (0.36% and around 1.9%), and bar diameters (Ø10 and Ø20 mm). The paper reports test results on mean and maximum crack spacing, as well as mean and maximum crack widths recorded using digital cameras and an electronic microscope. The test cracking characteristics were compared to the predictions by Eurocode 2 and Model Code 2010.

2 Experimental Program

The experimental program was aiming at evaluating the effect of section height, reinforcement ratio and bar diameter on cracking, deformation and tension stiffening behaviour of RC beams.

Seven RC beams of the rectangular section were tested under a four-point bending configuration. The beam cross-sections were designed as to have two different cases of section heights (200 and 600 mm), reinforcement ratios (0.36% and around 1.98% and bar diameters Ø10 and Ø20 mm). The beams of 200 mm height nominally had 210 mm width, 2000 mm span and 1000 mm pure bending zone, whereas the members of 600 mm height had 310 mm width, 4000 mm span and 1500 mm. The shear span was reinforced with stirrups to avoid shear failure. The clear cover from the longitudinal steel to the nearest concrete surface was 25 mm. The cross-section of the 200 mm and 600 mm beams is presented in Figs. 1 and 2, respectively.

The specimens were cast using an industrially produced C30/37 concrete mix of slump class S4. Main geometrical and material characteristics of the RC beams are given in Table 1, where *h* is the height of the beam; *b* is the width of the beam; *d* is the effective depth to the tensile steel area; a_2—is the effective depth to the compressive steel; A_{s1} is the area of the tensile steel; A_{s2} is the area of the compressive steel;

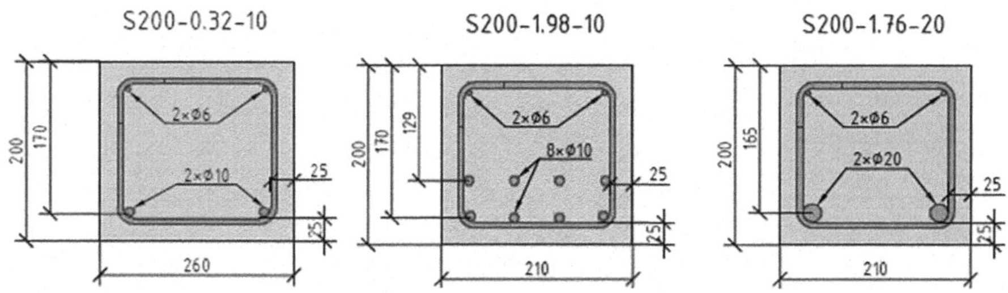

Fig. 1 Cross-sections of test RC beams having 200 mm section height

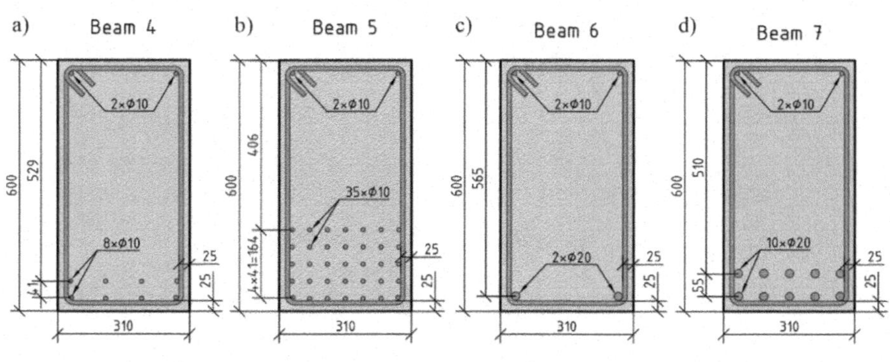

Fig. 2 Experimental test RC beams of 600 × 310 mm cross-section

Table 1 Main characteristics of experimental beams

No.	Beam	h (mm)	b (mm)	d (mm)	a_2 (mm)	A_{s1} (mm^2)	A_{s2} (mm^2)	ρ (%)	f_y (MPa)	E_s (GPa)	Age (days)	f_{cm} (MPa)
1	S200-0.32-10	214	262	184	28	155	57	0.32	500	200	41	50.4
2	S200-1.98-10	198	212	149	28	628	57	1.98			33	50.9
3	S200-1.76-20	201	216	162	28	628	57	1.76			33	50.9
4	S600-0.37-10	604	308	551	30	628	157	0.37			48	51.1
5	S600-1.81-10	605	312	489	30	2749	157	1.81			62	52.5
6	S600-0.36-20	606	311	565	30	628	157	0.36			47	51.1
7	S600-1.88-20	605	311	539	30	3142	157	1.88			69	52.5

ρ is the ratio of tensile reinforcement; f_y and E_s are the yielding strength and the elastic modulus of the bar reinforcement, respectively; f_{cm} is the cylinder strength of concrete. As the study was dedicated to the serviceability behaviour, the tests were stopped prior to yielding of tensile reinforcement.

3 Experimental Crack Patterns and Crack Spacing

Crack width was recorded on one surface of the beam throughout the pure bending zone using an electronic microscope and digital cameras. The images were captured by the digital cameras placed on a tripod 0.9–1.2 m from the test specimens. One camera was employed for 200 mm section beams and two cameras were used for 600 mm section beams. Examples of final crack patterns are given in Fig. 3. Mean experimental crack spacings are presented in Tables 2 and 3, along with the respective spacings calculated by Eurocode 2 and Model Code 2010. The calculated mean crack spacing for the both codes was determined as $s_{r,m}= s_{r,max}/1.7$ (Tables 4 and 5).

From the results of comparing experimental mean and maximum crack spacings with codes it can be seen that for the beams with lower reinforcement ratio Eurocode 2 and Model Code 2010 tend to give larger values.

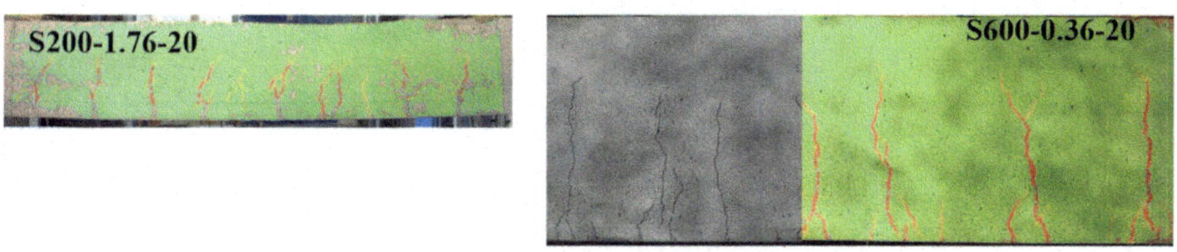

Fig. 3 Examples of crack patterns captured by digital cameras

Table 2 Experimental mean crack spacings and comparison with EC2

No.	Name of the beam	$s_{r,m,exp}$ (mm)	$s_{r,m,EC2}$ (mm)	$s_{r,m,EC2}/s_{r,m,exp}$	μ	COV
1	S200-0.32-10	111.1	152.8	1.38	1.05	0.21
2	S200-1.98-10	74.6	62.0	0.83		
3	S200-1.76-20	96.6	84.6	0.88		
4	S600-0.37-10	100.7	124.2	1.23		
5	S600-1.81-10	76.0	76.3	1.00		
6	S600-0.36-20	134.7	163.5	1.21		
7	S600-1.88-20	102.8	87.0	0.85		

Table 3 Experimental mean crack spacings and comparison with MC2010

No.	Name of the beam	$s_{r,m,exp}$ (mm)	$s_{r,m,MC2010}$ (mm)	$s_{r,m,MC2010}/s_{r,m,exp}$	μ	COV
1	S200-0.32-10	111.1	197.4	1.78	1.12	0.38
2	S200-1.98-10	74.6	53.2	0.71		
3	S200-1.76-20	96.6	83.9	0.87		
4	S600-0.37-10	100.7	142.3	1.41		
5	S600-1.81-10	76.0	61.9	0.81		
6	S600-0.36-20	134.7	202.3	1.50		
7	S600-1.88-20	102.8	79.4	0.77		

Table 4 Experimental maximum crack spacings and comparison with EC2

No.	Name of the beam	$s_{r,max,exp}$ (mm)	$s_{r,max,EC2}$ (mm)	$s_{r,max,EC2}/s_{r,max,exp}$	μ	COV
1	S200-0.32-10	162.4	259.8	1.60	1.21	0.27
2	S200-1.98-10	101.9	105.4	1.03		
3	S200-1.76-20	153.1	143.8	0.94		
4	S600-0.37-10	134.3	211.1	1.57		
5	S600-1.81-10	135.0	129.7	0.96		
6	S600-0.36-20	191.0	277.9	1.45		
7	S600-1.88-20	169.2	147.9	0.87		

Table 5 Experimental maximum crack spacings and comparison with MC2010

No.	Name of the beam	$s_{r,max,exp}$ (mm)	$s_{r,max,MC2010}$ (mm)	$s_{r,max,MC2010}/s_{r,max,exp}$	μ	COV
1	S200-0.32-10	162.4	335.6	2.07	1.29	0.44
2	S200-1.98-10	101.9	90.4	0.89		
3	S200-1.76-20	153.1	142.6	0.93		
4	S600-0.37-10	134.3	241.8	1.80		
5	S600-1.81-10	135.0	105.2	0.78		
6	S600-0.36-20	191.0	343.9	1.80		
7	S600-1.88-20	169.2	135.0	0.80		

4 Mean Crack Widths and Comparison to Code Predictions

In general, good agreement was obtained for mean crack widths' recordings obtained by the electronic microscope and digital cameras as shown Fig. 4. Further analysis will be based on the crack width values obtained by the microscope. Figure 5 shows the test mean crack width versus reinforcement strain graphs against the predicted

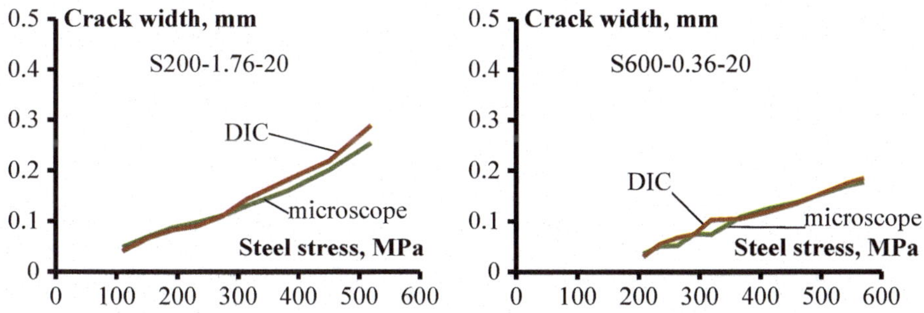

Fig. 4 Examples of mean crack width graphs recorded by digital cameras and an electronic microscope

values by Eurocode 2 and Model Code 2010. The values of experimental mean crack widths at 300 MPa reinforcement stress and statistical analysis of comparison with Eurocode 2 and Model Code 2010 are given in Tables 6 and 7.

From the results given in Tables 6 and 7 it can be seen that for the beams with lower reinforcement ratio Eurocode 2 and Model Code 2010 tend to give larger mean crack width values.

5 Maximum Crack Width and Comparison to Code Predictions

Figure 6 depicts maximum crack width—reinforcement strain diagrams recorded by the electronic microscope and digital cameras. Similarly to the case of mean crack width values, a good agreement can be stated. Figure 7 shows the test maximum crack width versus reinforcement strain graphs along with the predicted values by Eurocode 2 and Model Code 2010. As can be seen from Fig. 7, for the beams with high reinforcement ratio, the maximum crack width predictions by the Eurocode 2 and Model code 2010 were close to each other and to the tests. However, the predictions for the beams with low reinforcement ratio were not that accurate with the calculated values notably exceeding the test results. The exception were the early cracking stages when rather accurate prediction results were obtained as given in Tables 8 and 9.

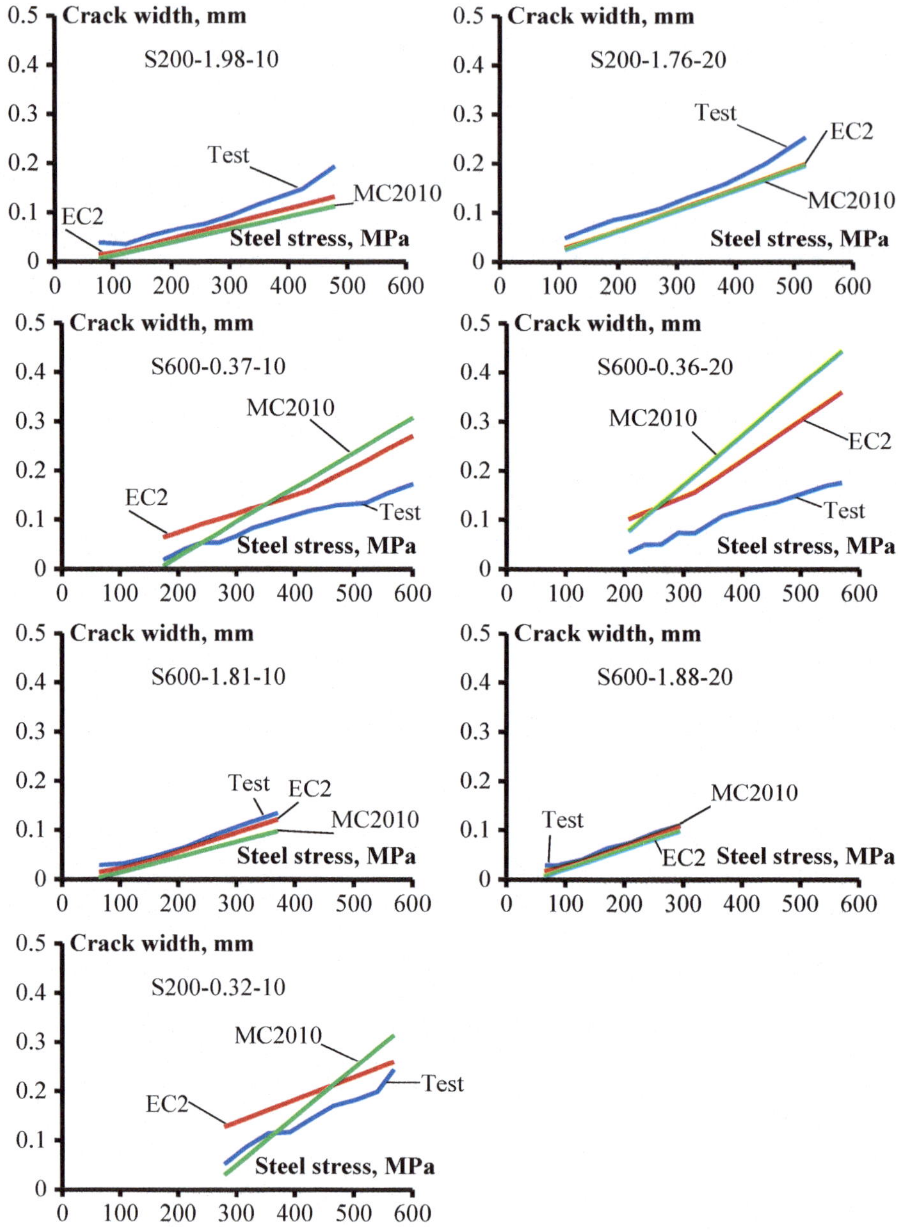

Fig. 5 Mean crack widths versus reinforcement stress

Table 6 Experimental mean crack widths at 300 MPa reinforcement stress and comparison with EC2

No.	Name of the beam	$w_{m,exp}$ (mm)	$w_{m,EC2}$ (mm)	$w_{m,EC2}/w_{m,exp}$	μ	COV
1	S200-0.32-10	0.073	0.138	1.89	1.30	0.40
2	S200-1.98-10	0.094	0.078	0.82		
3	S200-1.76-20	0.123	0.106	0.86		
4	S600-0.37-10	0.067	0.112	1.68		
5	S600-1.81-10	0.107	0.095	0.89		
6	S600-0.36-20	0.074	0.147	1.98		
7	S600-1.88-20	0.114	0.111	0.98		

Table 7 Experimental mean crack widths at 300 MPa reinforcement stress and comparison with MC2010

No.	Name of the beam	$w_{m,exp}$ (mm)	$w_{m,MC2010}$ (mm)	$w_{m,MC2010}/w_{m,exp}$	μ	COV
1	S200-0.32-10	0.073	0.051	0.71	1,09	0.55
2	S200-1.98-10	0.094	0.067	0.71		
3	S200-1.76-20	0.123	0.105	0.85		
4	S600-0.37-10	0.067	0.096	1.44		
5	S600-1.81-10	0.107	0.077	0.72		
6	S600-0.36-20	0.074	0.172	2.32		
7	S600-1.88-20	0.114	0.102	0.89		

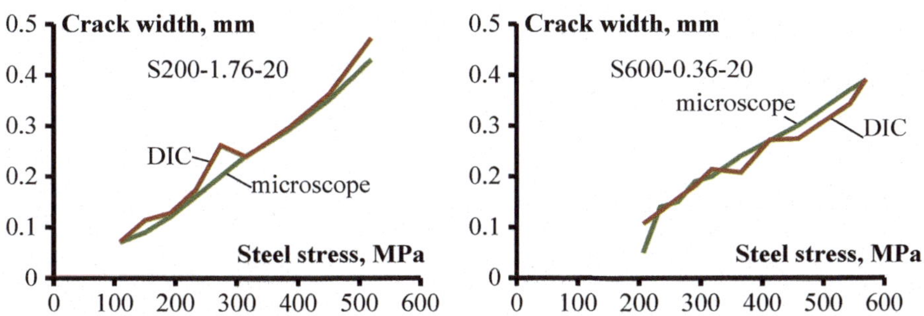

Fig. 6 Examples of maximum crack width graphs recorded by digital cameras and an electronic microscope

6 Concluding Remarks

The study carried out at Vilnius Gediminas Technical University (Vilnius Tech) reports test results on cracking of seven RC beams of the rectangular section. The

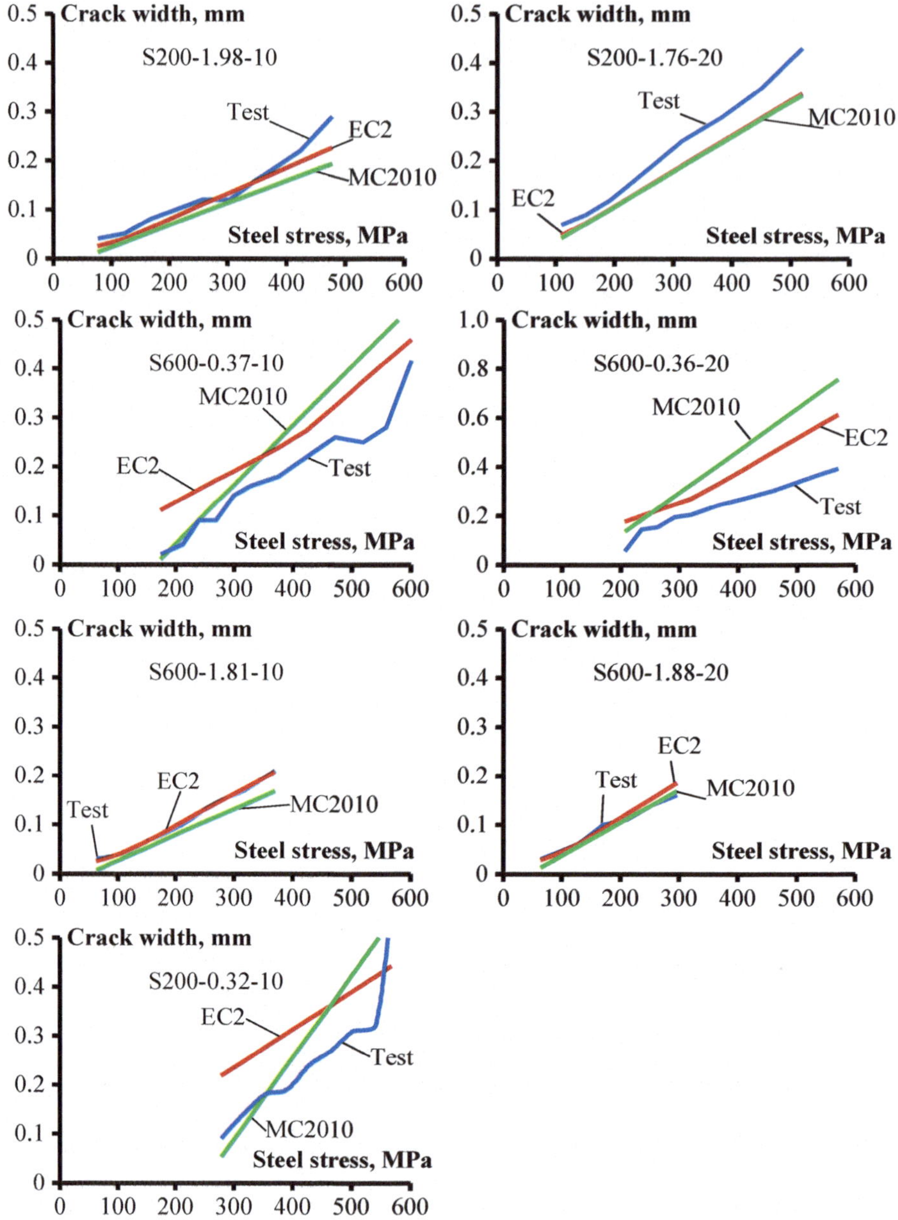

Fig. 7 Maximum crack widths versus reinforcement strain in the cracked section

Table 8 Experimental maximum crack widths at 300 MPa reinforcement stress and comparison with EC2

No.	Name of the beam	$w_{max,exp}$ (mm)	$w_{max,EC2}$ (mm)	$w_{max,EC2}/w_{max,exp}$	μ	COV
1	S200-0.32-10	0.118	0.234	1.99	1.24	0.30
2	S200-1.98-10	0.120	0.132	1.10		
3	S200-1.76-20	0.226	0.180	0.80		
4	S600-0.37-10	0.141	0.190	1.35		
5	S600-1.81-10	0.160	0.162	1.01		
6	S600-0.36-20	0.193	0.250	1.29		
7	S600-1.88-20	0.163	0.189	1.16		

Table 9 Experimental maximum crack widths at 300 MPa reinforcement stress and comparison with MC2010

No.	Name of the beam	$w_{max,exp}$ (mm)	$w_{max,MC2010}$ (mm)	$w_{max,MC2010}/w_{max,exp}$	μ	COV
1	S200-0.32-10	0.118	0.087	0.74	1.00	0.27
2	S200-1.98-10	0.120	0.113	0.94		
3	S200-1.76-20	0.226	0.178	0.79		
4	S600-0.37-10	0,141	0.163	1.16		
5	S600-1.81-10	0.160	0.131	0.82		
6	S600-0.36-20	0.193	0.293	1.51		
7	S600-1.88-20	0.163	0.173	1.06		

beams were subjected to a short-term loading under a four-point bending configuration. The experimental program was aiming at evaluating the effect of section height, reinforcement ratio, and bar diameter on cracking of RC beams. The beam cross-sections were designed to have two different cases of section heights (200 and 600 mm), reinforcement ratios (0.36% and around 1.9%), and bar diameters (Ø10 and Ø20 mm). The paper reports test results on mean and maximum crack spacings, as well as mean and maximum crack widths. These characteristics were compared to the predictions by Eurocode 2 and Model Code 2010. For the beams with high reinforcement ratio, the mean and maximum crack width predictions by the Eurocode 2 and Model code 2010 were close to each other and to the test results. However, the calculated values for the beams with low reinforcement ratio were not that accurate with the predictions at most load levels notably exceeding the test characteristics.

Acknowledgements The authors gratefully acknowledge the financial support provided by the European Social Fund under grant agreement with the Research Council of Lithuania (LMTLT) for the project No. 09.3.3-LMT-K-712-01-0145.

References

1. Kuuskoski, V.: Über die Haftung zwischen Beton und Stahl. Valtioneuvoston Kirjapaino, Helsinki (1950)
2. Rehm, G.: Über die Grundlagen des Verbundes zwischen Stahl und Beton. DAfStb Heft Nr. 138, Ernst & Sohn, Berlin (1961)
3. EN 1992-1-1: Eurocode 2: Design of Concrete Structures: Part 1-1: General Rules and Rules for Buildings. CEN European Committee for Standardization. Brussels, Belgium (2004)
4. fib (International Federation for Structural Concrete): Model Code for Concrete Structures 2010, p. 434. Wiley, Berlin (2013)
5. Rehm, G., Rüsch, H.: Versuche mit Betonformstählen Teil I (1963) Teil II (1963) Teil III (1964). Deutscher Ausschuss für Stahlbeton Heft 140 (1964)
6. Gilbert, R.I., Nejadi, S.: An Experimental Study of Flexural Cracking in Reinforced Concrete Members Under Short Term Loads. University of New South Wales, School of Civil and Environmental Engineering (2004)
7. Calderón Bello, E.: Estudio experimental de la fisuración en piezas de hormigón armado sometidas a flexión pura. Ph.D. thesis, Universidad Politécnica de Madrid, Madrid (2008)

Imposed Deformation Reduction in Semimassive Walls of RC Tanks by Internal Cooling

Mariusz Zych and Trong-Chuc Nguyen

Abstract Early-age cracks in the RC walls of the tanks very often cause leaks. In turn, limiting the value of imposed strains reduces the risk of cracking or the width of cracks if cracking is not to be excluded in general. In the paper the analysis of the impact of a cooling pipe system (CPS) on the changes of temperature and strains in a RC semimassive wall was presented. In the numerical calculations a model covering non-linear and non-stationary temperature field variations resulting from cement hydration, internal cooling and heat exchange with the surroundings was used. It was demonstrated that during concrete maturing CPS contributes not only to mean temperature changes reduction but also to the reduction of temperature gradients, which helps effectively limit or eliminate concrete cracking. The analysis of the so-called "self-equilibrated stress" inducing temperature indicated that the application of CPS contributes to the reduction to zero the positive temperature difference in the immediate surroundings of the cooling pipes and to the reduction of temperature extreme changes, i.e. positive changes in the wall interior and negative ones in its corners. As a result, the restrained part of the imposed strains, to a greater extent, may remain below the tensile strain capacity of concrete. This solution can protect the tank wall from cracking or significantly reduce it.

Keywords Cracks · Cooling pipe system · Wall · RC tank · Early-age concrete

1 Introduction

The period of concrete hardening, i.e. of cement hydration, is the time when significant imposed strains occur in massive and semimassive structures. According to

M. Zych (✉)
Department of Civil Engineering, Institute of Building Materials and Structures, Cracow University of Technology, Cracow, Poland
e-mail: mzych@pk.edu.pl

T.-C. Nguyen
Institute of Techniques for Special Engineering, Le Quy Don Technical University, Hanoi City, Vietnam

F. Kanavaris et al. (eds.), *International RILEM Conference on Early-Age and Long-Term Cracking in RC Structures*, RILEM Bookseries 31,
https://doi.org/10.1007/978-3-030-72921-9_9

the definition proposed by Falga [8], structures for which 2 $m^{-1} < m_c < 15$ m^{-1} are considered semimassive. This approach is based on determining the surface area-to-volume ratio $m_c = u_c/v_c$, where: u_c is the surface area of element that is exposed to air and v_c is the volume of the element. The number of variable parameters in the concrete mix as well as the variety of possible hydration conditions make it difficult to find optimal solutions. The issue of the complexity of the cement hydration process together with the influence of the currently used concrete admixtures was extensively presented by Aitcin and Flatt [1]. For example König et al. [13] described the early temperature changes in maturing concrete which result in self-equilibrating strains of a structural member as well as mean strains. The effect of excessive imposed strains in structures with a significant degree of external restraint are cracks, which are responsible for leakage in reinforced concrete (RC) tanks. Knoppik-Wróbel [12], proposing her own advanced numerical model for cracking risk assessment, verified it on semimassive wall members. This allowed a parametric analysis of the distribution of the degree of external restraint to be performed. Jeon [10] described the phenomenon of crack formation as a result of the release of the heat of hydration in members restrained by internal restraints. They proposed a hypoelastic model that can incorporate an incremental constitutive equation. The issue was discussed taking into account various mixtures and construction conditions. Seruga and Zych [20, 21] carried out extensive research studies on the material, temperature changes, strain and cracking during the execution of semimassive cylindrical and rectangular tanks. Further numerical analyses of these tanks carried out by Zych [25, 26] demonstrated the expected compliance of the advanced numerical model with the study results.

In the case of massive structures, one of the design solutions is the use of CPS, thanks to which the temperature rise resulting from cement hydration can be effectively lowered. However, the use of a CPS in mass concrete blocks also has many drawbacks such as local stress formation around large pipes, rapid reduction of concrete temperature can affect the strength development of concrete as presented by Bofang [5]. For a CPS to be efficient it is necessary to monitor the parameters affecting the working process of the CPS in concrete blocks [16, 22] such as: start and end time of water flow in the cooling pipe, water velocity, the type of water pipe, layout of the water pipe along the wall height (for example, in a concrete dam, the vertical and horizontal layouts are in the range of 1.5–3.0 m as proposed by Bofang [5]), water temperature in the cooling pipe, etc. Bofang [5] recommends that for a dam structure the height of the cooling area not be less than 0.4 times the height of the dam. Water temperature control in the pipes is very important to avoid a sudden drop in temperature in concrete blocks. In the studies by Bofang [5], Tang et al. [23], and Aniskin and Chuc [2] it is shown that the difference between water temperature and the initial temperature of concrete mixture should not exceed 10 °C.

The beneficial effect of internal cooling with air-filled prestressing ducts on the reduction of the risk of cracking in a massive wall was presented by Azenha et al. [3]. The authors demonstrated that in spite of the acknowledged limitations of air cooling, it may prove quite feasible in relatively cool climates and small lengths of embedment. Besides, the water pipe layout not only affects the cooling efficiency in concrete but also the economy of the project. In the case of a small water pipe, the temperature difference will be small, which can lower the maximum temperature in the concrete block significantly, thus reducing the formation of thermal stress.

This research paper analyses in detail the effect of CPS on temperature changes and imposed strains in a semimassive wall of a RC tank. A detailed analysis of the effect of CPS on changes in stresses was presented by the authors in Zych and Chuc [27].

2 Structure

The subject of analysis is one of the wall segments of a RC tank. This tank, due to the occurrence of cracks during the hardening of concrete and the resulting leaks, was the subject of the analysis presented by Zych [24]. Therefore, this tank was also a very good example for verifying the efficiency of internal cooling in order to reduce imposed strains. From the practical point of view, this has a large effect on the possibility of reducing the degree of reinforcement, which is usually very high in structures that are required to be watertight. The necessity of using a very high amount of reinforcement in a classic construction solution, has been experimentally confirmed in tests on a real object e.g. by Seruga and Zych [20]. The subject of the analysis is one of the wall segments (Fig. 1). This wall was constructed in stages, i.e. by executing subsequent segments. The second wall segment (which is being analyzed) that was 0.75 m thick, 1.4 m high and 30.3 m long was concreted in one cycle without a vertical construction joint in the middle of it. Despite the procedure of introducing horizontal and vertical construction joints in the rest part of the wall, numerous vertical cracks were observed on the outside and inside surfaces of the wall during concrete hardening (Fig. 1a, b).

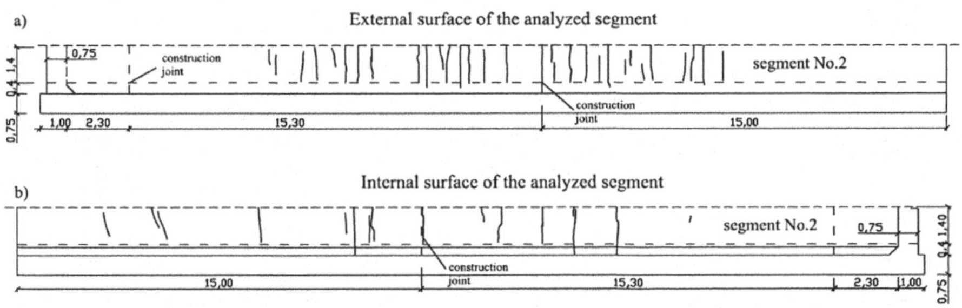

Fig. 1 Layout of cracks and their widths on the **a** outside and **b** inside segment surfaces

3 Numerical Model

The changes in the temperature field in the wall containing CPS were calculated by solving two basic differential equations, i.e. the Fourier equation taking into account the heat of hydration of cement and the equation taking into account the heat transfer between CPS and concrete, by following the principle of energy balance and described by Myers et al. [15] and Qiu et al. [18]. Convection is expressed by Newton's law of cooling as presented by Nguyen and Aniskin [17]. A detailed description of this model is presented by Zych and Chuc [27]. Its main simplification was to perform calculations in 2D. The 2D finite element model to determine temperature fields in mass concrete structures with CPS was presented in details by Lagundžija and Thiam [14]. Possible differences may result from the heating of the flowing water, therefore the obtained results in this respect should be treated as approximate. The tank was designed using C20/25 concrete. The concrete mix was used as specified in Seruga and Zych [21]. Segment No. 2 of this wall was made from CEM IIIA/42.5 cement. The following basic mechanical properties of concrete were tested: the secant modulus of elasticity, the mean value of concrete cylinder compressive strength and mean value of axial tensile strength of concrete [21]. In order to analyze the temperature changes in the early-age concrete hardening, it was necessary to know the amount of heat released in the concrete hardening process. Pursuant to Kiernożycki [11], the average values of clinker and GGBS content in the type of the cement CEM III A/42.5 was used, i.e. 55% clinker and 45% GGBS. Figure 2 illustrates the FEM mesh and the numbering of the nodes for which the results of the calculations are presented further in this paper. The calculations take into account both the influence of the changing ambient temperature as well as the thermal boundary conditions in line with the technology of the construction of this segment, i.e. the presence of formwork and the fact that the segment was built on the

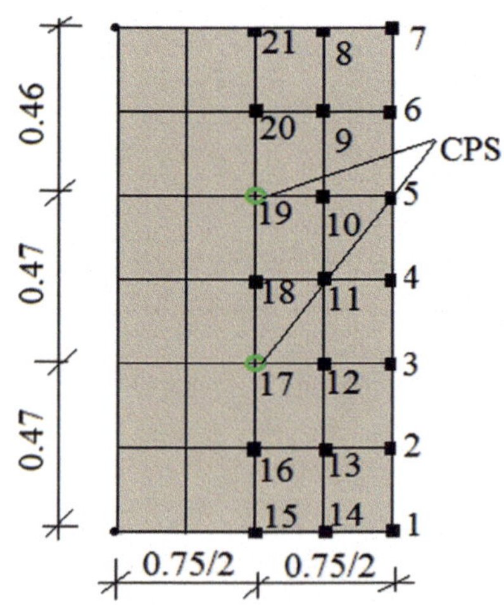

Fig. 2 Segment FEM mesh [m]

Table 1 Characteristics of the pipe cooling system [23]

Parameters	Values
Heat transfer coefficient on the border with concrete (W/(m^2 °C)	381.20
Specific heat (J/kg °C)	4200
The coefficient of thermal conductivity W(m °C)	0.64
Volume (m^3/h)	1.08
The water velocity in the pipe (m/s)	0.6
The water temperature in pipe (°C)	15
Water density (kg/m^3)	1000
Diameter (m)	0.025
Section area (m^2)	0.00008

foundation slab. It was assumed that the cross-section includes two cooling pipes at a height of 0.47 and 0.94 m. The CPS characteristics contained in Table 1 were used.

4 Results

4.1 Temperature Changes

In Fig. 3 examples of temperature changes $T_i(y, z, t)$ for both calculation variants are compared (i.e. without and with CPS). Figure 3a, b concern ordinate $y = 0.47$ m

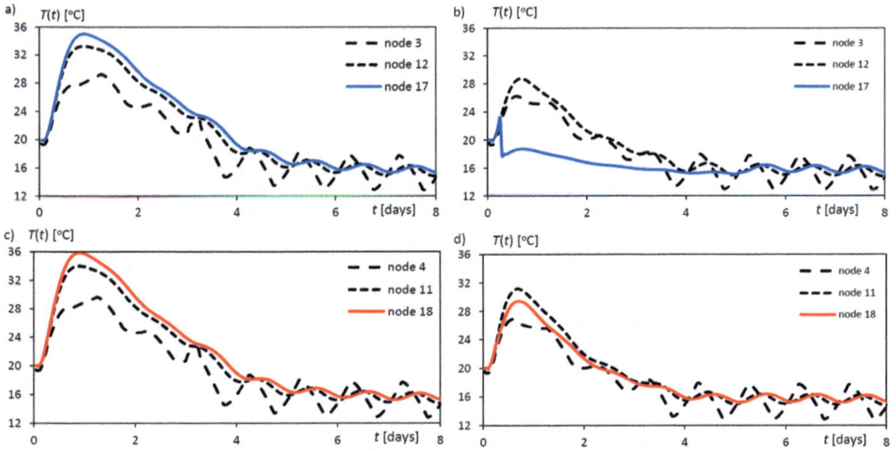

Fig. 3 Temperature changes in nodes on ordinate y = 0.47 m for: **a** variant without CPS, **b** variant with CPS and ordinate y = 0.7 m for: **c** variant without CPS, **d** variant with CPS

(measured from the wall base) on which in the variant with CPS the cooling pipe is located (cf. Fig. 2). The results are presented in the central node (node 17), in front of the extreme node (node 12) and on the wall surface (node 3). The comparison of Figs. 3a and b indicates a substantial reduction of the value of temperature changes $T_{17}(t)$ in the place of the location of the cooling pipes and slightly lower in the nodes more remote $T_{12}(t)$ and $T_3(t)$. The temperature changes shown in Fig. 3c, d concern the ordinate at mid-height of the wall (i.e. $y = 0.7$ m), and the ordinate is at an equal distance between ordinates $y = 0.47$ m and 0.94 m, on which the cooling pipes are located (cf. Fig. 2). The differences between the two variants observed on this ordinate are definitely smaller.

From a general comparison of temperature changes $T_i(y, z, t)$, between the two calculation variants it can be concluded that an effective impact of internal cooling, that is a reduction of the maximum temperature in each node of the section is clearly noticeable. For the presented nodes No 17, 12, 3 (i.e. $y = 0.47$ m) and nodes No 18, 11, 4 (i.e. $y = 0.7$ m) the reduction was 34, 14, 10% and 18, 8, 9%, respectively. In conclusion, the temperature reduction rate heavily depends on the position of the given node relative to the cooling pipe.

Most of the analytical (e.g. [4]) and standard (e.g. [9]) methods of determining thermal strain induced stresses in members restrained by external restraints based on the 1D model, i.e. averaging the temperature changes along the wall thickness. Therefore in the first stage the changes of mean temperatures $\Delta T_{i\_y\_mean}$ were determined for individual ordinates subtracting the current mean temperature from the initial temperature of concrete in the wall. The mean temperature changes $\Delta T_{i\_y\_mean}$ thus obtained for individual wall ordinates as a function of time for the variants without and with CPS are shown in Fig. 4. As is most clearly seen, the application of internal cooling results in a considerable reduction of temperature variations in the mid-wall and the temperature decrease in the wall section, in the variant with internal cooling, begins five hours earlier.

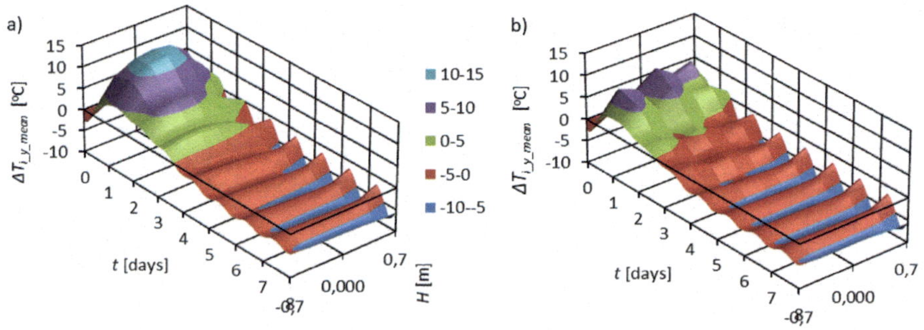

Fig. 4 Mean temperature variations $\Delta T_{i\_y\_mean}$ for individual wall ordinates in function of time for variants: **a** without CPS and **b** with CPS

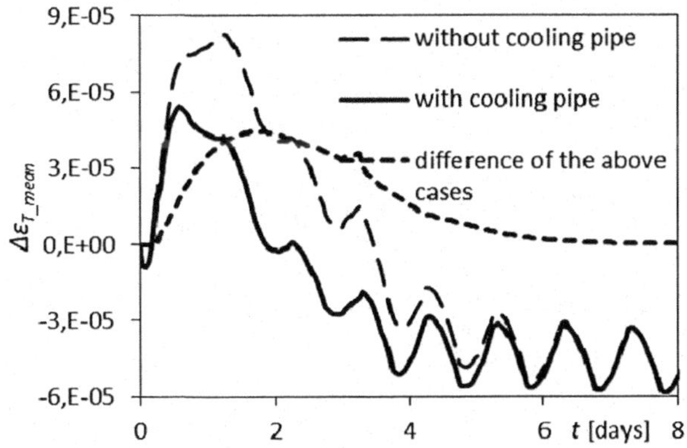

Fig. 5 Variations of free mean strain $\Delta\varepsilon_{T\_mean}$(t) in the cross-section of segment no 2

4.2 Mean Strains

Figure 5 illustrates changes of free mean strain $\Delta\varepsilon_{T\_mean}(t)$ of segment No 2 for both variants, without and with CPS. In the calculations the value of the coefficient of thermal expansion $\alpha_T = 1 \cdot 10^{-5}$/°C recommended by Eurocode 2 [6] was adopted. Major differences in free strains occur in the period of up to four days of concrete maturing, and their maximum value of about 50% is for the time $t = 2$ days. Free mean strain $\Delta\varepsilon_{T\_free\_mean}(t)$ is an important factor in the case of a high degree of restraint which, in turn, depends mainly on the type of external restraints and geometry of the member. In the case analyzed here both these conditions are extremely unfavorable as the external restraints are constituted by the slab of 0.75 m thickness, and the wall length to height *L*/*H* ratio is as high as 24. Despite, however, the final value of free strain, in both variants without and with CPS, in the case of nearly complete restraint of strains, concrete creep will additionally take place. Since concrete creep takes the highest value in the period of positive strains formation, the restrained part of imposed strain for the variant without CPS will be definitely higher. In the design practice [7] the period of positive imposed restraints increase is disregarded, therefore on the basis of the comparison of strains after reaching $T_{\max}$ only, it can be concluded that the internal cooling reduces the mean free strain of thermal shrinkage by ca. 20%.

4.3 Strain Gradient

Another component of strains, i.e. free strain gradient in the function of time $\Delta\varepsilon_{T\_free\_grad}(t)$ for both variants is shown in Fig. 6. Similarly to the case of free mean strain $\Delta\varepsilon_{T\_free\_mean}(t)$, free strain gradient $\Delta\varepsilon_{T\_free\_grad}(t)$ is a significant factor in the case of external restraints an unfavorable geometry of the member. In RC tank walls the geometry factor is of essential importance as mainly the gradient in

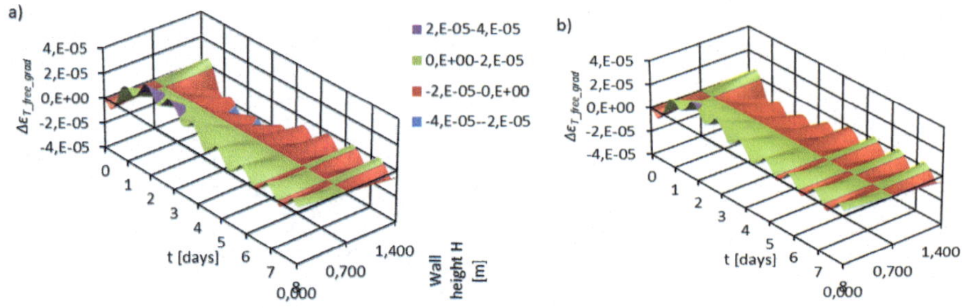

Fig. 6 Variations of free strain gradient $\Delta\varepsilon_{T\_free\_grad}(t)$ in vertical section of segment no 2 in the function of time for cases: **a** without CPS, **b** with CPS

vertical direction is analyzed. In some cases even if there are no external restraints, the substantial role is played by the self-weight of the member, which itself can effectively increase the restrained part of the strain gradient [19].

In the case analyzed here the external restraints, apart from free shortening of the segment, restrain the possibility of free rotation. In segment No 2 the effect of both the member's geometry and the monolithic connection with slab foundation is that the degree of restraint of rotation derived after JSCE [9] is 1.0, which means that 100% of strains resulting from temperature gradient are restrained strains that lead to the occurrence of stress gradients in the vertical section of the wall.

Figure 7 illustrates the gradients of restrained strains (which are 100% of the value of strain in unrestrained member) in the vertical section of segment No 2 for selected points in time. As in the case of free mean strains covering the period of both temperature increase and drops, from the presented values the time in which restrained strains causing tension occur cannot be deduced, because it depends on the full history of strains and changes of the mechanical properties of concrete over this period of time.

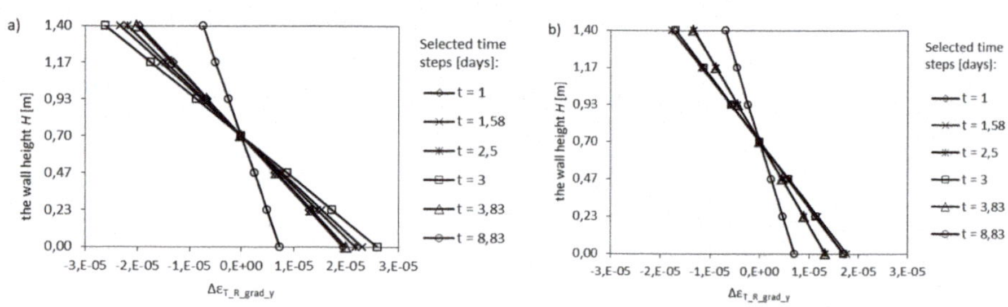

Fig. 7 Gradient of strain restrained part $\Delta\varepsilon_{T\_R\_grad\_y}(t)$ in vertical section of segment no 2 for cases: **a** without CPS, **b** with CPS

4.4 Self-equilibrated Strains

The last component of thermal strains $\Delta\varepsilon_{T\_eq}(t)$ concerns restrained self-equilibrated strains which regardless of the presence or lack of external restraints are an effect of temperature non-linear variations occurring in the section. These strains are 100% restrained part of the strain because it results directly from the design assumption of plane sections before and after the load has been applied to the member.

Figure 8 illustrates strains $\Delta\varepsilon_{T\_eq}(t)$ for wall variants without and with CPS. Definitely higher strains $\Delta\varepsilon_{T\_eq}(t)$ are recorded at the initial period of concrete maturing for the variant without CPS. From the comparison of the obtained values of $\Delta\varepsilon_{T\_eq}(t)$ with the tensile strain capacity of maturing concrete $\varepsilon_{ctu\_e\text{-}a} = 70\ \mu\varepsilon$ [4] it follows that in the variant without CPS the wall may undergo cracking at the bottom edge where the self-equilibrating temperature induced restrained strains are 75 $\mu\varepsilon$ after time $t = 18$ h. The restrained parts of mean strains, strain gradients and self- equilibrated strains occur at different periods of time, so the likelihood of cracking can be finally stated when the impact of other strains has been taken into account. Nevertheless, also in the case of RC tank walls it should be stated that the values of self-equilibrated strains are an important component and should not be disregarded in the assessment of the crack risk.

Figure 9 illustrates self-equilibrating strains $\Delta\varepsilon_{T\_eq}$ for selected points in time in the vertical section of segment No 2. As in the previously discussed strain components, i.e. mean strains and strain gradients that covered the periods of temperature increases and drops, at this stage of analysis it cannot be concluded on the time of the initiation of tensile stress inducing strains because it depends on the full history of strains and changes of the mechanical properties of concrete over this period of time.

The presented strain components of imposed strain and self-equilibrated strain occur simultaneously over the entire period of concrete maturing but their extreme values are shifted in time. Consequently, the following sub-section of the paper will deal with, inter alia, the combined effect of these strains for the entire period of concrete maturing.

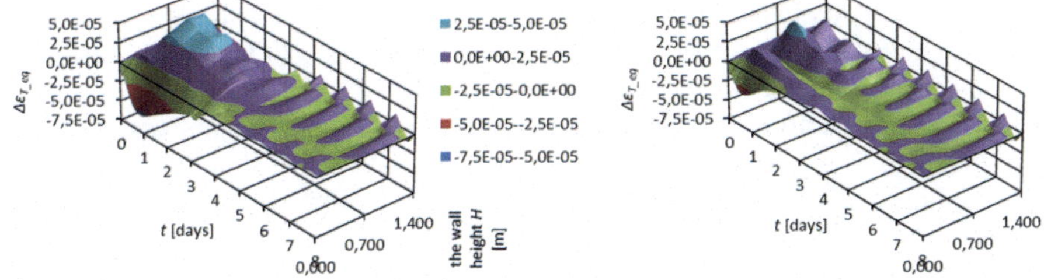

Fig. 8 Variations of self-equilibrating strains $\Delta\varepsilon_{T\_eq}(t)$ in vertical section of segment no 2 in the function of time for variants: **a** without CPS, **b** with CPS

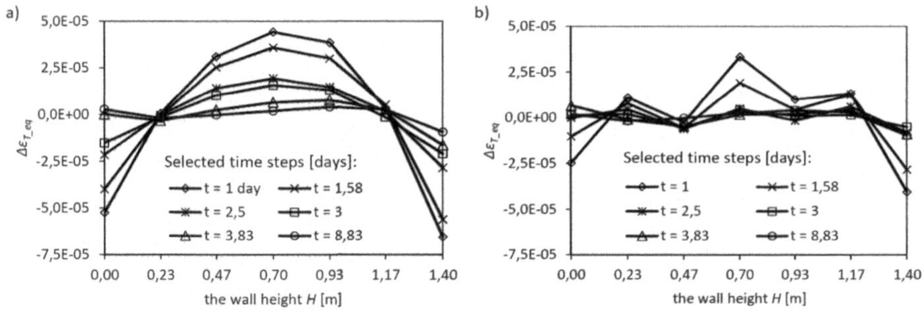

Fig. 9 Variations of self-equilibrating strains $\Delta\varepsilon_{T\_eq}$ in vertical section of segment no 2 for selected points of time for variants: **a** without CPS, **b** with CPS

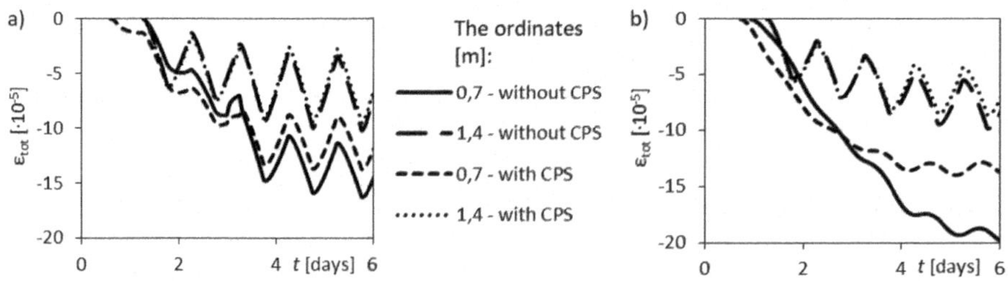

Fig. 10 Total strains **a** on the vertical edge of the wall, **b** along the wall axis

4.5 *Total Strains*

The total restrained strains resulting from changes in the temperature of the maturing concrete are the sum of the strains that are calculated above. Figure 10 presents a comparison of the total strains ε_{tot} calculated for selected points in the wall segment for the variants without CPS and with CPS. However, the presented results include only those strains which occurred due to the temperature drop at a given point of the cross-section. This approach is a conservative one, yet it is recommended by some guidelines. Moreover, due to the occurrence of shrinkage strains in practice that were not included in this analysis, the calculated values were not compared with the tensile strain capacity of the concrete. Figure 10a and b refer to the ordinates 0.7 and 1.4 m on the vertical edge of the wall and along the wall axis, respectively. In general, it can be stated that after 3 days, in the variant with CPS, the strains are smaller. It also means that the cooling intensity during the first 3 days should be limited by the possibility of the occurrence of premature tensile strains. By using CPS, a much greater reduction of strains was obtained for the central part of the wall, i.e. the place where the influence of ambient cooling is the smallest.

The strain ratio for the variants without and with CPS for analogous places in the wall cross-section is illustrated in Fig. 11. It demonstrates that CPS reduces strains in the central part of the wall by almost 50%, while at the remaining points this

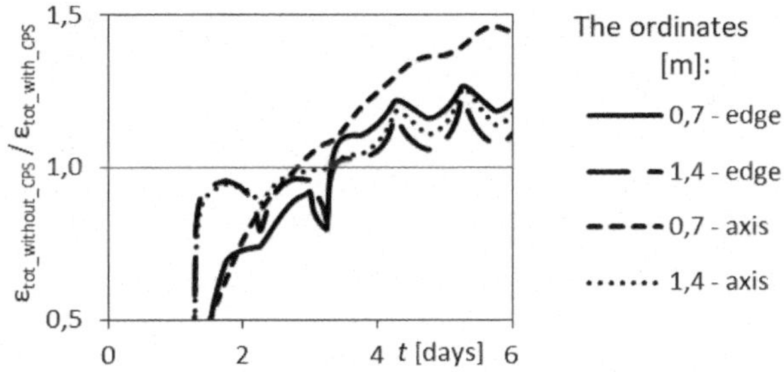

Fig. 11 Total strain ratio for variants without and with CPS

reduction is locally limited to 25%. It should be noted, however, that earlier cooling of the wall results in the occurrence of early-age strains which are larger at some initial stage, mainly in the wall axis. Despite the significant percentage differences observed between these strains (Fig. 11), their absolute values do not differ much (Fig. 10a, b).

5 Conclusions

The analysis of the impact of a cooling pipe system (CPS) on temperature changes in a semimassive wall of a RC tank provides a basis for both scientific conclusions suggesting further research aims and practical implications for design engineers:

- The novelty in this research paper is the analysis of the effect of CPS on the state of strains in a semimassive structure on the example of a RC tank wall.
- In RC semimassive tank walls internal cooling reduces very effectively the temperature rise of maturing concrete only in the immediate vicinity of the cooling pipes. The temperature variations reduction in the other areas is definitely smaller, which may result in locally higher self-equilibrated stresses. Therefore, cooling pipes' spacing should be determined by, inter alia, local temperature changes in the wall section.
- In the process of concrete maturing internal cooling contributes to a reduction of mean strain changes and strains gradient. In the analyzed semimassive wall of a RC tank these reductions are 19 and 35%, respectively, which enables concrete cracking to be limited or eliminated.
- The analysis of self-equilibrated strains indicates that the application of CPS contributes to the reduction to zero of the positive (compressive) strain difference in the immediate vicinity of cooling pipes. Moreover, a CPS helps reduce the extreme strain changes, i.e. positive (compressive) in the wall interior and negative (tensile) in its corners.
- The analysis presented herein provide a basis for further studies on the identification of the most effective/optimal engineering solution employing internal cooling in semimassive RC tank walls, including changes of key parameters such as cooling pipes spacing, cooling rate and duration. These parameters will be

determined by, inter alia, the wall geometry, composition of concrete mix, external conditions and expected results, i.e. complete elimination or merely limiting of the range of early-age concrete cracking.

- CPS results in a significant reduction of thermal strains after 3 days of concrete hardening. During the first 3 days, however, CPS causes greater increments of these strains. Hence, the cooling intensity is limited by the possible occurrence of premature tensile strains.

References

1. Aitcin, P.C., Flatt, R.J.: Science and Technology of Concrete Admixtures. Woodhead Publishing, Great Britain (2015)
2. Aniskin, N.A., Chuc, N.T.: Temperature regime and thermal stress in a concrete massif with pipe cooling. Power Technol. Eng. **52**(6), 638–643 (2019). https://doi.org/10.1007/s10749-019-01009-9
3. Azenha, M., Lameiras, R., de Sousa, C., Barros, J.: Application of air cooled pipes for reduction of early age cracking risk in a massive RC wall. Eng. Struct. **62–63**, 148–163 (2014)
4. Bamforth, P.B.: Control of Cracking Caused by Restrained Deformation in Concrete. CIRIA, London (2018)
5. Bofang, Z.: Thermal Stresses and Temperature Control of Mass Concrete. Butterworth-Heinemann, Great Britain (2014)
6. CEN (European Committee for Standardization): Design of concrete structures. General rules and rules for buildings. Part 1.1. Eurocode 2, CEN, Brussels, Belgium (2004)
7. CEN (European Committee for Standardization): Design of concrete structures. Liquid retaining and containment structures. Part 3. Eurocode 2, CEN, Brussels, Belgium (2006)
8. Falga, K.: Shrinkage stress and reinforcement in concrete structure, Monograph 295. Cracow University of Technology, Cracow, Poland (2004)
9. JSCE (Japan Society of Civil Engineers): Guidelines for concrete. No. 15: Standard specifications for concrete structures. Design. JSCE, Tokyo, Japan (2011)
10. Jeon, S.J.: Advanced assessment of cracking due to heat of hydration and internal restraint. ACI Mater. J. **105**(4), 325–333 (2008)
11. Kiernożycki, W.: Concrete Massive Structures. Polish Cement, Cracow, Poland (2003)
12. Knoppik-Wróbel, A.: Analysis of early-age thermal-shrinkage stresses in reinforced concrete walls. Doctoral thesis, Silesian University of Technology, Gliwice, Poland (2015)
13. König, G., Lieberum, K.-H., Tue, N.: Rißschäden an Betonbauteilen und eine Strategie zu ihrer Vermeidung [Crack damage in concrete elements and a strategy to avoid them] [In German]. Bauingenieur **4**, 155–161 (1996)
14. Lagundžija, S., Thiam, M.: Temperature reduction during concrete hydration in massive structures. Royal Institute of Technology Fortum Stockholm (Master of Science Project Stockholm, Sweden) 132p. (2017)
15. Myers, T.G., Fowkes, N.D., Ballim, Y.: Modeling the cooling of concrete by piped water. J. Eng. Mech. **135**(12), 1375–1383 (2009). https://doi.org/10.1061/(ASCE)EM.1943-7889.0000046
16. Nguyen, T.C., Huynh, T.P., Tang, V.L.: Prevention of crack formation in massive concrete at an early age by cooling pipe system. Asian J. Civil Eng. **20**(8), 1101–1107 (2019). https://doi.org/10.1007/s42107-019-00175-5
17. Nguyen, C.T., Aniskin, N.A.: Temperature regime during the construction massive concrete with pipe cooling. Mag. Civ. Eng. **89**(5), 156–166 (2019)

18. Qiu, Y., Zhang, G.: Stress and damage in concrete induced by pipe cooling at mesoscopic scale. Adv. Mech. Eng. **9**(2), 1–17 (2017).https://doi.org/10.1177/1687814017690509
19. Schlicke, D., Matiašková, L.: Advanced computational methods versus analytical and empirical solutions for determining restraint stresses in bottom-restrained walls. J. Adv. Concrete Technol. **17**(6), 335–349 (2019). https://doi.org/10.3151/jact.17.335
20. Seruga, A., Zych, M.: Thermal cracking of the cylindrical tank under construction. I: case study. J. Perform. Constr. Facil. **29**(4), 04014100 (2015). https://doi.org/10.1061/(ASCE)CF.1943-5509.0000581
21. Seruga, A., Zych, M.: Research on thermal cracking of a rectangular RC tank wall under construction. I: case study. J. Perform. Constr. Facil. **30**(1), 04014198 (2016). https://doi.org/10.1061/(ASCE)CF.1943-5509.0000704
22. Sun, J.C., Pang, Y.J., Zhao, W.Z.: FEM analysis of massive concrete pile using of cooling pipe in shahe bridge. Appl. Mech. Mater. **501–504**, 1359–1363 (2014). https://doi.org/10.4028/www.scientific.net/AMM.501-504.1359
23. Tang, H., Cai, D.S., Yang, L.: New planning of pipe cooling in temperature control for mass concrete. Appl. Mech. Mater. **300–301**, 1584–1588 (2013). https://doi.org/10.4028/www.scientific.net/AMM.300-301.1584
24. Zych, M.: Numerical analysis of cracking in hardening concrete of RC tank's wall. Tech. Trans. **20**(105), 227–249 (2008)
25. Zych, M.: Thermal cracking of the cylindrical tank under construction. ii: early age cracking. J. Perform. Constr. Facil. **29**(4), 04014101 (2015). https://doi.org/10.1061/(ASCE)CF.1943-5509.0000577
26. Zych, M.: Research on thermal cracking of a rectangular RC tank wall under construction. II: comparison with numerical model. J. Perform. Constr. Facil. **30**(1), 04014199 (2016). https://doi.org/10.1061/(ASCE)CF.1943-5509.0000703
27. Zych, M., Chuc, N.T.: Forthcoming. Early age cracking control in semimassive walls of RC tanks by internal cooling. J. Perform. Constr. Facil. (2021). https://doi.org/10.1061/(asce)cf.1943-5509.0001585

Initiation Period of Corrosion by Chloride Ion in Cracked Concrete Structures

Julio Torres and Javier Sánchez

Abstract In the current regulations there is no clear indication about what is the limit of crack opening that can produce a decrease in the service life of reinforced concrete structures, in terms of the arrival of chlorides to the reinforcement and the initiation of corrosion. The crack becomes a preferential way of entry of aggressive agents, especially chlorides in solution. In this work, the transport of chlorides in cracked concrete is studied, obtaining the coefficients of stationary and non-stationary diffusion, the period of initiation of the corrosion, as well as the decrease of this according to the width of the crack. Concrete samples have been manufactured with different cement compositions that, after their conditioning, have been cracked by indirect traction test. The method includes natural and accelerated chloride transport tests, the latter by applying an electric current, according to current Spanish regulations.

Keywords Diffusion coefficient · Chlorides · Concrete · Cracks · Initiation period

1 Introduction

The codes and standards used in the construction of structures take into account a maximum crack opening of the cover concete [1, 2], however, there is no criterion that links the crack opening with the durability of a structure exposed to a certain environment.

One of the most aggressive exposure classes for reinforced concrete structures are environments with chlorides [3–5]. Some researchers propose that for cracks less than 0.4 mm there is no significant increase in the entry of chlorides [6–8]. However, some researchers claim that the presence of cracks greater than 0.3 mm can reduce the life of a structure [9, 10]. In this work it is shown how, for small cracks there is an increase in the transport of aggressive agents to the rebar. The results of both accelerated and natural tests will be shown, using concretes with different contents of cement with cracks of different sizes.

J. Torres (✉) · J. Sánchez
Instituto de Ciencias de La Construcción Eduardo Torroja (CSIC), Madrid, Spain
e-mail: juliotorres@ietcc.csic.es

F. Kanavaris et al. (eds.), *International RILEM Conference on Early-Age and Long-Term Cracking in RC Structures*, RILEM Bookseries 31,
https://doi.org/10.1007/978-3-030-72921-9_10

Table 1 Concrete dosages used

	300 kg	360 kg	400 kg
Water (kg)	150	180	200
CEM I 52,5R (kg)	300	360	400
w/c effective	0.50	0.50	0.50
Gravel (4/12 mm) (kg)	800	743	707
Sand (0.3–4 mm) (kg)	1062	986	938
Fine sand (0–0.3 mm) (kg)	99	92	87
Additive (% weight of cement)	1.20	1.80	2.80

2 Methodology

2.1 *Materials*

Table 1 shows the dosages used. With each dosage, 4 cylindrical samples of 75 mm diameter and 150 mm height were produced, and 9 prismatic samples of 50 × 50 × 250 mm with a steel reinforcement bar of 8 mm diameter. All of them have been cured for 28 days in a curing chamber at 25 °C and 90% humidity.

2.2 *Cracking*

Disc samples. Five 20 mm thick discs are cut from each cylindrical sample. 60 discs have been obtained, of which 4 discs from each dosage (12 in total) do not crack (reference). The remaining 48 samples are cracked in a universal testing machine (INSTRON 8801 model) applying a compression load in two opposite generatrixes by displacement control (see Fig. 1-left). To facilitate the cracking, a silicone strip is applied around each disc to avoid total breakage. The crack width after load removal is between 0.03 and 0.7 mm. The final sample relationship is summarized in Table 2.

Fig. 1 Cracking system of concrete discs and prismatic samples used. Mitutoyo Fisurometer used for crack width measurement

Table 2 Summary of tested discs according to cement content and crack width

300 kg		360 kg		400 kg	
Sample	Crack width (mm)	Sample	Crack width (mm)	Sample	Crack width (mm)
REF 300	0.00	REF 360	0.00	REF 400	0.00
300-1	0.06	360-1	0.03	400-1	0.07
300-2	0.09	360-2	0.05	400-2	0.12
300-3	0.40	360-3	0.18	400-3	0.25
300-4	0.45	360-4	0.35	400-4	0.30
		360-5	0.45	400-5	0.30
		360-6	0.50	400-6	0.40
		360-7	0.70		

Table 3 Summary of tested prismatic specimens according to cement content and crack width

300 kg		360 kg		400 kg	
Sample	Crack width (mm)	Sample	Crack width (mm)	Sample	Crack width (mm)
300-1	0.05	360-1	0.05	400-1	0.05
300-2	0.20	360-2	0.20	400-2	0.25
300-3	0.50	360-3	0.50	400-3	0.50

Prismatics samples. 9 prismatic samples of the 12 total samples with embedded steel are cracked through a 3-point bending test with displacement control in the same testing machine defined above (Fig. 1-right). Crack sizes between 0.05 and 0.5 mm have been obtained. The crack widths obtained are summarized in Table 3.

Crack width measures. All cracked samples are measured with a Mitutoyo portable Fisurometer with an 8x magnification lens, it has a metric scale with an error of ±0.01 mm, with which it is measured in three separate zones of the crack, obtaining the average. The crack regions chosen are those with the widest crack width (see Fig. 1-center).

2.3 Accelerated Tests

The samples without steel, cracked and not cracked (references) are tested according to the UNE 83987 standard "*Concrete durability. Test methods. Measurement of chloride diffusion coefficient in hardened concrete. Multiregime method*". This accelerated method is based on the migration of the chloride ion through the concrete by the application of a 12 V electric field [11]. The concrete disc, after its conditioning, is inserted between two solutions: (i) a solution of 1 M NaCl and (ii) a solution of distilled water. The chloride ions migrate by the effect of the electrical field from the first one, through the concrete, to the second solution of distilled water. The electrical field is generated between two metal electrodes inserted one in each solution (Fig. 2). The increase of chloride ions in the distilled water solution is quantified by measuring the electrical conductivity of the solution. Chloride ions pass through the concrete in two phases, (i) through a non-stationary state, in which the chloride ions can react with components of the concrete and therefore their passage is slower, (ii) through a stationary state, in which the effect of the chloride reaction is considered to be negligible.

The diffusion coefficients obtained through the method described in the standard [11] are those corresponding to the stationary period (Ds) and the non-stationary period (Dns). The stationary diffusion coefficient, Ds, is calculated through the following equation:

$$D_s = \frac{A \cdot R \cdot T \cdot l}{z \cdot S \cdot F \cdot Cl \cdot \gamma \cdot \Delta\phi_{ss}} \quad (1)$$

where, Ds is the coefficient of chloride diffusion in steady state, in cm^2/s, A is the slope of the linear regression equation of the stationary period (mol/s), R is the constant of the perfect gases, in cal/molK (19872), T is the average temperature of the anolite during the test, in K, l is the thickness of the sample, in cm, z is the valence of the chloride ion ($z = 1$), S is the surface area of the sample exposed in the chloride solution, in cm^2, F is Faraday's constant, in cal/Vmol (23060), Cl^- is the initial concentration of chloride in the catholyte, in mol/cm^3, ϒ is the activity coefficient of the catholyte solution (0.657) and $\Delta\Phi$ss is the weighted average effective voltage during the stationary period, in V.

The non-stationary diffusion coefficient, Dns, according to Eq. (2):

$$D_{ns} = \frac{e^2}{2 \cdot t_{paso} \cdot \varphi} \quad (2)$$

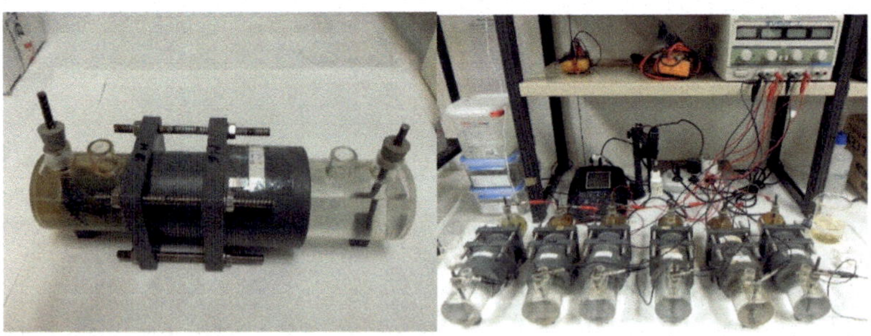

Fig. 2 Multiregime test according to UNE 83987

where Dns is the coefficient of chloride diffusion in the non-stationary state, in cm^2/s, e is the thickness of the specimen, in cm, t is the passage time, in s, φ is the electrical field acceleration factor (39.61 for 20 °C).

Once the diffusion coefficients are obtained, the equations proposed in the regulations can be applied, in this case Fick's second law, to try to obtain the initiation period for corrosion of the reinforcements.

Calculation of the initiation period through the diffusion coefficient. The initiation period due to the penetration of chlorides, is governed by a diffusive process through the concrete that can be modeled through the Eq. (3).

$$d = K_{cl}\sqrt{t} \tag{3}$$

where d is the depth (mm), t is the time (years) and K_{Cl} is the chloride penetration coefficient.

This period of time is necessary so that the concentration of chlorides at the height of the reinforcement, takes a critical value that produces the depassivation of the steel and begins the corrosion. To calculate this time, Eq. (4), if aging is not taken into account, thus, considering a constant diffusion coefficient, the chloride penetration coefficient, K_{Cl}, is calculated according to Eq. (5).

$$t = \left(\frac{d}{K_{Cl}}\right)^2 \tag{4}$$

$$K_{Cl} = \alpha\sqrt{12D_w}\left(1 - \sqrt{\frac{C_{th} - C_b}{C_s - C_b}}\right) \tag{5}$$

where α is the unit conversion factor, its value is 56157, D_w is the diffusion coefficient, for a crack width (cm^2/s), Cth is the critical chloride concentration, (% weight of cement), Cs is the surface chloride concentration (% weight of cement) and Cb is the content of chlorides in raw materials (% weight of cement).

In this study, it has been taken as diffusion coefficient, Dns and Ds, obtained in the multi-regime test of cracked concrete. It has been considered that passivity of reinforcement exists for Cth contents up to 0.6%, according to the Spanish regulations [1] for passive reinforcements. The surface concentration, Cs, which depends on the conditions of exposure and, in the absence of experimental values, values are taken according to the Spanish standards, these values in % of cement are summarized in Table 4. The content of chlorides, Cb, in the raw materials used is taken as zero.

The value of the cover of the reinforcement used for the calculation of the time of the initiation period, depth (d), is also referenced in the normative according to the type of cement, marine environment and the useful life of the project, the summary of these values is observed in Table 4.

Table 4 Surface concentration (Cs) in % in weight of cement and minimum concrete cover according to marine environment and dosage for 50 years service life

Marine environment IIIa (Spanish regulations EHE-08) or XS1 (EN 206), [1]		
Kg cement/m^3 (kg)	C_s	Concrete cover (mm)
300	1.092	45
360	0.894	45
400	0.802	45

2.4 Natural Tests

These tests are based on exposing, the cracked zone of the samples, to a marine environment through a continuous dissolution of 30 g/l of NaCl during 1190 days. During the duration of the test, a measurement of corrosion potential and corrosion intensity is made every 5 days. With these measurements, the active or passive state of the steel is checked and the end of the initiation period is determined, which is reached when the steel has a corrosion intensity higher than 0.2 $\mu A/cm^2$ during several consecutive measurements [12]. The corrosion intensity is obtained by measuring the resistance to polarization (Rp). This electrochemical method is used to calculate the corrosion speed according to the UNE 112072 standard [13].

Calculation of the initiation period through natural tests. The initiation period is defined as the time from the manufacture of the concrete until the aggressive agents of the environment reach the reinforcing steel, in this case the chloride ion breaks the passive layer and corrosion begins. In the natural test the initiation period ends when several consecutive measurements of the corrosion intensity are above the limit value set at 0.2 $\mu A/cm^2$ [12].

To choose which diffusion coefficient is correct to use to obtain the initiation period through the accelerated test, the values obtained from diffusion coefficients in that test are compared with those obtained in the natural test. For this purpose, the natural diffusion coefficient in cracked samples is obtained by clearing it from Eq. (5), with the help of Eq. (4). The values of the concentration of chloride in the reinforcement, C_{th}, and of the surface concentration, Cs, both in % in weight of cement, have been obtained from the natural test through an analysis of the concrete sample in both areas by potentiometric analysis, according to the UNE 80213 standard, Cements Test Methods. Potentiometric determination of chlorides [14].

3 Results

3.1 Accelerated Tests

The results of the multiregime tests for each concrete as a function of the cement content are shown in Fig. 3. There are two phases in the testing of the uncracked

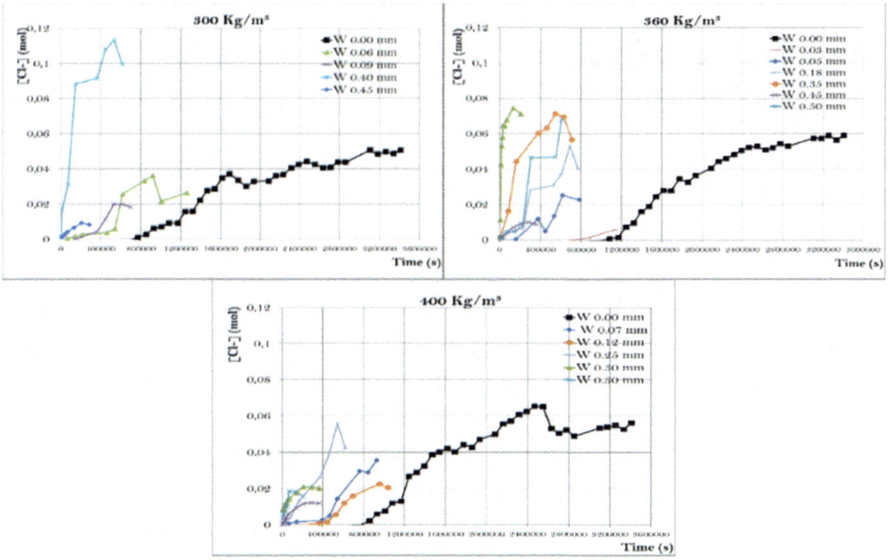

Fig. 3 Variation of chloride concentration with time in the multiregime tests

samples (in black in the figure), a first stage in which the arrival of the chloride in the deposit is not detected initially without chlorides (non-stationary phase). This is because the chloride is passing through the thickness of the sample and reacting with the hydrated phases of the cement. And a second stage, called stationary, in which the chloride already appears in the anolite, increasing its concentration gradually and regularly, during the first hours. In the case of cracked samples, the first stage corresponding to the non-stationary state is practically non-existent. This indicates that the chloride is passing a freely into the anolite, independent of the size of the crack. This may indicate that independently of the crack size, the diffusion coefficient of the chloride ion that should be used for the calculation of the initiation period in cracked concrete is the stationary state.

Stationary and non-stationary diffusion coefficients increase linearly with crack width independent of cement content and therefore the data can be taken as the same population, see Fig. 4. If the non-stationary coefficients are observed there seems to be a critical crack, between 0.2 and 0.3 mm, from which the values increase substantially. This performance allows us to deduce that there is a first stage in which the chloride diffuses within the crack, its width is critical, and when it arrives at a critical size the coefficients do not depend on the crack width and the chloride diffuses freely. In the stationary coefficients there is no such strong change in the tendency, producing an increase in the diffusion coefficient practically linear with the width of the crack. The relationship between diffusion coefficients and crack width is represented by adjusting the trend line to these representations, resulting in the following Eqs. (6) and (7).

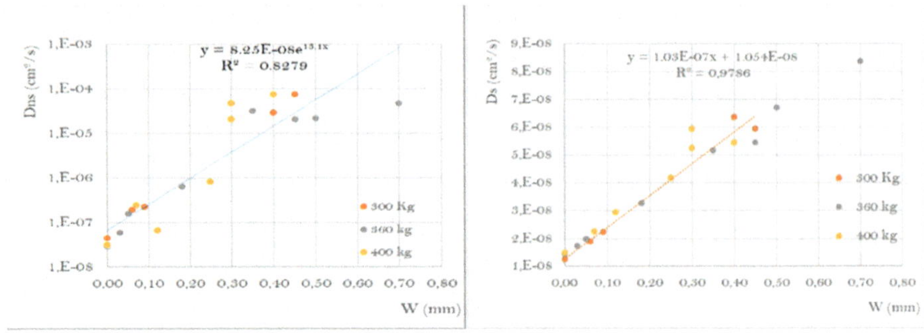

Fig. 4 Non-stationary and stationary diffusion coefficients as a function of the crack width for the different dosages

$$D_s = 1.03 \times 10^{-7} \cdot w + 1.54 \times 10^{-8} \tag{6}$$

$$D_{ns} = 8.25 \times 10^{-8} \cdot e^{13.1.w} \tag{7}$$

3.2 *Natural Tests*

Table 5 shows the results of the initiation period, the surface concentration of chlorides, C_s, the concentration of chlorides in the concrete in contact with the reinforcement, C_{th}, and the diffusion coefficient calculated through Eqs. (4) and (5), D_{real}, as a function of the crack width and cement content in the concrete. As shown in Table 5 the initiation period is reduced with the crack width as expected.

Table 5 Summary of results obtained in the natural test

Kg cement/m^3	Crack width, W, (mm)	Time initiation period (year)	Cs (% cement weight)	Cth (% cement weight)	Diffusion Coefficient, Dreal, (cm^2/s)
300	0.05	1.189	1.770	1.600	4.043E–08
	0.20	0.425	1.560	1.200	1.816E–08
	0.50	0.304	1.040	0.800	2.535E–08
360	0.05	0.518	1.330	0.990	1.195E–08
	0.20	1.112	1.270	1.100	2.179E–08
	0.50	0.304	1.190	0.890	2.097E–08
400	0.05	0.425	1.380	0.900	7.411E–09
	0.25	0.304	1.210	0.742	8.144E–09
	0.50	0.304	1.280	0.959	2.121E–08

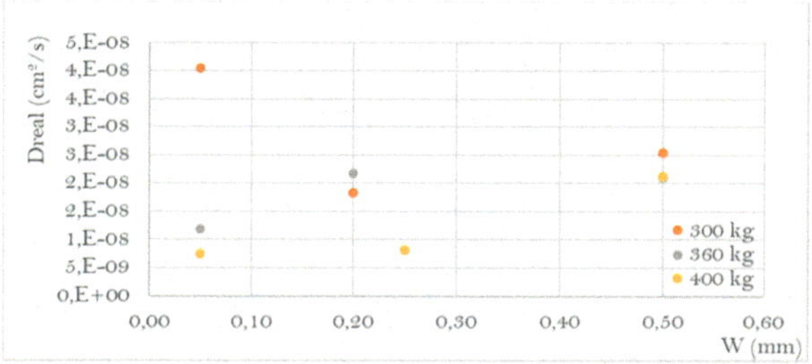

Fig. 5 Diffusion coefficients in the natural test

The diffusion coefficients depend on the crack width, as it occurred in the stationary and non-stationary diffusion coefficients in the accelerated test, see Fig. 5. This increase of diffusion coefficients is supported by the decrease of the initiation period time. The difference is between 28–74% between smaller crack width specimens with respect to those with larger crack widths. These variations with crack width are not as clear in the natural tests as in the accelerated tests because in the natural test there is no current to force the chloride to move.

4 Discussion

4.1 Selection of the Coefficient for the Calculation of Service Life

In view of the results, a first question is which of the coefficients obtained in the accelerated test is the best adapted to a natural test and therefore the most appropriate to apply to the calculation of the service life. Normally the one used is the non-stationary D_{ns} [15–18] since it takes into account both transport and reaction. In reality, a stationary regime is never reached due to the fact that the chloride advance front will always imply a new reaction with the hydrated phases of the cement. However, in the case of cracked elements it seems that a stationary regime is produced since the cracked area does not represent a physical impediment to the mobility of chlorides, as can be seen in Fig. 6. The calculations of the initiation period will be made with the stationary diffusion coefficient calculated with the accelerated test.

4.2 Calculation of the Initiation Period

In this study, it is assumed as stationary diffusion coefficient, Ds, the value obtained in the multiregime test of the cracked concrete, Table 6. The critical chloride concentration, Cth, is the value of 0.6% in weight of cement, which summarizes the EHE-08 for rebars. The surface concentration, Cs, which depends on the exposure conditions

Fig. 6 Diffusion coefficients in the natural and accelerated tests

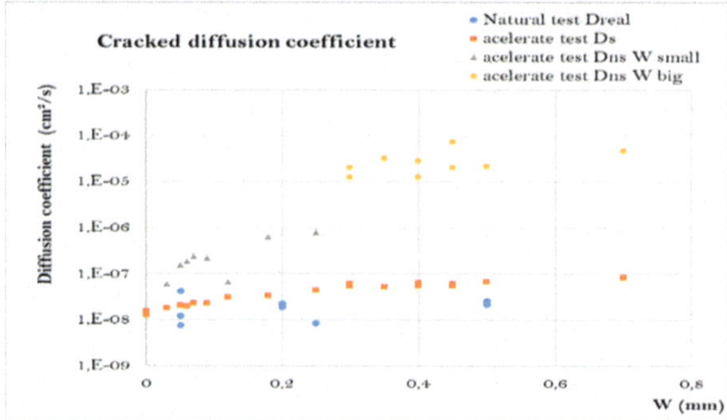

Table 6 Diffusion coefficients of samples tested by the accelerated test

Kg cement/m³ 300 kg								
W (mm)	0.00	0.06	0.09	0.40				
Ds (cm²/s)	1.2E–08	1.9E–08	2.2E–08	6.3E–08				
Kg cement/m³ 360 kg								
W (mm)	0.00	0.03	0.05	0.18	0.35	0.45	0.50	0.70
Kg cement/m³ 400 kg								
W (mm)	0.00	0.07	0.12	0.25	0.30	0.30	0.40	
Ds (cm²/s)	1.4E–08	2.2E–08	2.9E–08	4.1E–08	5.9E–08	5.2E–08	5.4E–08	

and which, in the absence of experimental values, is taken from Table A.9.4. of Annex 9 of EHE-08 [1], these values in % of cement are summarized in Table 4.

The value of the concrete cover used for the calculation of the initiation period, depth (d), is also referenced in the EHE-08 according to the type of cement, marine environment and the useful life of the project, the summary of these values is observed in Table 4. The results of the duration of the initiation periods according to the crack width, for the considered environment, is observed in Fig. 6. Two aspects are emphasized from the results: (1) that the useful life decreases with the crack width as it was deductible from the previous Figures and (2) that although the coefficients are very similar in the three concretes, the result of its service life changes. This is due to the difference in surface concentration indicated by the EHE 08 model: the higher the Cs (concrete with lower cement content), the shorter the service life of the calculation.

Based on Eqs. (4) and (5), substituting the stationary diffusion coefficients obtained in the multiregime test, the relationship between the initiation periods of cracked and non-cracked concrete is obtained. This relationship helps to estimate the loss of service life according to the width of the crack, which is important when you want to know the time until corrosion in cracked concrete.

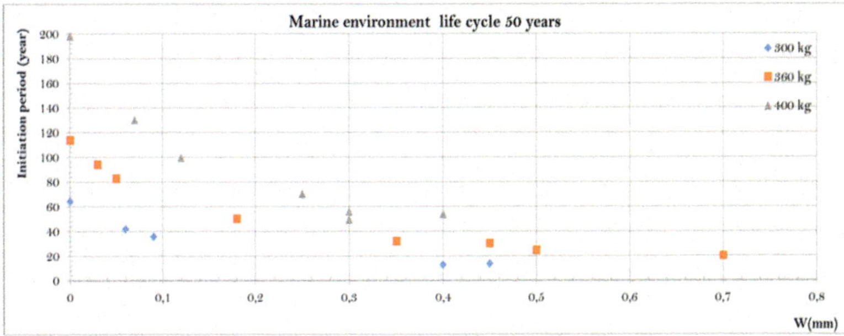

Fig. 7 Variation of the initiation period in marine environment IIIa according to crack width for 50 years of service life

In Fig. 7, the adjustment of the experimental data is observed, which relates the variation of the initiation period (t_W/t_0) with the crack width, regardless of the dosage in the environment in which the concrete is exposed.

A structure that has a maximum crack width allowed in the standard of 0.4 mm, has a reduction of its initiation period and therefore of its service life by 74%. The crack width is a basic variable to take into account to determine the useful life of a cracked element.

5 Conclusions

The main conclusions obtained in this study are summarized below:

(1) It is possible to use the accelerated test referred to in the regulations for the calculation of the diffusion coefficients of cracked concrete.
(2) The relationship between the chloride ion diffusion coefficient in cracked and non-cracked concrete (CEM I) is independent of the cement content of the concrete perhaps because the a/c ratio is the same in all three dosages.
(3) The relationship between the crack width and the stationary diffusion coefficient found fits the Eq. (6).
(4) Applying these values to the formulas of EHE 08 and assuming the rest of the values of the constituent parameters of these equations, it results that the service life in cracked elements decreases exponentially (see adjustment Fig. 8).

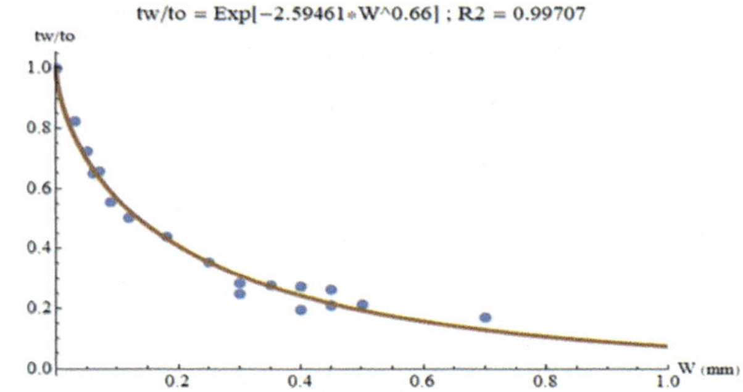

Fig. 8 Variation of initiation period with crack width

References

1. EHE–08 Instrucción de hormigón estructural—Ministerio de Fomento. Spain. (2008)
2. Model Code 2010. fib (2012)
3. Mohammed, T., Otsuki, N., Hamada, H.: Corrosion of steel bars in cracked concrete under marine environment. J. Mater. Civ. Eng. **15**(5), 460–469 (2003)
4. Okada, K.: Chloride Corrosion of Reinforcing Steel in Cracked Concrete. ACI Publication, (1980)
5. De Schutter, G.: Quantification of the influence of cracks in concrete structures on carbonation and chloride penetration. Mag. Concr. Res. **51**(6), 427–435 (1999)
6. Schießl, P., Raupach, M.: Laboratory studies and calculations on the influence of crack width on chloride-induced corrosion of steel in concrete. Mat. J. **94**(1), 56–61 (1997)
7. Nürnberger, U.: Chloride corrosion of steel in concrete, part 2. Betonwerk Fertigteil-technik **10**, 697–704 (1984)
8. Hartl, G., Lukas, W.: Investigations on the penetration of chloride into concrete and on the effect of cracks on chloride-induced corrosion of reinforcement. Betonwerk Fertigteil-technik **7**, 497–506 (1987)
9. Shaikh, F.U.A.: Effect of cracking on corrosion of steel in concrete. Int. J. Concr. Struct. Mater. **12**(1), 3 (2018)
10. Otieno, M.B., Alexander, M.G., Beushausen, H.D.: Corrosion in cracked and uncracked concrete–influence of crack width, concrete quality and crack reopening. Mag. Concr. Res. **62**(6), 393–404 (2010)
11. UNE 83987.: Concrete durability. Test methods. Measurement of chloride diffusion coefficient in hardened concrete. Multiregime method (2014)
12. Vennesland, Ø., Raupach, M., Andrade, C.: Recommendation of Rilem TC 154-EMC: "Electrochemical techniques for measuring corrosion in concrete" measurements with embedded probes. Mater. Struct. **40**(8), 745–758 (2007)
13. UNE 112072.: Laboratory measurement of corrosion speed using the polarization resistance technique (2011)
14. UNE 80213.: Methods of cement testing. Potentiometric determination of chlorides
15. Sergi, G., Yu, S.W., Page, C.L.: Diffusion of chloride and hydroxyl ions in cementitious materials exposed to a saline environment. Mag. Concr. Res. **44**(158), 63–69 (1992)
16. Saetta, A., Scotta, R., Vitaliani, R.: Analysis of chloride diffusion into partially saturated concrete 24. ACI Mater. J. **90**, 441–451 (1993)
17. Djerbi, A., Bonnet, S., Khelidj, A., Baroghel-Bouny, V.: Influence of traversing crack on chloride diffusion into concrete. Cem. Concr. Res. **38**, 877–883 (2008)
18. Jang, S.Y., Kim, B.S, Oh, B.H.: Effect of crack width on chloride diffusion coefficients of concrete by steady-state migration tests. Cem. Concr. Res. **41**, 9–19 (2011)

Investigation on Crack Control and Crack Pattern Analysis of Self-compacting Concrete Exposed to Standard Fire Exposure

Mervin Ealiyas Mathews, N. Anand, A. Diana Andrushia, and Tattukolla Kiran

Abstract Development of thermal cracks is inevitable in concrete structures under extreme fire conditions. Intense heat flux on the exposed surface and the thermal incompatibility between the mortar and aggregate phase results in micro-cracks. The geometrical properties of cracks such as width, length, density and pattern are to be quantified accurately to adopt suitable control methods. This study aims to analyse the crack pattern and crack control measures on Self-Compacting Concrete (SCC) exposed to standard fire temperature. SCC mixes were developed to achieve a strength of 30MPa using Fly Ash (FA), Ground Granulated Blast Furnace Slag (GGBFS) and Expanded Perlite Aggregate (EPA). The rheological behaviour of the developed mixes was confirmed as per European Federation of National Associations Representing for Concrete (EFNARC 2005) guidelines. After the curing process, specimens were exposed to elevated temperatures for different durations (30, 60, 90, and 120 min) following the International Organization for Standardization ISO 834 standard fire curve. A mortar combination of Cement-Perlite Plaster (CPP) was considered as the sacrificial layer (protection layer) over the concrete surface to understand the efficiency of perlite in controlling the crack growth. GGBFS-SCC specimens exhibited severe surface cracking than FA-SCC specimens. The protected (CPP) specimens showed better resistance on mitigating the surface cracks even at 120 min of heating. Attempts have been made to quantify the cracks on the concrete surface with the help of Image-Processing Tool (IPT). Scanning Electron Microscopy (SEM) analyses have been carried out to analyse the micro-cracks in the fire-damaged concrete.

Keywords Self-compacting concrete · Elevated temperature · Thermal cracks · Image processing · EFNARC

M. E. Mathews (✉) · N. Anand · A. D. Andrushia · T. Kiran
Karunya Institute of Technology and Sciences, Coimbatore, India

F. Kanavaris et al. (eds.), *International RILEM Conference on Early-Age and Long-Term Cracking in RC Structures*, RILEM Bookseries 31,
https://doi.org/10.1007/978-3-030-72921-9_11

1 Introduction

Concrete has many versatile characteristics and advantages in contrast to other construction materials. Some of them are compressive strength, inherent fire resistance, cost-effectiveness in implementing the construction projects [1]. However, the cracks in concrete persist in its plastic and hardened form due to various causes, as shown in Fig. 1. The crack formation and propagation adversely affect the durability of concrete [2].

According to the American Concrete Institute, the term durability is stated as the ability to remain serviceable, retain quality and stability even after exposed to critical environmental conditions or accidental events [3]. The intensity of surface crack generation and propagation on hardened concrete due to unexpected fire is larger than the cracks formed by other durability problems. As the concrete becomes more brittle after the exposure of fire, these cracks gain more importance in reducing the service life of concrete structures [4]. The effect of fire-induced damages ultimately affects the durability of concrete structures. The intense irregular formation of surface cracks causes a discontinuity in the particle phase and reduce the local stiffness [5]. Proper, timely assessment and preventive measures to reduce the crack induced damages that will be useful to avoid structural failure [6].

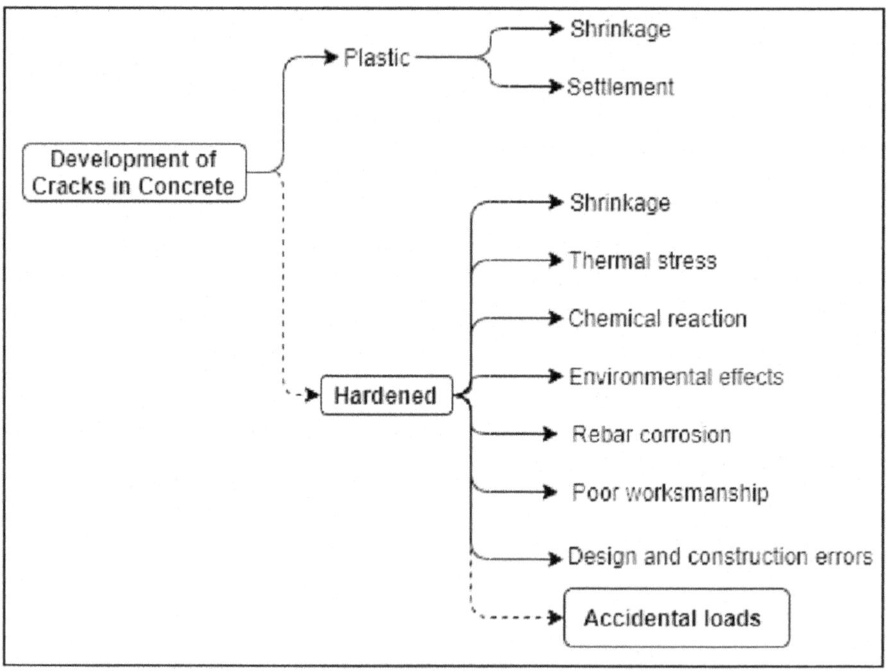

Fig. 1 Development of cracks in concrete as per ACI

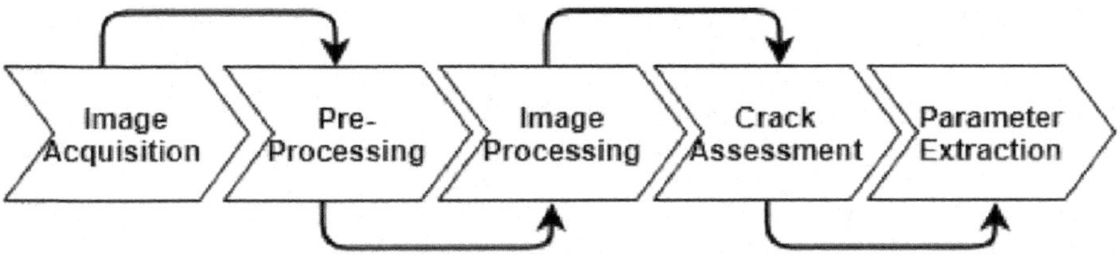

Fig. 2 Structural flow of crack detection and analysis based on IPT

Destructive Testing (DT) and Non-Destructive Testing (NDT) are the methods for crack detection and assessment with the help of human visual observations and inspecting tools thereby the surface deformities can be assessed and evaluated [7]. The reduction in the load-carrying capacity and durability level of the concrete structure can be predicted by evaluating the crack properties such as type, length, width, and density of surface cracks [8]. The human inspective methods of surface crack evaluation can be done quickly and instantly, but the concerns about accuracy and localized evaluation restriction exist, also its objective-based deterioration analysis is complicated [9]. Automatic surface defect analysis is being developed with the help of various image processing tools (IPT) to overcome the drawbacks of manual inspection [10]. This will enhance the accuracy of damage assessment, allows objective deterioration evaluation and the process can be executed with high efficiency. Infrared (IR), ultrasonic, laser, thermal and radiographic methods are other assessment adopted for crack defects identification [9].

The crack detection and analysis based on IPT is having great interest among researchers because of its simplicity and reliability [11, 12]. Algorithmic, morphological, percolation and practical based methods are the four approaches generally adopted in IPT procedures [13]. The typical image processing difficulties are noise, shades, blemish and sharpness, which are expected while extracting the results. Figure 2 explicit the structural flow of crack detection and analysis based on IPT.

The main objective of the present work is to assess and analyse the effect of temperature on the crack intensifying nature of SCC specimens developed with FA, GGBFS and EPA. Manual inspection and observations were made using inspecting tools, and the mean width of the visually largest crack is noted. IPT based crack detection and quantification were used to analyse the damage level of the fire exposed concrete. Scanning Electron Microscopy (SEM) based microstructure analysis was carried out to understand the particle phase discontinuities in the fire affected concrete. The outcome of the results was used to develop a crack control mortar combination of CPP over the concrete surface to understand the efficiency of perlite in controlling the crack growth.

2 Experimental Program

The experimental investigation focused on analysing the effect of temperature on the formation of surface cracks on the SCC specimen developed with FA and GGBFS. To understand the effectiveness of perlite in regulating the crack growth, a mortar combination of CPP was wrapped over the concrete surface. The maximum temperature simulates the heat effect of a real fire varied from Room Temperature (RT) to 1029 °C following ISO 834 fire rating curve. Siliceous aggregates are used for the development of concrete mixes.

2.1 Mix Proportion

Four different SCC mixes were developed, and the mix proportioning is stated in Table 1. Ordinary Portland Cement (OPC) of 53-grade was used for the investigation. The slump flow test assessed the consistency and workability of fresh SCC. The flow spread diameter values were measured between 660 and 680 mm, and it satisfies the permissible limit stipulations of EFNARC 2005 [14].

Mineral Admixtures

Class-F FA and GGBFS are the industrial by-products, used as a replacement material for cement in the development of SCC mix. The chemical and physical properties of materials are given in Table 2.

Aggregates

Crushed granite stones of maximum 12 mm size and Manufactured Sand (M-Sand) with a maximum particle size of 4.75 mm were used as coarse and fine aggregate respectively for Mix 1 and Mix 3. Mix 2 and Mix 4 consist of 2.5% fine aggregate is replaced with EPA content to understand the fire resistance capability of perlite in the SCC core structure. The optimum level of EPA is achieved as 2.5% by different trial on fresh concrete. This mix satisfied the strength requirement of hardened concrete at elevated temperature. The properties of EPA is given in Table 2.

Table 1 Mix proportions of SCC

Description	Cement/Mineral admixtures	Aggregates (coarse/fine; EPA)	Water	Chemical admixtures (SP/VMA)
	(kg/m^3)			
Mix 1	320/144(FA)	804/963	218	0.80/0.10
Mix 2	320/144(FA)	804/939; 24 (EPA)	218	0.90/0.15
Mix 3	320/144(GGBFS)	804/963	218	0.90/0.15
Mix 4	320/144(GGBFS)	804/939; 24 (EPA)	218	1.0/0.20

Table 2 Characteristics of the materials

Material	Chemical composition (%)						Physical property	
	CaO	SiO_2	Al_2O_3	Fe_2O_3	MgO	K_2O	Specific gravity	Fineness (m^2/kg)
OPC	61.33	18.78	8.75	3.40	2.25	2.20	3.15	335
FA	2.25	58.50	28.25	3.45	0.30	1.26	2.60	306
GGBFS	37.34	37.73	14.42	1.11	8.71	–	2.90	441
EPA	0.6–2.40	72–76	11–16	0.6–1.40	–	2–5	0.055–0.30	–

Water

Potable water is used throughout the experimental investigation for developing, casting and curing.

2.2 *Details of Test Specimens*

Test specimens with a diameter of 150 mm and a height of 63.50 mm are used for the investigation. It is decided to conduct a surface crack assessment on cylindrically shaped samples. As the cylindrical shape undergoes comparatively more uniform heating for varying heating cycles, it is preferable. Also, the formation of surface cracks on the exposed face will be ideally distributed across the cross-section, but in case of square or rectangular specimens, the crack distribution is more seen in corner regions due to the heat flux causing high-stress concertation in that region. The heating procedure is carried out on two different specimen conditions. The unprotected specimens are taken for heating after the curing period without plastering. The protected (with CPP layer) specimens were cast with plaster of OPC and EPA mortar (1:4). Both the unprotected and protected specimens were dried for 24 h before taking it for heating in the furnace. The view of the test specimens and heating of specimen in the furnace is shown in Fig. 3.

2.3 *Test Methodology and Heating Procedures*

The test methodology is as follows:

- Preparation of test specimens (FA; FA/CPP; FA/EPA; FA/EPA/CPP; GGBFS; GGBFS/CPP; GGBFS/EPA and GGBFS/EPA/CPP) [Details of test specimens id are mentioned in Table 3].
- Curing process for 28 days.
- Specimens kept for drying in laboratory conditions at RT (for 24 h).
- Selected batch of specimens are wrapped with CPP layer.

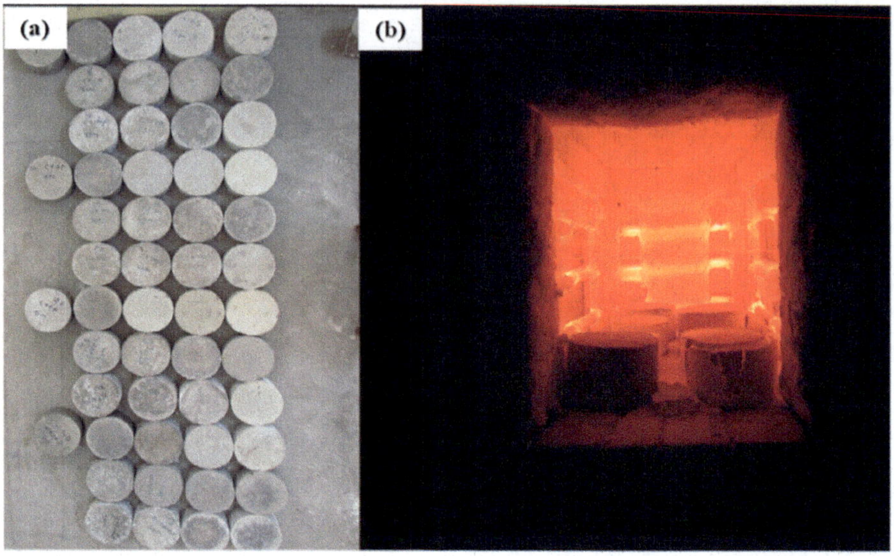

Fig. 3 **a** View of the test specimens. **b** Heat test of specimen in computerised electrical furnace

- Heating of specimens following ISO 834 fire rating curve (for 821 °C at 30, 925 °C at 60, 986 °C at 90, and 1029 °C at 120 min) [15].
- Crack detection and analysis.

The content of moisture inside the specimen was measured by taking the difference between the actual weight and the weight of the specimens after heat exposure, and it was noted that moisture loss was varying between 1.5 and 6.5%.

The importance of simulating the real effect of fire was of great concern in this study. The rate of heating and duration of the heating pattern was followed similar to the standard fire curve (ISO 834) with the help of an integrated microprocessor-based control panel in the electric furnace. The duration considered for the study is 30, 60, 90, and 120 min. Once the target temperature is reached, the furnace turns off automatically. After that, the specimens were taken out, and the necessary natural cooling is provided.

3 Test Results and Discussions

3.1 Effect of Elevated Temperature and Mineral Admixtures on Crack Development

The temperature and duration of the fire exposure have a significant effect on the surface crack formation and its growth. Figure 4 shows the captured surface images

Table 3 Measured crack width of SCC specimens at different durations of heating

Specimen id	Crack width (mm)				Remarks
	W_{min}	W_{max}	W_{avg}	$W_{std.dev}$	
FA/CPP 0M	0.00	0.00	0.00	0.00	FA-SCC Specimens with CPP layer
FA/CPP 30M	0.00	0.00	0.00	0.00	
FA/CPP 60M	0.02	0.04	0.03	0.01	
FA/CPP 90M	0.04	0.06	0.05	0.01	
FA/CPP 120M	0.18	0.20	0.19	0.07	
FA/EPA 0M	0.00	0.00	0.00	0.00	FA-SCC Specimens with EPA as filler material
FA/EPA 30M	0.02	0.04	0.03	0.01	
FA/EPA 60M	0.09	0.12	0.11	0.04	
FA/EPA 90M	0.16	0.20	0.18	0.05	
FA/EPA 120M	0.26	0.28	0.27	0.06	
FA/EPA/CPP 0M	0.00	0.00	0.00	0.00	FA-SCC Specimens with EPA as filler and CPP layer
FA/EPA/CPP 30M	0.00	0.00	0.00	0.00	
FA/EPA/CPP 60M	0.01	0.02	0.02	0.01	
FA/EPA/CPP 90M	0.07	0.08	0.08	0.03	
FA/EPA/CPP 120M	0.09	0.12	0.11	0.05	
GGBFS/CPP 0M	0.00	0.00	0.00	0.00	GGBFS-SCC Specimens with CPP layer
GGBFS/CPP 30M	0.01	0.02	0.02	0.01	
GGBFS/CPP 60M	0.02	0.04	0.03	0.01	
GGBFS/CPP 90M	0.13	0.16	0.15	0.04	
GGBFS/CPP 120M	0.24	0.26	0.25	0.06	
GGBFS/EPA 0M	0.00	0.00	0.00	0.00	GGBFS-SCC Specimens with EPA as filler material
GGBFS/EPA 30M	0.11	0.12	0.12	0.05	
GGBFS/EPA 60M	0.20	0.22	0.21	0.08	
GGBFS/EPA 90M	0.29	0.30	0.29	0.10	
GGBFS/EPA 120M	0.31	0.36	0.33	0.12	
GGBFS/EPA/CPP 0M	0.00	0.00	0.00	0.00	GGBFS-SCC Specimens with EPA as filler and CPP layer
GGBFS/EPA/CPP 30M	0.00	0.00	0.00	0.00	
GGBFS/EPA/CPP 60M	0.01	0.02	0.02	0.01	
GGBFS/EPA/CPP 90M	0.15	0.18	0.17	0.05	
GGBFS/EPA/CPP 120M	0.18	0.20	0.19	0.07	

and the thermal crack width of different unheated and heated SCC specimens. It was noted that as the rate and duration of heat exposure increases, the crack width also increases significantly. The magnified visuals of measuring crack width are enclosed within the actual view of the surface images. The crack width is measured using the Elcometer. The density of the surface crack increased drastically as the duration of heating increases from 30 to 120 min.

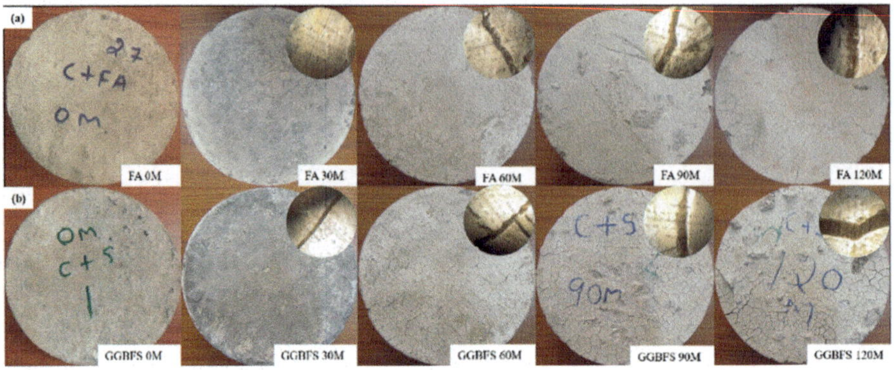

Fig. 4 View of thermal crack width (Unheated and heated SCC specimens). **a** FA-SCC specimens. **b** GGBFS-SCC specimens

Figure 5 shows the plot showing the duration of heating and the average crack width of SCC specimens developed with FA and GGBFS. The data is formulated by recording several numbers of crack width measured with mean values. It is observed from the plot that GGBFS-SCC specimens has comparatively larger surface thermal cracks compared to FA-SCC specimens. The 30 min heated FA-SCC specimens (FA 30M) had 0.02 mm of average crack width ($W_{avg.}$), but in case of GGBFS-SCC specimens (GGBFS 30M) heated for 30 min, the $W_{avg.}$ were found to be 0.10 mm.

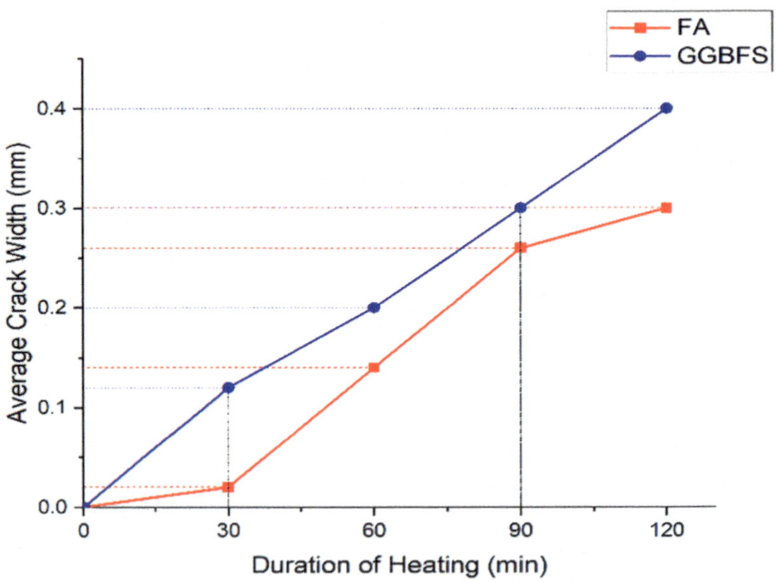

Fig. 5 Average crack width of FA-SCC and GGBFS-SCC specimens

The $W_{avg.}$ values observed for 120 min heated specimens of FA (FA 120M), and GGBFS (GGBFS 120M) are 0.29 mm and 0.39 mm respectively.

3.2 *Effect of CPP Technique on Surface Crack Control of SCC*

In the previous phase of the study, an analysis carried out regarding the effect of elevated temperature and the variation in the crack intensity of SCC specimens with different admixtures. Subsequently, in the second phase of the investigation, authors have attempted to develop a crack control method to reduce the effect of heat directly falling on the SCC surface by incorporating the EPA as a protective coating material. The study has been carried out in three different ways such that by adding EPA at an optimum level (2.5% of fine aggregate content) as internal filler material along with the powder (binder) content, as a mortar combination of Cement-Perlite Plaster (CPP) was considered as the sacrificial layer (protection layer) over the concrete surface and at last a combination of the EPA as filler and CPP layer.

Table 3 illustrates the summary of results recorded in terms of minimum, maximum, average and standard deviation of crack width measured using Elcometer on the surface of FA-SCC specimens and GGBFS-SCC specimens exposed to elevated temperature. The significant surface crack formation was not observed up to 30 min of heating. Notable surface rupture was noticed between 30 and 120 min of heating. It is noted that in the test specimens without protective CPP layer exposed to 120 min of heating has a significant effect on wider surface crack development. The protected specimens with CPP layer in both the FA and GGBFS-SCC specimens (i.e., FA/EPA/CPP, FA/CPP, GGBFS/EPA/CPP & GGBFS/CPP) were resistant to the expansion of crack even at 120 min. But as noted in the previous section, significant surface cracks were observed due to heat even at 60 min of heating for both the specimens FA and GGBFS, without EPA and CPP. This clearly shows the effectiveness of the protection technique using EPA mineral as a sacrificial material.

3.3 *SEM Analysis*

SEM analysis is carried out to understand the effect of damage incurred inside the SCC core due to elevated temperature exposure. As visual observations help identify the surface damage concerning thermal cracks, SEM analysis is helpful to understand the morphological changes that happened inside the SCC structure before and after heating. The cracks and voids increases, as the exposure duration increases for FA and GGBFS specimens. Also, the CPP layer had protected specimens from the severe effect of fire. Figure 6 shows the SEM images of 120 min heat-exposed specimens (a) FA-SCC specimens, (b) GGBFS-SCC specimens.

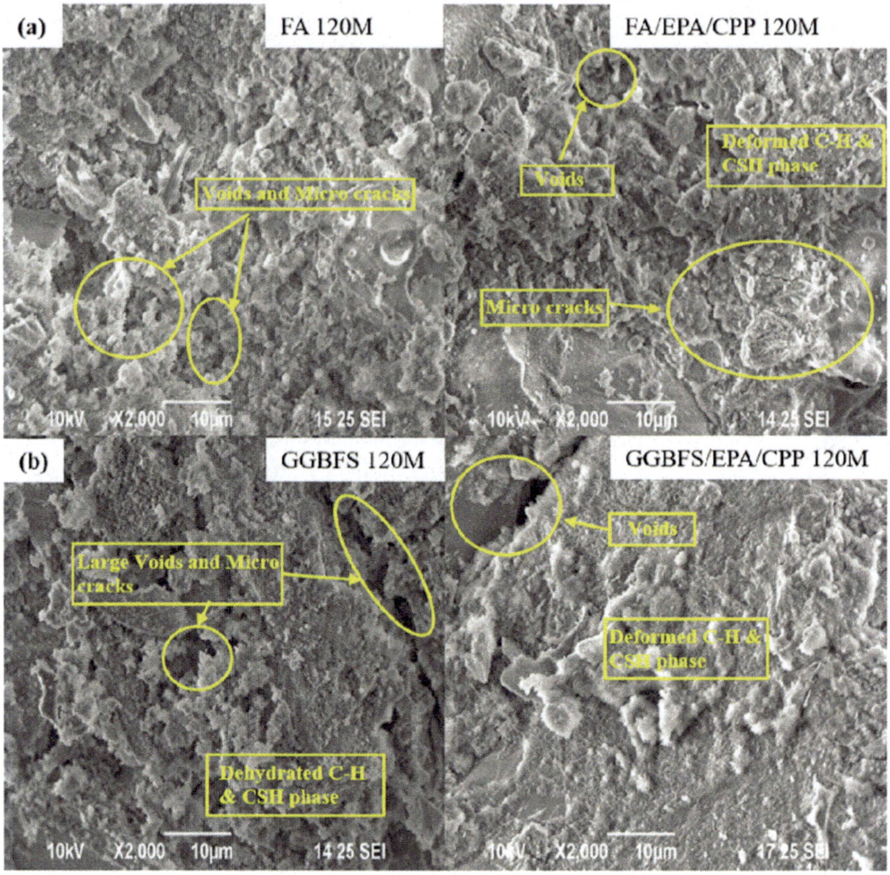

Fig. 6 SEM images of heat-exposed specimens. **a** FA-SCC specimens. **b** GGBFS-SCC specimens

3.4 IPT Analysis

The input images are collected from real-time experiments, i.e., FA and GGBFS based SCC exposed to 120 min duration. In both the cases, CPP and EP/CPP type specimens are used for the analysis. The image collection setup consists of the Canon SX510 HS camera for taking a photograph with optical image stabilizer of intelligent IS for sharp photos. It has 30x optical zoom with 24 mm wide lens in a mini body. It has 60x Zoom Plus HS System with 12.1 Megapixel CMOS having DIGIC 4 facility and can produce quality capture even in low light. The image analysis experiment is performed with the help of MATLAB 2018 software.

The image-based damage analysis consists of two significant steps, namely damage detection and quantification. To identify the damages clearly, the input images are applied with the prerequisites process of background estimation and noise removal. The damage detection is done through the method specified in A DA

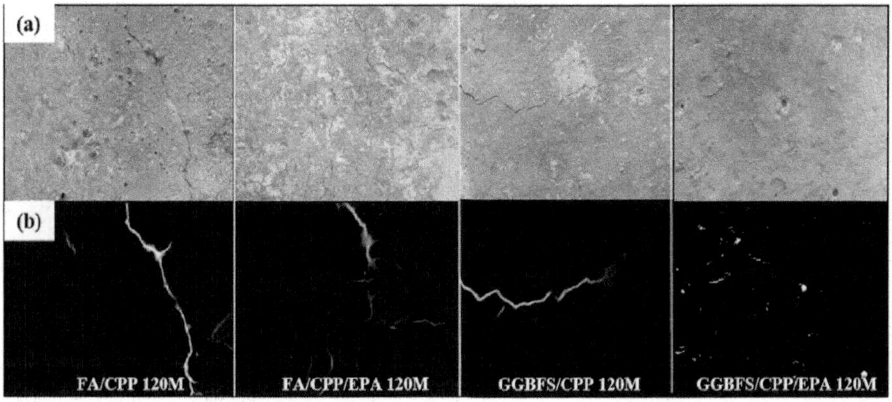

Fig. 7 Image analysis based damage detection of heat exposed FA and GGBFS SCC specimen. **a** Raw images. **b** Processed images

[16]. Initially, the input images are decomposed into higher and lower sub-bands. The uneven backgrounds are removed with the help of column filtering. Ripplet Coefficient Variance Parameter (RCVP) is leveraged to separate the crack pixel from the noisy pixels in the sub-bands. The directional properties of ripplet transform are used to identify the weak cracks. All the cracks are identified in the reconstructed images. Figure 7 shows the input images of FA and GGBFS based concretes and their corresponding damages.

The damage quantification is done after the crack detection process. The morphological information of cracks is used to quantify the cracks. The morphological operations such as closing, opening, pruning and thinning are used to determine the crack properties. Thinning process is used to perform skeletonization with respect to branch point processing [17]. The isolated pixels which are related to noises are removed by the cleaning process. Let $m(x, y)$ be the skeleton point index as given in Eq. (1), which regulates the displacement of the point in the detected cracks from the input image [18].

$$m(x, y) = \begin{cases} 1, (x, y) \in skeleton \\ 0, otherwise \end{cases} \tag{1}$$

The length of the cracks can be obtained from the skeleton points using the Eq. (2).

$$z_l = \int m(x, y)dz \tag{2}$$

where dz represents pixel level, finite crack length at the skeleton line. The particular distance between two adjacent points is calculated using the Eq. (3),

$$m\left((x_{i+1} - x_i)^2 + (y_{i+1} - y_i)^2\right)^{1/2} \quad (3)$$

Sebastien demonstrated the crack detector for inputs. Thus, the crack properties of crack length and crack width are obtained [19]. The damage quantification is calculated through the total crack length attribute. The input images are resized into 512 × 5122. The total crack length of FA-CPP and GGBFS-CPP based concretes are 143.23 and 872.3 pixels. It is evident from the analysis that, concrete with GGBFS exhibited significant damage in terms of thermal crack.

4 Conclusion

The main objective of this research was about to investigate the effective crack control measures that can be adopted on SCC specimens exposed to standard fire based on the analysis of crack pattern. The surface cracks developed due to the high-temperature exposure will lead to severe durability problems and thereby reduces the overall service life of the structure. Therefore, crack assessment and control are the control factors in the proper health monitoring of SCC structures.

Here, an extensive surface crack property has been carried out using conventional crack property surveying tool (Elcometer) and the IPT. The properties of ripplet transform were used in the IPT approach to assessing the fire induce crack parameters. Also, the SEM analysis was incorporated into the study to understand the internal deformation incurred due to high-temperature exposure. The outcome of the study clearly shows the effectiveness of perlite as a protective material in the form of internal filler as well as a mortar combination of CPP layer. The development of surface cracks generation and propagation has been significantly delayed on the protected SCC specimens compared to the unprotected. The novel method of assessing the heat-induced cracks and the crack control method suggested in work can be effectively implemented on the structures subjected to accidental fires.

References

1. Aboalarab, L.: The effect of crack opening size and repair methods on corrosion of steel reinforcement in concrete (2019)
2. 224, A.C., 224, A.C.: Causes, evaluation and repair of cracks in concrete structures, ACI 224.1R-93. Man. Concr. Pract. (1998)
3. ACI 201.2R-01: Guide to durable concrete reported by ACI Committee 201. ACI Comm. **201** (2008)
4. Mathews, M.E., Nandhagopal, M., Anand, N., Arulraj, G.P.: Versatility at its best: an integrated review on development of self-compacting concrete. Int. J. Sci. Technol. Res. **8**, 513–519 (2019)
5. Thanaraj, D.P., Anand, N., Prince Arulraj, G., Zalok, E.: Post-fire damage assessment and capacity based modeling of concrete exposed to elevated temperature. Int. J. Damage Mech. (2020). https://doi.org/10.1177/1056789519881484

6. Barris, C., Torres, L., Vilanova, I., Miàs, C., Llorens, M.: Experimental study on crack width and crack spacing for Glass-FRP reinforced concrete beams. Eng. Struct. (2017). https://doi.org/10.1016/j.engstruct.2016.11.007
7. Dhital, D., Lee, J.R.: A fully non-contact ultrasonic propagation imaging system for closed surface crack evaluation. Exp. Mech. (2012). https://doi.org/10.1007/s11340-011-9567-z
8. Shan, B., Zheng, S., Ou, J.: A stereovision-based crack width detection approach for concrete surface assessment. KSCE J. Civ. Eng. (2016). https://doi.org/10.1007/s12205-015-0461-6
9. Fujita, Y., Hamamoto, Y.: A robust automatic crack detection method from noisy concrete surfaces. Mach. Vis. Appl. (2011). https://doi.org/10.1007/s00138-009-0244-5
10. Mohan, A., Poobal, S.: Crack detection using image processing: a critical review and analysis. Alexandria Eng. J. (2018). https://doi.org/10.1016/j.aej.2017.01.020
11. Manjurul Islam, M.M., Kim, J.M.: Vision-based autonomous crack detection of concrete structures using a fully convolutional encoder–decoder network. Sensors (Switzerland) (2019). https://doi.org/10.3390/s19194251
12. Valença, J., Dias-da-Costa, D., Gonçalves, L., Júlio, E., Araújo, H.: Automatic concrete health monitoring: assessment and monitoring of concrete surfaces. Struct. Infrastruct. Eng. (2014). https://doi.org/10.1080/15732479.2013.835326
13. Wang, P., Huang, H.: Comparison analysis on present image-based crack detection methods in concrete structures. In: Proceedings—2010 3rd International Congress on Image and Signal Processing, CISP (2010). https://doi.org/10.1109/CISP.2010.5647496
14. EFNARC: The European guidelines for self-compacting concrete: specification, production and use (2005)
15. ISO 834-1: Fire-resistance tests—elements of building construction—part 1: general requirements. ISO Stand (1999)
16. Andrushia, A.D., Anand, N., Arulraj, G.P.: A novel approach for thermal crack detection and quantification in structural concrete using ripplet transform. Struct. Control Heal. Monit. (2020). https://doi.org/10.1002/stc.2621
17. Arena, A., Delle Piane, C., Sarout, J.: A new computational approach to cracks quantification from 2D image analysis: application to micro-cracks description in rocks. Comput. Geosci. (2014). https://doi.org/10.1016/j.cageo.2014.01.007
18. Ji, A., Xue, X., Wang, Y., Luo, X., Xue, W.: An integrated approach to automatic pixel-level crack detection and quantification of asphalt pavement. Autom. Constr. (2020). https://doi.org/10.1016/j.autcon.2020.103176
19. Drouyer, S.: An "all terrain" crack detector obtained by deep learning on available databases. Image Process. Line. (2020). https://doi.org/10.5201/ipol.2020.282

Long-Term Experimental Campaign on RC Shrinkage Cracking: Conceptualization, Planning and Experimental Procedures

José Gomes, José Granja, Carlos Sousa, Cláudio Ferreira, Dirk Schlicke, Rui Faria, and Miguel Azenha

Abstract Numerical and analytical simulation of cracking in reinforced concrete (RC) structures requires laboratory testing for validation of assumptions and calibration of modelling strategies. The scale, complexity and innovation of these experiments can present a challenge, even for experienced professionals, and a significant amount of time, effort and money might be inadequately invested on unsuitable test setups, or poorly planned experimental campaigns. This paper deals with the lessons learned from a complex long-term experimental campaign on RC slabs subjected to flexural cracking and restrained shrinkage. It intends to serve as a steppingstone for future works of the same nature. The final version of the restraining device is described and decisions on tested specimens, test conditions, sensing and material characterization are explained based on the available resources. The most relevant tasks during preparation of the long-term experimental campaign are described in detail. Finally, the hard-ships and problems faced during implementation of the experimental program, and the way they were overcome, are addressed. The application of concepts in line with Building Information Modelling (BIM) methodologies, such as object-oriented modelling and process maps, for design of the test setup and task scheduling, was of crucial importance for the success of the experimental campaign.

Keywords Service life behaviour · Experimental testing · Restrained shrinkage · BIM · Reinforced concrete slabs

J. Gomes · J. Granja · M. Azenha (✉)
School of Engineering, University of Minho, Guimarães, Portugal
e-mail: miguel.azenha@civil.uminho.pt

C. Sousa · C. Ferreira · R. Faria
Faculty of Engineering, University of Porto, Porto, Portugal

D. Schlicke
Graz University of Technology, Graz, Austria

F. Kanavaris et al. (eds.), *International RILEM Conference on Early-Age and Long-Term Cracking in RC Structures*, RILEM Bookseries 31,
https://doi.org/10.1007/978-3-030-72921-9_12

1 Introduction

One of the main features of concrete that designers have to deal with is the time-dependent nature of its behaviour. This can either have an adverse effect on the structural performance (e.g., when self-induced stresses or loss of prestressing are caused by restrained shrinkage or creep effects, respectively), or a beneficial effect when stresses are redistributed or decreased due to creep [1, 2]. The study of interactions that occur in the context of the aforementioned time-dependent properties in the process of crack development is of paramount importance to provide better methodologies for crack width prediction in RC structures.

The main challenge in studies of this nature is the experimental validation of assumptions and models. In that regard, several experimental studies on this subject have been performed over the last years. Gilbert and Nejadi [3] performed an experimental campaign of long-term tests on RC beams and slabs in order to assess the effect of internal restraint on crack development. Different Variable Restraint Frame (VRF) testing systems, which enable the testing of restrained RC ties, have been proposed by Cusson and Hoogeven [4], Faria et al. [5] and Schlicke et al. [6]. More recently, Gomes et al. [7, 8] proposed a test setup for simultaneous application of axial restraint and vertical loads in slab-like specimens. By providing information that is complementary to [8], this paper describes in detail the conceptualization, planning and experimental procedure of the latter, which was performed by this research team, and it is aimed to be a guideline for future experimental programs of the same nature.

Section 2 reports the development of the experimental setup, devised to axially restrain a slab-like specimen subjected to vertical loading, while in Sect. 3 the preparation and conduction of the experimental program is addressed.

2 Description of the Restraining Device

2.1 Overview

The main requirement of the restraining device is the ability to provide a controlled axial restraint level to a pre-defined control region of a slab-like test specimen, while simultaneously allowing it to endure bending and shear stresses. The underlying operating principle can be explained with support of Fig. 1. It relies on a simply supported RC specimen (thus representing the region of positive bending moments of a real continuous slab) in which the supports are embedded at mid-height (and shrouded by the longitudinal reinforcement), and the longitudinal displacement of one of the supports is controlled by the user with a hydraulic driven force. Even though detailed information on sizing and requirements can be found in [7], they are generally addressed in the following paragraphs.

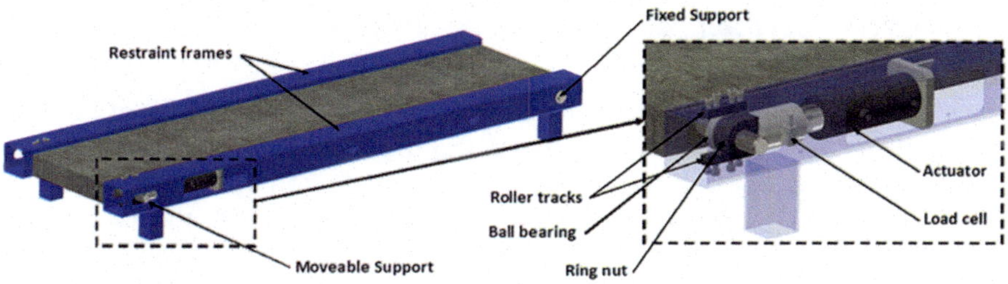

Fig. 1 Schematic view of the restraining device

The design of the test specimens is based on a continuous solid slab supported by transverse beams at every 4 m. The slab was designed for the ultimate limit state, considering permanent loads of 2.0 kN/m^2 (besides self-weight) and live loads of 3.2 kN/m^2. Such approach resulted in a 0.10 m thick slab (C20/25), longitudinally reinforced for bending with ϕ8//0.10 (S400C) and with a secondary transverse reinforcement of ϕ6//0.25, providing adequate service life behaviour without consideration of the effects of axial restraint (which was considered a design starting point).

The design of the restraining device was performed by taking into account the following assumptions: (i) the test specimen is 2.6 m long, with a free span of 2.4 m to match the distance between zero bending moments of the slab (~3/5 of the slab's effective span); (ii) the cross section of the test specimen is 0.1 $\times$ 0.5 m^2 to ensure a width-to-depth ratio of 1/5 and, consequently, a slab-type behaviour; (iii) even though the test specimen is subjected to positive bending moments only, no reinforcement curtailment is considered, and the same reinforcement is attributed to the top and bottom layers (for simplification of execution and analysis); (iv) the test specimen will not endure restraining forces higher than concrete's axial cracking force $N_{cr} = A_c \times f_{ctm}$.

The restraining device consists of two reaction frames supporting the test specimen, each one housing a hydraulic actuator and a load cell connected to one of the supports. The test specimen is supported by two solid steel rods, with a diameter of 40 mm, embedded at mid-height. One steel rod—the fixed support—is inserted in the 40 mm holes located in the lateral side of the reaction frames; the other steel rod—the moveable support—is inserted in the ring nuts (connecting the support to the load cells and actuators) and in the radial ball bearings (which carries the vertical load). The radial ball bearing is located between two roller tracks whose position is adjusted with 4 socket head cap screws, being relieved from top roller track immediately after casting to enable longitudinal sliding. The axial restraint of the test specimen is performed with the two double effect hydraulic actuators (with a 100 kN capacity in compression, each), connected in parallel to a manual hydraulic pump. The axial load applied to the test specimen is monitored with two resistive load cells (measurement range of 100 kN, each), positioned between the actuator and the ring nut.

2.2 Object Oriented Modelling of the Restraining Device

Given the complexity of the devised test setup, with several elements of different nature interacting with each other, the modelling of the restraining device was performed with an object-oriented approach, in which the model is composed by objects that, besides the geometric parameters, are attributed with other type of data (e.g., information about manufacturer, cost and technical features) and procedures (namely spatial relations with adjacent objects and behaviour rules). Such approach enables the development of complex models in periods of time that would not be possible with conventional computed aided design (CAD) software, and reduces the risk of mistakes through automatic detection of incompatibilities.

The model for the restraining device was developed in AutoDesk Inventor Professional 2018. It is composed by several assemblies, which includes the steel components of the reaction frames and support system, actuators and load cells, wood components for formwork and propping system, and the test specimen. Figure 2 shows an exploded view of the restraining device and subsequent stages of the test setup assembly, automatically generated from the final model.

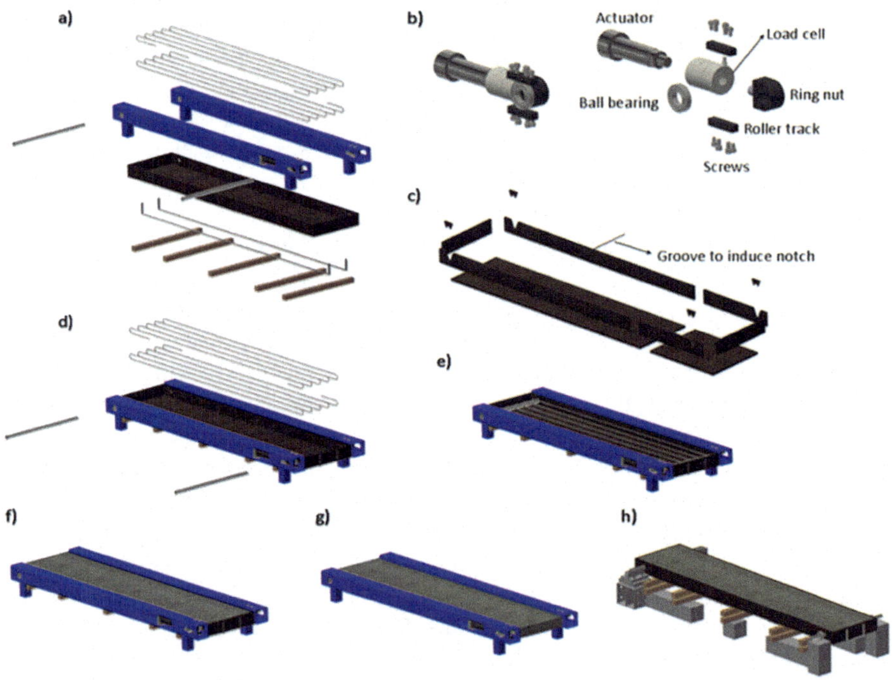

Fig. 2 Exploded view of test setup: **a** pre-assembled restraining system; **b** components inside the restraint frame; **c** pre-assembled formwork; **d** positioning and propping of formwork; **e** positioning of longitudinal rebars and steel rods; **f** casted test specimen; **g** test specimen after formwork removal; **h** support and formwork propping systems for unrestrained specimen

The STP file format for interoperability of the general industry [9] was used to order the production of these components, allowing for a straightforward process. The model contains information regarding the geometry of the different components and the spatial relation between them, as well as material information and object standardized references. The components inside the restraint frame, presented in Fig. 2b (assembled and isolated), were defined based on the requirements described in Sect. 2.1; specific information regarding the brand and commercial references was also included in the model. For the remaining components, only information about the material was attributed, besides the geometry and spatial relation with other components.

The formwork was designed to facilitate its extraction from the climatic chamber and to avoid damaging the test specimens during its removal. It is divided longitudinally into two parts (connected by two metallic plates with a 5×20 mm^2 cross-section), and it is composed by eight 18 mm thick plywood panels, connected with metallic L-shaped brackets screwed to the bottom and lateral sides (Fig. 2c). A PVC groove with triangular cross-section (13 mm wide and 8 mm deep) is glued to the bottom of the formwork, to induce a notch on the test specimen and consequent cracking at mid-span. After the formwork is assembled, it is positioned and fixed through five timber slats bolted to the bottom face of the restraint frames (Fig. 2d). The longitudinal rebars are then duly positioned inside the formwork. The notches on the lateral panels of the latter enable the insertion of the steel rods, as described in Sect. 2.1, through the reinforcement, between top and bottom rebars (Fig. 2e). The negative of the notch that is not filled by the supporting rod is sticked to the formwork with insulating tape. The position of the moveable support is adjusted and fixed with the roller tracks and head cap screws before casting the specimen. Immediately after casting (Fig. 2f), the top roller tracking is relieved to enable longitudinal displacement. After the curing period, the formwork is removed by unscrewing the metallic plates and L-shapes, as well as the timber slats (Fig. 2g). Alternative support and formwork propping systems were devised to account for simultaneous testing of unrestrained specimens under similar conditions (Fig. 2h).

3 Long-Term Experimental Campaign

3.1 Experimental Strategy

In order to address the effects of both degree of restraint and level of flexural loading, the experimental campaign involved the simultaneous testing of four distinct situations, as presented in Fig. 3. Two test specimens were subjected to external axial restraint and different levels of vertical load: SLAB1 was submitted to the quasi-permanent load combination, while SLAB2 was merely subjected to its self-weight. One test specimen, SLAB3, was subjected to the same vertical loading as SLAB1, but without any external axial restraint. A fourth situation, with no restraint forces

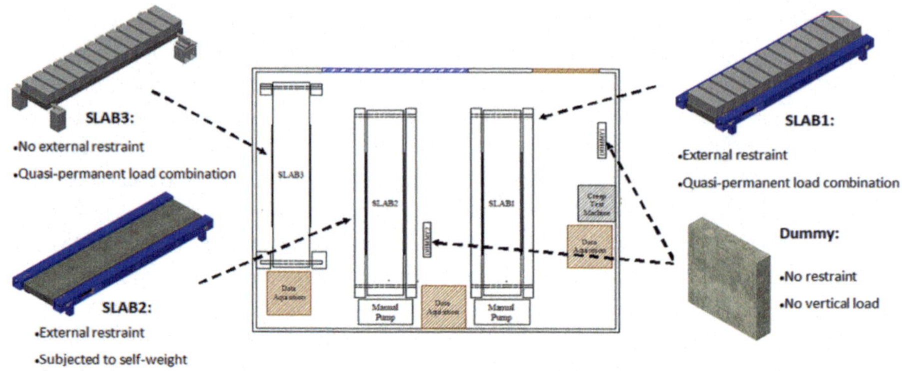

Fig. 3 Arrangement of test specimens in the climatic chamber

nor vertical loading (allowing to determine the imposed deformation that former slabs are subjected to) was simulated with two dummy specimens, DUMMY1 and DUMMY2, with the same cross section and subjected to the same drying history as former slabs, 0.5 m long (positioned vertically) and with no reinforcement.

The experimental program was conducted over a period of 18 months, in a climatic chamber with constant environmental conditions (T = 20 °C and RH = 60%). It was complemented with companion specimens for measurement of RH profiles and characterization of the following concrete properties, at different ages: compressive strength, tensile strength, E-modulus and creep in compression.

The structural behaviour of the test specimens was assessed, during the entire experimental campaign, with continuous monitoring of vertical displacements at mid-span, average in-plane deformations in the control region, steel strains at both top and bottom reinforcement and crack width development.

3.2 Preparation of the Experimental Campaign

The preparation for the long-term experimental campaign involved several tasks at different stages: it included minor enhancements to the test setup based on the pilot test outcomes, the construction of a large-scale climatic chamber, instrumentation of test specimens and elaboration of a micromanagement plan for casting day.

In order to increase efficiency and minimize errors, these tasks were carefully planned with process maps, in which both structured workflows for predicted scenarios and protocols to respond to dynamic situations were considered. A process map visually shows, through different flowcharts shapes, the steps of a work activity, the actors that are involved in carrying out those activities and information flows between them. The standard Business Process Model and Notation (BPMN), developed by the Object Management Group (OMG), is often used to create process maps, as it provides a notation that is readily understandable by all users and creates a standardized bridge between process design and process implementation [10].

The most relevant preparation activities will be addressed in the following subsections, apart from the pilot test, whose detailed description can be found elsewhere [7] and will not be discussed here for the sake of brevity.

Construction of climatic chamber. The climatic chamber, depicted in Fig. 4, was specifically constructed for this experimental campaign to eliminate the influence of temperature and ambient RH variations in the results, thus simplifying the analysis. Given the large area that was needed to include all test specimens, as well as the available funds for the project, the following cost-efficient solution was envisaged: a room with a net area of 3.6×4.9 m^2, built with double layer wood panels and 5 cm thick XPS modules. A moveable roof made of a sandwich panel was considered to enable the removal of test specimens at the end of the experimental campaign, thus allowing it to be reused for other projects. The environmental conditions were kept constant with an air conditioning device, a humidifier and a dehumidifier working simultaneously. This solution provided a test room with constant T and RH of 20 °C $\pm$ 0.5 °C and 60% $\pm$ 5%, respectively. A restricted access policy was implemented to avoid unnecessary entrances from people outside the research group, and a double layer laminated glass (2.0×2.1 m^2) allowed to visualize the test setup from outside. These measures minimized environmental contamination and human-caused technical issues.

Instrumentation and setup of the longitudinal strain measurement system. For assessment of the structural behaviour of the test specimens, the following sensors were used: (i) vibrating wire strain gauges for measurement of dummy specimens' deformations; (ii) electrical resistance strain gauges and optical fibre Bragg grating sensors for measurement of strains in the longitudinal rebars; (iii) hand-held USB microscope for measurement of crack widths; (iv) thermo-resistance PT100 for measurement of the temperature inside the test specimens and (v) linear variable differential transducers (LVDTs) for measurement of deflections and longitudinal average strains in the control region. The following paragraphs describe the devised measurement system for assessment of the longitudinal strains in the control region, which was gradually improved until the presented final version. Detailed information regarding the remaining measurement devices can be found elsewhere [8].

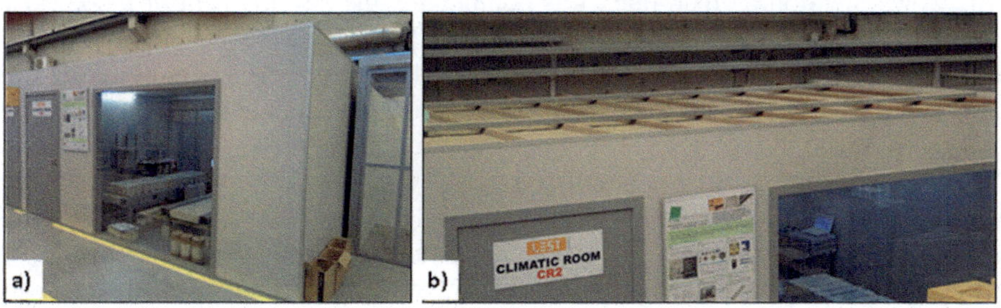

Fig. 4 Climatic chamber: **a** view from the outside; **b** roof detail

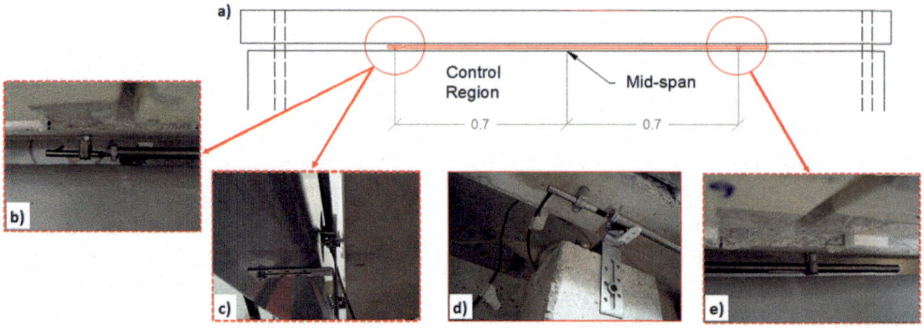

Fig. 5 Longitudinal strains measurement system (final version): **a** outline; **b** detail of LVDT; **c** detail of roller support; **d** setup for unrestrained specimen; **e** fixed extremity

After application of the restraint load, significant cracking was expected at both extremities due to stress concentration near the steel rods. For that reason, the measurement of the average axial strains (for control of the restraining condition) was limited to the 1.4 m long central region. For assessment of the longitudinal strains in this control region, a measurement system with one LVDT on each side of the test specimen was used, as illustrated in Fig. 5.

For installation of this measurement system, two hexagonal nuts M6 were pre-embedded on each lateral side of the test specimens, at mid-height and spaced 0.7 m from the mid-span. The average longitudinal deformation was measured with one LVDT on each side of the slab, fixed to a metallic clamp screwed to the pre-embedded nut on one extremity and in contact with a 12 mm diameter steel tube, which was fixed to a metallic clamp screwed to the other extremity. Near the point of contact between the LVDT and the steel tube, the latter was supported by a roller fixed to the metallic frame of the restraint device (SLAB1 and SLAB2) or to a concrete block (SLAB3).

Micromanagement plan for the casting day. Considering the limited space inside the climatic chamber, the number of sensors installed and the number of concrete specimens for testing and material characterization, a micromanagement plan for the casting day was developed to ensure that all test specimens were cast in a small-time period, during which the fresh concrete age can be considered the same. It is worth mentioning that, due to its complexity, the development of a dedicated process map [11] for this activity, which involved 14 people from the research team, was of paramount importance for its success. The roles of each person during this task, and the interaction between them, are described in detail in the next paragraphs and schematized in Fig. 6.

Persons identified with the numbered circles 01–05 were responsible for casting the test and dummy specimens. Two vibrating needles were available inside the climatic chamber to be used by persons 01 and 02, which were responsible for vibrating the concrete. Persons 03–05 were responsible for filling the formworks with concrete buckets. The interchange of the concrete buckets between inside and outside the climatic chamber was assured by persons 06 and 07, while persons 08–10 were

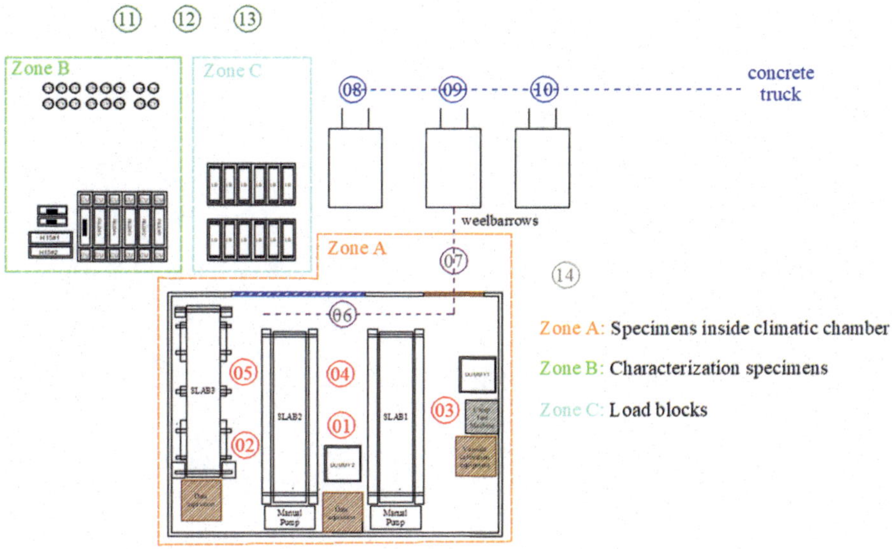

Fig. 6 Micromanagement plan for the casting day

responsible for the transportation of concrete on wheelbarrows from the concrete truck and for filling the empty buckets. The specimens for concrete characterization and load blocks were cast outside the climatic chamber by persons 11–13. Person 11 was in charge of vibrating the concrete with a vibrating needle, and persons 12–13 of filling the moulds with concrete buckets. Person 14 was responsible for the photographic recording of the casting (Fig. 7).

Right after casting, all specimens were duly sealed with plastic film, and characterization specimens were taken to the climatic chamber. The casting took place on April 17, 2018, mixing of concrete started at 13:54 and the concrete truck arrived at the casting site at 14:35. A slump test was performed to check the consistency class (120 mm slump—class S3). The casting of the test specimens started at 14:50 and ended at 15:40.

3.3 Experimental Campaign and Control of Restraining Condition

The test specimens were left undisturbed until formwork removal. After a curing period of 7 days, all test specimens were carefully demoulded and exposed to the climatic chamber environment. The formwork was gradually removed by unscrewing the metallic L-shaped brackets first (to reduce friction between concrete and formwork), and then removing the timber slats. During this process, 3 LVDTs

Fig. 7 Casting of the long-term experimental campaign: **a** arrival of concrete truck; **b** filling the wheelbarrows; **c** transportation of concrete buckets; **d** casing the slab specimens; **e** vibrating slab specimens; **f** casting load blocks and characterization specimens; **g** vibrating load blocks and characterization specimens; **h** sealing the test specimens; **i** sealing the characterization specimens

were positioned on the top surfaces, at mid-span, to measure the vertical deflection (Fig. 8a).

The formwork removal was immediately followed by installation of LVDTs for measurement of vertical displacements, as well as the lateral measurement system for assessment of the longitudinal strains in the control region. Early measurements of longitudinal deformations in the control region showed that the devised measuring system was being affected by friction effects (recorded deformations developed in steps, instead of showing a continuous variation). Even though this issue was mitigated at 22 days of age, by reducing the distance between the measurement system and the lateral side of the slab (final form is presented in Fig. 5), the initial intent of providing full compensation of the average longitudinal strain in the control region by adapting the axial force, was withdrawn. Instead, the restraining procedure for SLAB1 and SLAB2 was performed by controlling an expected restraint force throughout time. The expected restraint force was estimated before each increment of axial load by performing viscoelastic analyses based on the age-adjusted elastic modulus (AAEM) method [12] and experimental measurements of free shrinkage and crack width: the AAEM method was applied to determine the axial load increment needed to fully compensate the longitudinal deformation, considering the recorded free shrinkage strains, while the recorded crack widths were used to update the average deformation in the cracked zone.

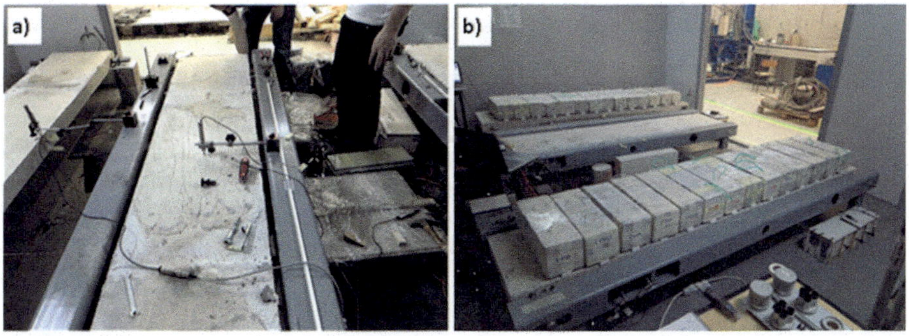

Fig. 8 Key instants of long-term experimental campaign: **a** formwork removal; **b** application of vertical loads

At the age of 50 days, SLAB1 and SLAB3 were subjected to a distributed load of 2.96 kN/m^2. In a passive restraint system, the imposition of external bending may result in the decrease of the axial restraint force due to widening and formation of cracks [13, 14]. However, the devised restraint system was not under closed loop displacement control (the restraint force depends on the oil pressure in the hydraulic actuators), and the restraint force decrease had to be estimated (using the AAEM method) and applied manually in SLAB1, over 4 stages. At each stage, the load blocks were strategically placed to induce ~25% of the final bending moment increment, followed by a reduction of ~25% of the calculated restraint force decrease. It is worth noting that the load blocks were adequately separated from each other (thus avoiding significant arch effects), and supported by $20 \times 40 \times 480$ mm^3 XPS stripes to enable concrete drying from the slabs' top surface (Fig. 8b).

At the end of the fourth stage, it was observed that the calculated restraint force decrease was overestimated (no new cracks were formed), and the axial load in SLAB1 was progressively increased until a situation of near full restraint (measured longitudinal strain in the control region close to 0 μm/m). From 60 days onwards, the axial force on both SLAB1 and SLAB2 was kept constant.

The presentation and analysis of the results obtained in this experimental campaign are out of the scope of this paper and can be found elsewere [8].

4 Conclusions

This paper deals with the preparation and conduction of a long-term experimental campaign on slabs subjected to restraining shrinkage and vertical loads, with special focus on design and planning approaches. The following conclusions are drawn.

The restraining device proposed in this work was capable of imposing and controlling a pre-defined level of axial load on the test specimens, while allowing them to endure shear and bending stresses. The object-oriented modelling approach allowed to develop the test setup without the need of substantial corrections after deployment of a pilot test.

The use of process maps at micromanagement level was of paramount importance for the successful casting of several test specimens, highly instrumented, in a very confined space, in a short period of time and without equipment failure.

Initial difficulties regarding the measurement system for longitudinal strains in the control region were overcome with an alternative plan to restrain the test specimens, based on different measurement procedures. Even though those measurements were not meant for redundancy, it shows the importance of having a wide range of measurement systems in large scale experiments.

Acknowledgements Funding provided by the Portuguese Foundation for Science and Technology (FCT) to the Research Project IntegraCrete (PTDC/ECM-EST/1056/2014 POCI-01–0145-FEDER-016841) and to the PhD Grant SFRH/BD/148558/2019 is acknowledged. This work was also financially supported by UIDB/04029/2020—ISISE and by UIDB/04708/2020—CONSTRUCT, funded by national funds through the FCTMCTES (PIDDAC). The authors also acknowledge the contribution of Ali Edalat-Behbahani, Amin Abrishambaf, António Matos, César Gonçalves, Fernando Pokke, José Ferreira, Marco Jorge, Meera Ramesh, Ricardo Cruz, Sérgio Soares and Sónia Soares during the casting of the experimental campaign.

References

1. Azenha, M.: Numerical simulation of the structural behaviour of concrete since early ages. Ph.D. Thesis, University of Porto, FEUP (2009)
2. Leitão, L.: Análise termo-higro-mecânica das tensões autoinduzidas em estruturas de betão. Ph.D. Thesis, University of Porto, FEUP (2018)
3. Gilbert, R.I., Nejadi, S.: An experimental study of flexural cracking in reinforced concrete members under sustained loads. In: UNICIV Report. School of Civil & Env. Eng., Sydney (2004)
4. Cusson, D., Hoogeveen, T.: An experimental approach for the analysis of early-age behaviour of high-performance concrete structures under restrained shrinkage. Cem. Concr. Res. **37**(2), 200–209 (2007)
5. Faria, R., Leitão, L., Teixeira, L., Azenha, M., Cusson, D.: A structural experimental technique to characterize the viscoelastic behavior of concrete under restrained deformations. Strain **53**(1) (2017)
6. Schlicke, D., Hofer, K., Tue, N.V.: Adjustable restraining frames for systematic investigation of cracking risk and crack formation in reinforced concrete under restrained conditions. In: Serdar, M., et al. (eds.) Advanced Techniques for Testing of Cement-Based Materials, pp. 211–239. Springer International Publishing (2020)
7. Gomes, J., Azenha, M., Granja, J., Faria, R., Sousa, C., Zahabizadeh, B., Edalat-Behbahani, A., Schlicke, D.: Proposal of a test set up for simultaneous application of axial restraint and vertical loads to slab-like specimens: sizing principles and application. In: SynerCrete'18: Interdisciplinary Approaches for Cement-Based Materials and Structural Concrete: Synergizing Expertise and Bridging Scales of Space and Time, RILEM, Funchal, Madeira Island (2018)
8. Gomes, J.G., Granja, J., Sousa, C., Zahabizadeh, B., Faria, R., Schlicke, D., Azenha, M.: A new test setup for simulation of the combined effect of bending and axial restraint in slab-like specimens. Eng. Struct. **225** (2020)

9. ISO: ISO 10303: Industrial automation systems and integration—product data representation and exchange (1994)
10. OMG: Business Process Model and Notation (BPMN). In. Edited by (OMG) OMG (2013)
11. Gomes, J., Azenha, M.: Process map for casting a long-term experimental campaign on RC shrinkage cracking. Dataset (2021). https://doi.org/10.5281/zenodo.4474539
12. Bazant, Z.P.: Prediction of concrete creep effects using age-adjusted effective. J. Am. Concr. Inst. **69**(4), 212–217 (1972)
13. Camara, J., Luís, R.: Structural response and design criteria for imposed deformations superimposed to vertical loads. In: The Second fib Congress: 2006, Naples, Italy, pp. 1–12 (2006)
14. Gomes, J., Carvalho, R., Sousa, C., Granja, J., Faria, R., Schlicke, D., Azenha, M.: 3D numerical simulation of the cracking behaviour of a RC one-way slab under the combined effect of thermal, shrinkage and external loads. Eng. Struct. **212** (2020)

Strength Behaviour of a High-Performance Concrete Under Drying

Ismail Yurtdas, Nicolas Burlion, Jianfu Shao, and Alex Li

Abstract Cement-based materials usually undergo drying which influences their mechanical behaviour. This study investigates the evolution of mechanical properties of a high-performance self-compacting concrete subjected to drying at an early age and after long-term maturation. Uniaxial compression and bending tests are performed at different drying times. The evolution of the mechanical properties obtained is controlled by a competitive effect between material strengthening (due to capillary depression, disjoining pressure, hygral gradients, and hydration if it still occurs) and drying-induced micro-cracking (due to material heterogeneity and differential shrinkage). The competitive effect shows a great dependence on maturation level. Such couplings have to be taken into account for a reliable durability analysis.

Keywords High-performance concrete · Maturation · Drying · Micro-cracking · Mechanical properties

1 Introduction

Concrete structures are subjected to drying due to the hygral imbalance between the early age concrete and environmental conditions. Shrinkage is a macroscopic consequence of drying and is induced by variation of capillary depression, disjoining pressure and surface tension [1, 2]. The drying which is non-uniform because of the low permeability of concrete and structure geometry creates hygral gradients that lead to a differential shrinkage producing tensile stress and surface micro-cracks as soon as the tensile strength is reached [1–6]. Micro-cracking also occurs at the interface of the hydrated cement paste/stiff inclusion because of a retracting matrix

I. Yurtdas (✉) · A. Li
MATIM, Université de Reims Champagne-Ardenne, 51687 Reims, France
e-mail: ismail.yurtdas@univ-reims.fr

N. Burlion · J. Shao
UMR 9013-LaMcube, CNRS, Centrale Lille, Université de Lille, 59655 Lille, France

F. Kanavaris et al. (eds.), *International RILEM Conference on Early-Age and Long-Term Cracking in RC Structures*, RILEM Bookseries 31,
https://doi.org/10.1007/978-3-030-72921-9_13

and inclusions [7–9]. Moreover, drying at an early age generates hydration gradients in the material [10]. The earlier the drying the stronger the hydration gradients become. Therefore, the interaction between drying, hydration and mechanical behaviour will greatly influence the durability of concrete structures. The previous studies on cement-based materials [3, 5, 11, 12] showed the occurrence of a competitive effect between strengthening of material and induced micro-cracking during drying. The strengthening of material is due to a variation in capillary suction, disjoining pressure and surface tension which create an isotropic pre-stressing in the material. This strengthening is amplified by hygral gradients [1–6] which lead to the confinement of the sample's core. The drying-induced micro-cracking comes from the heterogeneities of the material and differential shrinkage. However, if the hydration has not yet reached stabilisation, the strength increase due to the hydration process (which slows down and then stops in time from the outside to the inside of the material because of drying) has also to be considered in this competitive effect.

In an earlier work [13], the effect of drying on the evolution of the mechanical behaviour of a high-performance concrete has been investigated after a long-term maturation period. The objective of the present work is to perform a comparative study between the effects of drying at early age and after long-term maturation on the evolution of strength behaviour. This behaviour is evaluated by means of uniaxial compression and bending tests carried out at different times. These two tests also made it possible to show which is more sensitive to drying-induced micro-cracking. In the first part, the setting up of the experimental investigation is described. In the second part, the results obtained are presented and analysed by putting forward the coupling effects between drying, hydration and mechanical properties.

2 Experimental Investigation

The experimental study was set up to follow the evolution of the hygro-mechanical behaviour immediately after material casting. A high performance self-compacting concrete was used to perform the study (Table 1). The gravel (4/10 mm) was crushed while the sand (0/4 mm) was rolled, the two aggregates being silicocalcareous.

Table 1 Composition of concrete

Component	Quantity (kg/m^3)
Gravel (4/10)	793.9
Sand (0/4)	980.5
Cement	397.0
Limestone filler	109.2
Superplasticizer	6.4
Water	170.2
Water/binder (w/b)	0.41

Cement CEM I 52.5 N CP2 and limestone filler were used to manufacture the high performance self-compacting concrete. The high range water reducer admixture was a polycarboxylate modified superplasticizer. Cylindrical (Ø160 × h320 mm^3) and prismatic (40 × 40 × 160 mm^3) samples were manufactured to conduct the investigation.

One day after casting, the samples were separated into two series: the first series were stored in lime-saturated water at 21 °C while the second series were submitted to air-drying in a controlled environment with temperature = 21 ± 1 °C and relative humidity = 60 ± 5%. The comparison of the strength behaviour of these two series made it possible to evaluate the effect of air-drying at an early age with regard to water maturation. The testing programme was composed as follows:

- uniaxial compression tests on cylindrical samples (Ø160 × h320 mm^3),
- three-point bending tests on prismatic samples (40 × 40 × 160 mm^3) and uniaxial compression tests on cubic samples (40 × 40 × 40 mm^3) by using the half-prisms obtained after the bending tests,
- measurement of the loss in mass of samples with time,
- determination of porosity.

The effect of air-drying on strength behaviour was also tracked 75 days after casting. After this long-term maturation period in lime-saturated water, the effect of hydration on strength [12, 13] and endogenous shrinkage [14] can be considered as slight. The tests carried out on prismatic samples of the first series at different times of drying enabled comparison of the effect of drying at an early age with that occurring after an optimal maturity.

The compression tests on cylindrical samples (Ø160 × h320 mm^3) were performed with a loading velocity of 10 kN/s (EN 12390-3). To ensure a perfect transmission of the load and thus to minimise the bending effect, all the samples were surfaced and a rotating plate system was used. Three-point bending tests were carried out on prismatic samples (40 × 40 × 160 mm^3) with a loading velocity of 0.05 kN/s (EN 196-1). After the failure of the sample in the flexural test, the two parts of each prism were subjected to compressive stress with the help of a device consisting of two steel plates 40 mm wide, with a loading velocity of 2.4 kN/s (EN 196-1). The top of the device is equipped with a rotating plate while the bottom of the device is fixed. The loaded fraction of the half-prism sample is a cube of 40 × 40 × 40 mm^3. For each type of test and conditioning, at least two samples were tested. However, only the average values will be presented for reasons of clarity.

3 Results

3.1 *Effect of Drying at Early Age*

Figure 1 presents the evolution of the uniaxial compressive strength of cylindrical samples and cubic samples (previously tested in flexion) as a function of time and conditioning mode. The air-drying begins one day after casting. The strength of water-stored samples increases in time due to hydration, the rate of increase being higher up to 14 days. Then, it tends towards a stabilisation for large cylindrical samples but continues to increase slightly for small cubic samples. The increase in strength reaches 55% and 72% for large cylindrical samples and small cubic samples respectively. The porosity was also determined on water-stored samples at different times of mechanical tests by the difference between saturated and dried (at 60 °C) state. The porosity obtained between 4 and 75 days of maturation varies between 12.2 and 12.8%. This variation is non-significant and means that the microstructure is well formed after 4 days of casting. It does not enable the variation in compressive strength to be captured.

The compressive strength of air-dried samples also increases in the same proportion as that of saturated samples and then begins to decrease but remains clearly higher than the one day compressive strength (non-dried samples): the increase reaches 40 and 57% at 14 days then drops to attain 36 and 40% at 75 days for large cylindrical samples and small cubic samples respectively compared to the strength after one day.

The increase in compressive strength of the air-dried samples can be attributed to the hydration which slows down in time by faster drying and to the beneficial effect of capillary depression and disjoining pressure. The decrease in strength that follows shows the harmful effect of drying-induced micro-cracking which prevails

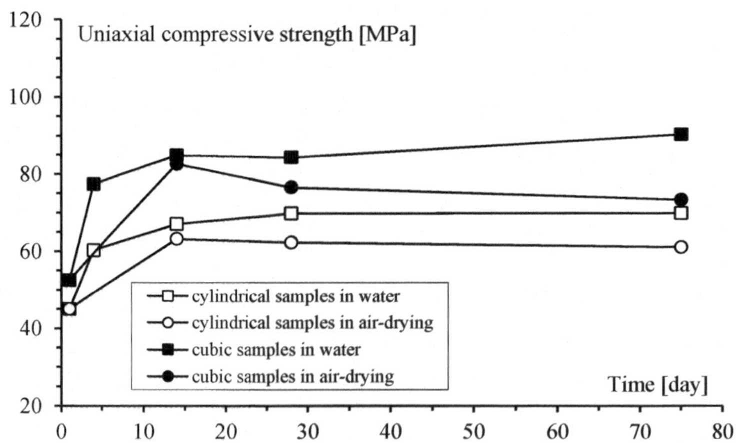

Fig. 1 Uniaxial compressive strength of cylindrical (Ø160 × h320 mm^3) and cubic (40 × 40 × 40 mm^3) samples versus time and conditioning mode

over the strengthening effect. This evolution is similar to that shown by the results of Price [12] on an ordinary strength concrete. Moreover, compared to large cylindrical samples, the increase then the decrease in the compressive strength of small cubic samples with time of drying with regard to one day strength is more pronounced. This could be caused by their faster drying which also leads to a smaller degree of hydration from the outside to the inside of the material.

Figure 1 also highlights the influence of a scale (size) effect on uniaxial compressive strength. Indeed, the compressive strength of cubic samples is higher than that of cylindrical samples for the two conditioning modes. Various reasons could explain this difference in strength: length-to-width (*l/w*) ratio whose effect decreases with the increase of material strength, smaller critical dimension of the prismatic samples which is conducive to contact between the coarse aggregates (it would take only four aggregates with a maximum size of 10mm in a line to span the sample) thus inducing a more marked opposing force of aggregates to the mechanical loading compared to cylindrical samples, probable effect of the difference between the loading rate of cubic (1.5 MPa/s) and cylindrical (0.5 MPa/s) samples.

Figure 2 shows the evolution of the bending strength of prismatic samples as a function of time and conditioning mode. As indicated previously, the air-drying begins one day after casting. The bending strength of the saturated samples is again higher than for the air-dried samples. The increase in bending strength of the air-dried samples due to hydration and variation of capillary depression and disjoining pressure is countered by drying-induced micro-cracking which reduces the bending strength although this still remains slightly higher than the one day strength.

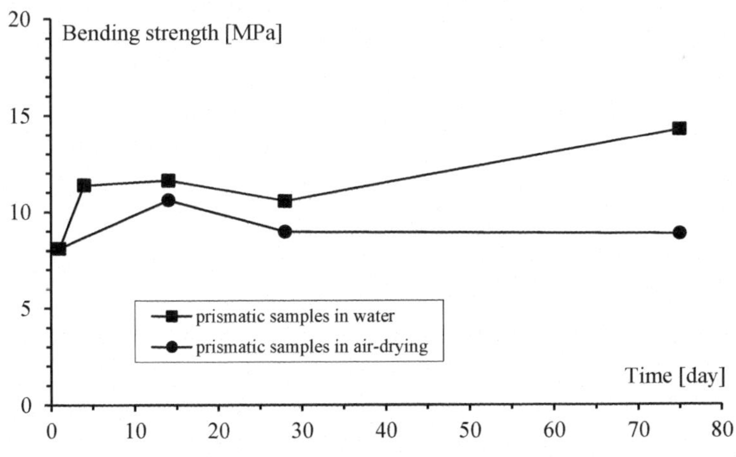

Fig. 2 Bending strength of prismatic samples (40 × 40 × 160 mm^3) versus time and conditioning mode

3.2 Effect of Drying After Long-Term Maturation

Figure 3 shows the evolution of the compressive strength of cubic samples versus time of drying one day and 75 days after casting. Time 0 corresponds to the beginning of drying process. Figure 4 gives the same evolution versus loss in mass. Note that the tests performed on saturated samples after 75 days of maturation did not show significant variation of strength due to continuation of hydration [13]. Therefore, any strength evolution after this long-term maturation can be attributed mainly to the effect of drying unlike the strength evolution of air-dried samples one day after casting.

The strength of mature samples increases with time of drying by about 22% and does not show a decrease, contrary to the case of early age drying for which strength first rises by 57% then decreases to about 44% compared to non-dried samples. Therefore, for mature material, the strengthening effect is dominant compared to the effect of drying-induced micro-cracking. Such an increase in strength has also been reported by other authors for cement-based materials whether or not the sample drying was homogeneous [3, 5, 11, 15, 16]. However, it has also been observed that the effect of drying-induced micro-cracking can become preponderant after sufficient air-drying generating a slight strength decrease following an initial higher increase [12].

Note that when the samples of low permeability, like those of the present study, are tested close to saturation, this will bring about an interstitial over-pressure which amplifies opening and propagation of micro-cracks. The presence of water also has a lubricating effect. The influence of these two effects decreases with drying.

Figure 5 displays the evolution of the bending strength of prismatic samples versus time of drying one day and 75 days after casting. Time 0 corresponds to the beginning

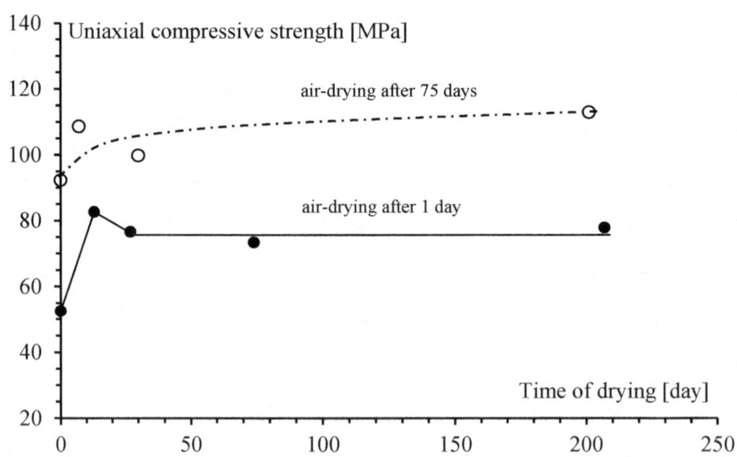

Fig. 3 Uniaxial compressive strength of cubic samples ($40 \times 40 \times 40\ mm^3$) versus time of drying

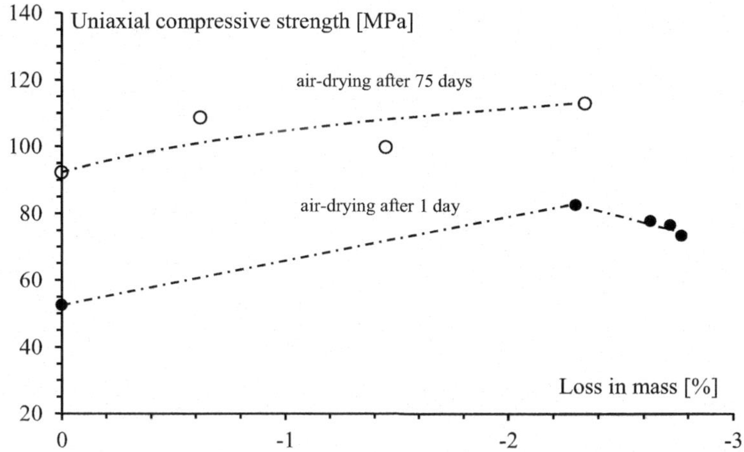

Fig. 4 Uniaxial compressive strength of cubic samples (40 × 40 × 40 mm^3) versus loss in mass

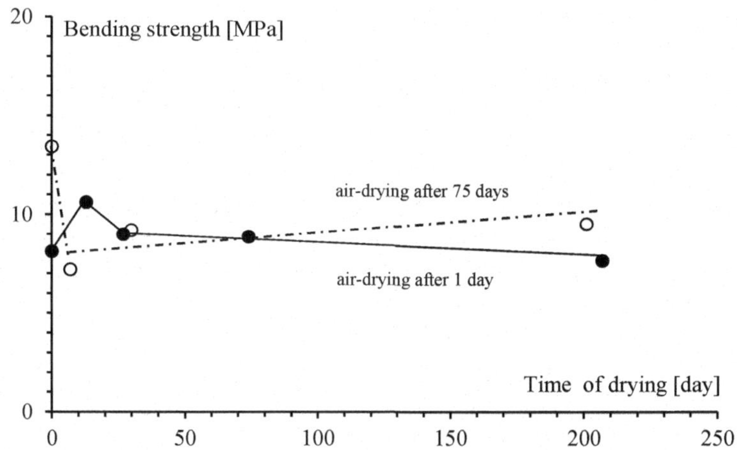

Fig. 5 Bending strength of prismatic samples (40 × 40 × 160 mm^3) versus time of drying

of the drying process. Figure 6 gives the same evolution versus loss in mass. Air-drying of mature samples provokes a drastic decrease of 44% in bending strength at the beginning of drying. Then, the bending strength begins to increase with drying to have a final net decrease of 31%. This evolution of bending strength occurs under the same above cited competitive effect between material strengthening and drying-induced micro-cracking. Thus, after a significant reduction due to drying-induced micro-cracking, the bending strength increases gradually without reaching the value of the water saturated samples. This increase illustrates the effect of capillary suction and disjoining pressure on the evolution of bending strength. However, the bending strength of the air-dried samples one day after casting increases by 31%. Then, it

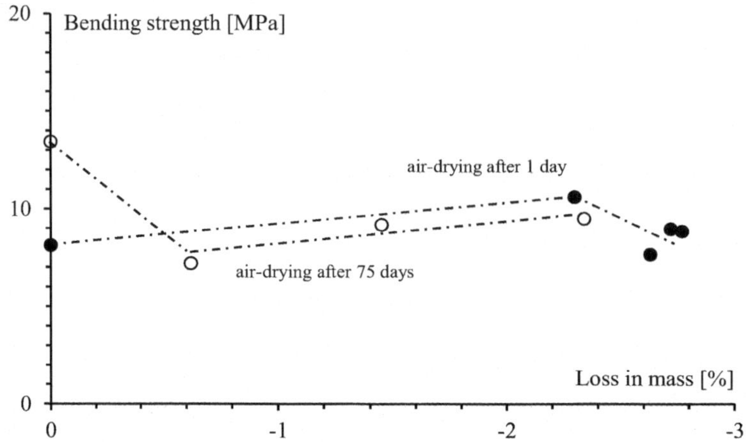

Fig. 6 Bending strength of prismatic samples (40 × 40 × 160 mm^3) versus loss in mass

decreases due to drying-induced micro-cracking and reaches a final drop of 6% compared to non-dried samples. Notice that the creep of matrix which depends on maturity level would contribute to the evolution of bending strength as it reduces the shrinkage stresses in aggregate/cement paste interfacial zone and then the occurrence of supplementary micro-cracks [6, 9]. The results obtained also highlight that the drying-induced micro-cracking is more easily detected by the measurement of the bending strength. Indeed, in the flexural tensile test, the external zone (outer fibre of the beam) weakened by the induced micro-cracking is directly subjected to the loading contrary to a uniaxial compression test.

The bending strength evolution reported here is comparable to that obtained by Walker and Bloem [17] who carried out their tests under non uniform drying conditions on mature concretes with w/c = 0.42 and 0.45 and containing air-entraining admixture. They note first a decrease of the bending strength that can reach 40% after 4 days of drying then register a gradual increase of the bending strength up to the end of their study, i.e. 32 days, which results in a drop of 19% compared to the bending strength of non-dried samples. Moreover, Kanna et al. [18] report an increase of 6% in bending strength of a mortar with w/c = 0.45 cured 7 days in water then subjected to 21 days of drying at 50% of relative humidity. Okajima et al. [15] who tested a mature mortar with w/c = 0.65 at various uniform internal humidities do not observe any significant bending strength variation at 60% relative humidity compared to saturated case. However, the bending strength increases for more severe conditions of drying. Moreover, the results of Pihlajavaara [16] who tested a mature mortar with w/c = 0.50 at various uniform internal humidities show an increase of about 28% at 60% relative humidity. All these results illustrate the importance of the competitive effect in the evolution of mechanical properties with drying.

4 Conclusions

This experimental study highlights the effect of drying at an early age (one day) and after long-term maturation (75 days) on mechanical properties of a high-performance self-compacting concrete. The evolution of mechanical properties depends on the competitive effect between material strengthening due to capillary depression, disjoining pressure, hygral gradients and hydration (if it still takes place), and drying-induced micro-cracking due to material heterogeneity and differential shrinkage. Thus, in the case of early age drying during which hydration still occurs, the uniaxial compressive strength and bending strength increase due to the strengthening of the material. However, these strengths begin to decrease when the drying-induced micro-cracking effect overcomes the material strengthening effect. After long-term maturation, the uniaxial compressive strength shows only an increase with air-drying whereas the bending strength drastically decreases at the beginning of drying then slightly increases. Therefore, the drying-induced micro-cracking effect is very quickly detected on the bending strength of a mature material, the subsequent limited strength increase showing the beneficial strengthening effect. The results obtained clearly show the effect of maturation in the evolution of mechanical properties with drying, which has to be considered in durability analysis.

References

1. Bazant, Z.P., Wittmann, F.H.: Creep and Shrinkage in Concrete Structures. Wiley , London (1982)
2. Acker, P.: Comportement mécanique du béton: apport de l'approche physico-chimique. PhD thesis of Ecole Nationale des Pont et Chaussées, Rapport de Recherche LPC n°152 Paris (1988)
3. Popovics, S.: Effect of curing method and moisture condition on compressive strength of concrete. ACI J **83**(4), 650–657 (1986)
4. Baroghel-Bouny, V.: Caractérisation des pâtes de ciment et des bétons. PhD thesis of Ecole Nationale des Ponts et Chaussées, Paris (1994)
5. Bartlett, F.M., MacGregor, J.G.: Effect of moisture condition on concrete core strengths. ACI Mater J **91**(3), 227–236 (1994)
6. Neville, A.M.: Properties of Concrete, 4th edn. Longman Group (1995)
7. Hearn, N.: Effect of shrinkage and load-induced cracking on water permeability of concrete. ACI Mater J **96**(2), 234–241 (1999)
8. Bisschop, J., Pel, L., van Mier, J.G.M.: Effect of aggregate size and paste volume on drying shrinkage microcracking in cement-based composites. In: Ulm, F.-J., Bažant, Z.P., Wittmann, F.H. (eds.) Creep, Shrinkage & Durability Mechanics of Concrete and Other Quasi-Brittle Materials, pp. 75–80. Elsevier, Amsterdam (2001)
9. de Sa, C., Benboudjema, F., Thiery, M., Sicard, J.: Analysis of micro-cracking induced by differential drying shrinkage. Cem. Concr. Comp. **30**, 947–956 (2008)
10. Khelidj, A., Loukili, A., Bastian, G.: Etude expérimentale du couplage hydro-chimique dans les b&tons en cours de maturation: Incidence sur les retraits. Mater. Struct. **31**, 588–594 (1998)
11. Yurtdas, I., Burlion, N., Skoczylas, F.: Experimental characterisation of the drying effect on uniaxial mechanical behaviour of mortar. Mater. Struct. **37**, 170–176 (2004)
12. Price, W.H.: Factors influencing concrete strength. J. ACI **47**(2), 417–432 (1951)

13. Yurtdas, I., Burlion, N., Shao, J.F., Li, A.: Evolution of the mechanical behaviour of a high performance self-compacting concrete under drying. Cem. Concr. Comp. **33**, 380–388 (2011)
14. Baroghel-Bouny, V., Godin, J.: Experimental study on drying shrinkage of ordinary and high-performance cementitous materials. In: Baroghel-Bouny, V., Aïtcin, P.-C. (Eds) Shrinkage of Concrete, pp. 215–232. RILEM Publication PRO 17, Paris (2000)
15. Okajima, T., Ishikawa, T., Ichise, K.: Moisture effect on the mechanical properties of cement mortar. Transact. Jpn. Concrete Inst. **2**, 125–132 (1980)
16. Pihlajavaara, S.E.: A review of some of the main results of a research on the ageing phenomena of concrete, effect of moisture conditions on strength, shrinkage and creep of mature concrete. Cem. Concr. Res. **4**, 761–771 (1974)
17. Walker, S., Bloem, D.L.: Effects of curing and moisture distribution on measured strength of concrete. Proc. Highw. Res. Bd. **36**, 334–346 (1957)
18. Kanna, V., Olson, R.A., Jennings, H.M.: Effect of shrinkage and moisture content on the physical characteristics of blended cement mortars. Cem. Concr. Res. **28**, 1467–1477 (1998)

The Cracking Sensitivity of a Na-Geopolymer

Heng Sounean, Hannawi Kinda, and Darquennes Aveline

Abstract This research study aims to better understand the mechanical behavior of a Na-geopolymer used to immobilize the nuclear waste (MgZr alloys). Based on a large experimental campaign that has been carried out at LGCGM, the mechanical properties of Na-geopolymer are investigated as well as its cracking sensitivity at early age. This last depends on several parameters studied in the present project: tensile strength, Young's modulus, and delayed deformations such as autogenous shrinkage and basic creep. All of these parameters are characterized experimentally under autogenous condition. These results obtained for a Na-Geopolymer are compared to those of a Portland cement-based mortar. From restrained ring tests, it appears that the cracking sensitivity is lower for Na-Geopolymer. This behavior is mainly due to a low Young's modulus, a fast stabilization of autogenous shrinkage and a high creep capacity.

Keywords Autogenous shrinkage · Basic creep · Cracking · Mechanical performances · Na-geopolymer

1 Introduction

Geopolymer named by Joseph Davidovits in the late 1970s [1] is an inorganic material made by mixing aluminosilicates sources with an alkaline solution. Geopolymers structure consists of tetrahedral frameworks of AlO_4- and SiO_4- by linking together with oxygen bridges and with alkali metal ions balancing the charge associated with

H. Sounean (✉) · H. Kinda · D. Aveline
Laboratoire de Génie Civil et Génie Mécanique (LGCGM), Institut National des Sciences Appliquées de Rennes, Rennes, France
e-mail: sounean.heng@insa-rennes.fr

H. Kinda
e-mail: kinda.hannawi@insa-rennes.fr

D. Aveline
e-mail: aveline.darquennes@insa-rennes.fr

F. Kanavaris et al. (eds.), *International RILEM Conference on Early-Age and Long-Term Cracking in RC Structures*, RILEM Bookseries 31,
https://doi.org/10.1007/978-3-030-72921-9_14

tetrahedral Al [2, 3]. The general empirical formula of geopolymer is Mn{-(SiO_2)z-AlO_2}n, wH_2O [2], where M is alkali metal cation (normally is Na^+ or K^+), and n is a degree of polycondensation with z = 1, 2, 3. Since several years, more and more research works focus on the design and the behavior of this inorganic material. Indeed, it presents several advantages: a low CO_2 release during its production [4–6], a high compressive strength, and a good chemical and fire resistance [7, 8].

The present study focuses on a Na-geopolymer designed to immobilize nuclear wastes such as Mg-Zr alloys fuel cladding used in the UNGG reactors (Uranium Natural Graphite Gas) in the framework of the DECIMAL project (Phenomenological Description Corrosion And Its Impact On The Durability Of Encapsulated Magnesium Wastes In Hydraulic Binders). In an aqueous media, Mg can corrode rapidly and lead simultaneously to hydrogen (H_2) release. But Rooses et al. [9] have shown that the corrosion of Mg-Zr immobilized in a Na-geopolymer matrix is low in comparison to an immobilization into a Portland cement matrix and further studies are being performed in the DECIMAL project to better describe the corrosion mechanisms of Mg alloys in this Na-matrix [10]. However, the chemical reactions driving geopolymerization are very complex and depend on many factors such as aluminosilicate source, curing time and temperature [11–13], nature and concentration of alkaline activator solution [14–16], as well as Si/Al/cation ratios [17, 18]. These parameters also affect the mechanical behavior of geopolymers at the macroscale. In the present study, the behavior of a Na-geopolymer is analyzed in autogenous condition to assess its cracking risk when its delayed deformations are restrained by an external element.

2 Materials and Curing Condition

To design the Na-geopolymer, a metakaolin is used as an aluminosilicates source. Its chemical composition is given in Table 1. Its area surface and specific density are equal to 17 m^2/g and 2.4 g/cm^3 respectively.

The metakaolin-based geopolymer mortar -named MGEO hereafter- presents the following formulation: $1Na_2O–3.96SiO_2–1Al_2O_3–12.5H_2O$. The alkaline activator solution is prepared by mixing sodium silicate solution (SiO_2 = 27.5 wt.%, Na_2O = 8.3 wt.%, H_2O = 64.2 wt.%, SiO_2/Na_2O = 3.42) with sodium hydroxide pellets (>98 wt.% purity) and distilled water (conductivity = 25.43 µs/cm at 20 °C, pH = 7.66). Moreover, sodium fluoride (1.25 M) used as corrosion inhibitor is also added, as well as quartz sand (0.315/1.6 mm). Furthermore, ordinary Portland cement mortar (CEM I 52.2 N) is used as a reference material and named hereafter MCEMI. It is mixed

Table 1 Chemical composition of metakaolin

Chemical	SiO_2	Al_2O	$K_2O + Na_2O$	Fe_2O_3	TiO_3	CaO + MgO
Metakaolin	55%	40%	0.8%	1.4%	1.5%	0.3%

with tap water (conductivity = 558.9 μs/cm at 20 °C, pH = 8.06) and normalized sand (0/2 mm). The two studied mortars show the same paste volume and liquid to binder ratio (L/B = 0.508).

All the specimens are demolded after 24 h and stored at 20 ± 1 °C. The autogenous condition is obtained thanks to a double layer of aluminum foil. For mechanical tests, a plastic foil is also added between the adhesive aluminum foils and specimen. To control the autogenous condition, specimens' mass is regularly measured during all the tests.

3 Experimental Method

3.1 Mechanical Properties

To determine the tensile strength, a three-point flexural test is performed on three prismatic specimens (4 × 4 × 16 cm) with a loading rate of 0.01 mm/s. The tests are performed on 1, 3, 7, 28, and 90 days old specimens. After this test, specimens are loaded again to determine the compressive strength with a loading rate equal to 2.4 kN/s according to EN 196-1.

Young modulus test is carried out on three cylindrical specimens (ϕ11 × 22 cm) at 3, 7, 28, and 90 days. Specimens are loaded with a rate of 0.5 MPa/s (NF EN 12390-4) until a value equal to 30% of their compressive strength determined previously on specimens having the same dimensions.

3.2 Autogenous Shrinkage

The autogenous shrinkage is measured on three prismatic specimens (4 × 4 × 16 cm) at their central axis by means of two inserts placed on each sample's extremities (top and bottom) with a displacement comparator from 1 to 40 days, as well as their mass loss.

In parallel, internal relative humidity evolution and water porosity are also measured to understand the autogenous shrinkage (amplitude and kinetics) as capillary depression is one of the main driving mechanisms responsible of this deformation. The internal relative humidity is monitored with a thermo-hygrometer sensor -having a precision of ±2%—placed weekly into the middle of cylindrical specimens (ϕ50 × 70 mm) thanks to a drilled hole (ϕ5 × 35 mm). To determine water porosity, three cylindrical 28 days old specimens (ϕ40 × 60 mm) are water-saturated under vacuum for 44 h. Their mass is measured in air (M_{air}) as well as in water (M_w) thanks to the hydrostatic weighing. After, specimens are dried in an oven at 40 °C until to reach a constant mass value (M_{dry})—a mass variation inferior to 0.05% between to two measurements (24 h) is acceptable. Based on these measurements, bulk density

(ρ_{app}) and water porosity (ϕ) are calculated using the Eqs. (1) and (2) respectively:

$$\rho_{app} = \frac{M_{dry}}{M_{air} - M_w} \rho_w \tag{1}$$

$$\phi = \frac{M_{air} - M_{dry}}{M_{air} - M_w} \rho_w \tag{2}$$

With ρ_W the water density (1 g/cm^3).

3.3 Basic Creep

Compressive basic creep test is carried out on two prismatic specimens (7 × 7 × 28 cm) using hydraulic jacks. The total deformation is measured on each specimen side via two glued balls with a distance of 20 cm. A load equal to 20% of compressive strength previously determined on cylindrical specimens (ϕ11 × 22 cm) is applied on 90 days old specimens. In parallel, autogenous shrinkage is also determined on unloaded samples (7 × 7 × 28 cm).

3.4 Restrained Ring Test

A ring test (Fig. 1) is performed to evaluate the cracking time of the studied mortars. Its steel ring is characterized by an inner (R_2) and outer (R_1) radius equal to 100 mm and 125 mm respectively. The outer radius (R_0) of the mortar ring cast around the steel ring is equal to 160 mm. The specimens are demolded after 24 h, and sealed

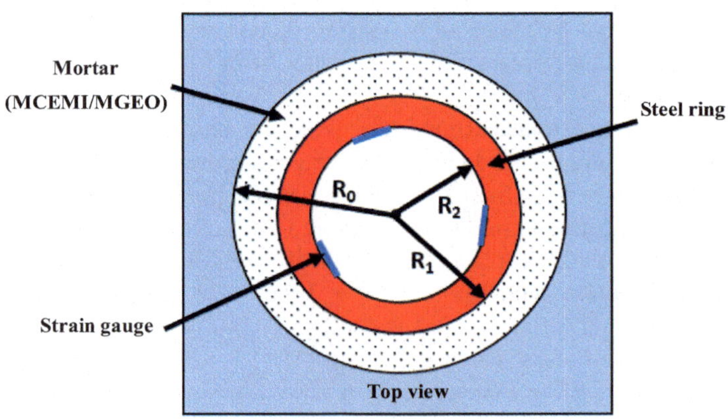

Fig. 1 Ring test setup

with a double layer of aluminum foils to obtain autogenous condition. The deformations of the steel ring are monitored thanks to three strain gauges glued at the mid-height of the inner surface of the steel ring (Fig. 1) and connected to the acquisition system. Moreover, the ambient temperature is also monitored to quantify the thermal deformation.

4 Experimental Results

4.1 Mechanical Properties

Figure 2 presents the flexural and compressive strength evolution for MCEMI and MGEO. According to Fig. 2a, the flexural strength of MGEO is lower than that of MCEMI for all the tested ages. It indicated that MGEO has lower tensile strength. It is probably related to the fact the geopolymer-based materials present an intrinsic brittleness due to their highly cross-linked framework [19]. The compressive strength of MGEO increases rapidly during the three first days (Fig. 2b). After 3 days, it evolves slowly and its value is close to the compressive strength of MCEMI. A similar evolution was found by Rocha et al. [20]. Figure 3 shows that the Young's modulus is lower for MGEO than for MCEMI at all the tested ages. This last behavior presents an advantage to limit the cracking risk of the geopolymer when its delayed deformations are restrained.

4.2 Autogenous Shrinkage

Figure 4 presents the autogenous shrinkage evolution of MGEO and MCEMI measured from 1 day. During the first 10 days, the autogenous shrinkage of MGEO evolves more rapidly. After, it slightly increases until 22 days. Beyond it tends to a constant value close to that obtained for MCEMI. For example, this deformation is equal to about 255 μm/m for both mixtures at 22 days. Notice that the mass loss (because the tightness is not perfect after several weeks—unwanted drying) is slightly higher for MGEO at this age: 0.018% for MGEO and 0.007% for MCEMI. Moreover, the final deformation amplitude for MCEMI is quite similar than that found by Olivier [21] (230 μm/m at 28 days for a mortar with W/C ratio equal to 0.5).

Water porosity

As one of the main mechanisms driving autogenous shrinkage for cementitious materials is the capillary depression, the water porosity was measured for the both studied mortars and the results at 28 days are presented in Table 2. It appears that MGEO is characterized by a more important porosity. In a next step, the pore size distribution

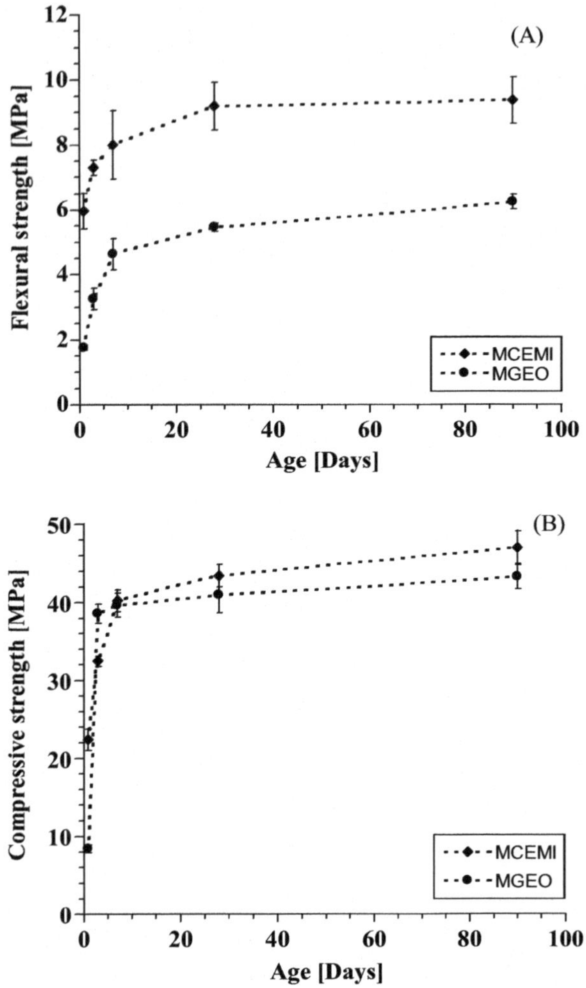

Fig. 2 Flexural (**a**) and compressive (**b**) strength for MCEMI and MGEO at 1, 3, 7, 28, and 90 days

will be characterized to better understand the effect of porosity on the autogenous deformation.

Internal relative humidity

On Fig. 4, the internal relative humidity evolution is also presented for both studied mortars. This parameter varies around a constant value (±95%) during the 40th days for MGEO, while it decreases progressively for MCEMI (89.6% at 43 days). Ma et al. [22] found a similar result for a fly-ash based geopolymer. They explain that by the decreasing of salt content (Na^+ and silicate species) in pore solution. Moreover, they assume that autogenous shrinkage of fly-ash based geopolymer is due to the continuous reorganization and polymerization of the gel structure instead of the self-desiccation process available for a classic cementitious material.

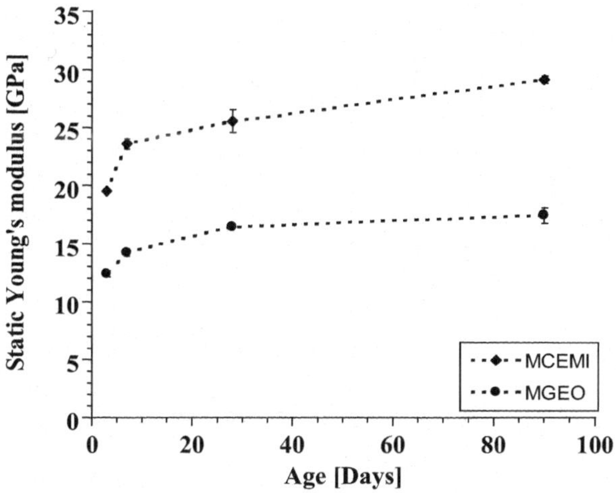

Fig. 3 Young's modulus for MCEMI and MGEO at 3, 7, 28, and 90 days

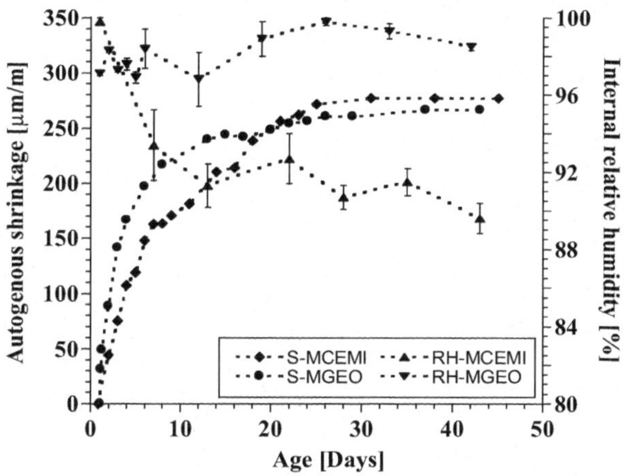

Fig. 4 Evolution of internal relativity humidity (RH) and autogenous shrinkage (S) for MCEMI and MGEO

Table 2 Bulk density (ρ_{app}) and water porosity (ϕ) for MCEMI and MGEO at 28 days

Test	MCEMI	MGEO
ρ_{app} (g/cm^3)	1.989 ± 0.003	1.813 ± 0.002
ϕ (%)	20.9 ± 0.2	26.2 ± 0.5

4.3 Basic Creep

The specific basic creep is illustrated in Fig. 5 for both studied mortars loaded at 90 days. It is higher for MGEO. Thus, this mortar is characterized by a higher capacity to relax internal stresses—a positive behavior to reduce the cracking risk when the

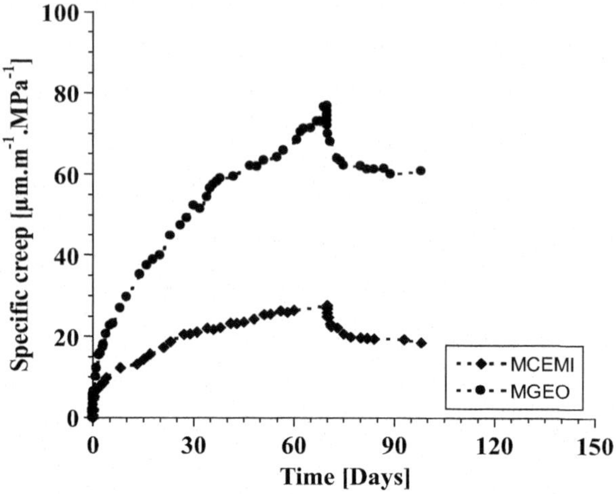

Fig. 5 Specific basic creep of MCEMI and MGEO, loading at age 90 days

delayed deformations are restrained. The specimens were unloaded at 70 days. The residual deformation is also more important for MGEO. Notice that the mass loss at the test end is equal to 0.23% and 0.074% for MGEO and MCEMI respectively.

4.4 Restrained Ring Test

To determine the cracking sensitivity, a restrained ring test was performed. The deformation evolutions of the steel ring surrounded by MCEMI and MGEO during the first 30 days are showed in Fig. 6. During these tests, the average ambient temperature is equal to 18.6 ± 1 °C and 19.7 ± 1 °C respectively. The deformation increases more rapidly for MGEO at early age, but their amplitude from 5 days is quite similar. At about 25 days, MCEMI cracks whereas no crack is observed for MGEO beyond 30 days. Based on these results, it appears that MGEO presents a lower cracking sensitivity when its delayed deformations are restrained.

5 Conclusion

The main objective of this experimental campaign is to study the delayed behavior of a Na-geopolymer and its cracking risk under autogenous condition. This material presents a low Young's Modulus and a low tensile strength. Even though the lower stiffness can reduce the internal stresses amplitude, a lower tensile strength may cause a higher cracking risk. However, the delayed deformation results show that Na-geopolymer is characterized by a rapid stabilization of autogenous shrinkage reaching a constant after 22 days and a high specific basic creep. These both results

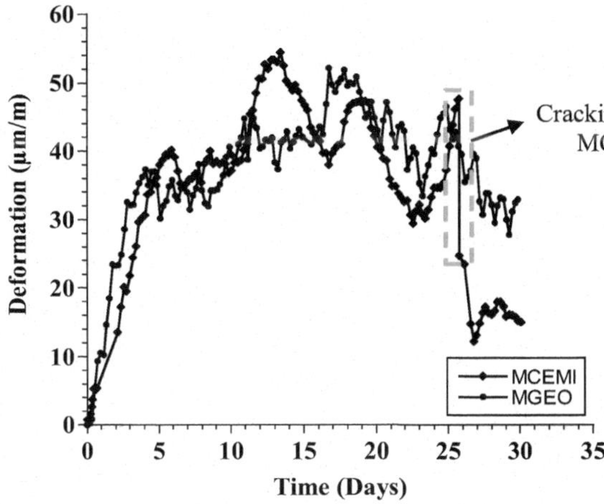

Fig. 6 Steel ring deformations due to the delayed deformations of MGEO and MCEMI under autogenous condition

are interesting to limit cracking sensitivity when the delayed deformations of the geopolymer are restrained. Thanks to a ring test coupling all these material parameters, the cracking sensitivity of the geopolymer was studied. This test has confirmed that the studied geopolymer presents a lower cracking sensitivity in comparison to a classic Portland cement mortar.

Acknowledgements Work performed within the Investments for the future program of the French Government and operated by French National Radioactive Waste Management Agency (Andra).

References

1. Davidovits, J.: The need to create a new technical language for the transfer of basic scientific information. Conference 1982, Transfer and Exploitation of Scientific and Technical Information, vol. 1, pp.19–23. Luxemburg (1982)
2. Davidovits, J.: Geopolymer: inorganic polymeric new materials. J. Therm. Anal. **37**, 1633–1656 (1991)
3. Duxson, P., Luckey, G.C., Separovic, F., Van Daventer, J.S.J.: Effect of alkali cations on aluminum incorporation in geopolymer gels. Ind. Eng. Chem. Res. **44**, 832–839 (2005)
4. Duxson, P., Provis, J.L., Luckey, G.C., Van Daventer, J.S.J.: The role of inorganic polymer technology in the development of green concrete. Cem. Concr. Res. **37**, 1590–1597 (2007)
5. Provis, J.L., Van Daventer, J.S.J.: Geopolymers: Structures, Processing Properties and Industrial Applications. Boca Raton, Boston, New York, Washington (2009)
6. McLellan, B.C., William, R.P., Lay, J., van Riessen, A., Corder, G.D.: Costs and carbon emission for Geopolymer pastes in comparison to ordinary Portland cement. J. Clean. Prod. **19**(9–10), 1080–1090 (2011)

7. Barkarev, T.: Resistance of geopolymer materials to acid attack. Cem. Concr. Res. **35**, 658–670 (2005)
8. Fernández-Jiménez, A., Palomo, A.: New cementitious materials based on alkali-activated fly ash: performance at high temperatures. J. Am. Ceram. Soc. **91**(10), 3308–3314 (2008)
9. Rooses, A., Lambertin, D., Chartier, D., Frizon, F.: Galvanic corrosion of Mg-Zr fuel cladding and steel immobilized in Portland cement and geopolymer at early ages. J. Nucl. Mater. **435**, 137–140 (2013)
10. Boubon, R., Deschanels, X., Cabié, M., Rébiscoul, D.: Evolution of corrosion products formed during the corrosion of MgZr alloy in poral solutions extracted from Na-geopolymers used as conditioning matrix for nuclear waste. Materials **13**, 4958 (2020)
11. Van Jaarsveld, J.G.S., Van Daventer, J.S.J., Luckey, G.C.: The effect of composition and temperature on the properties of fly as- and kaolinite-based geopolymers. Chem. Eng. J. **89**, 63–73 (2002)
12. Rovananik, P.: Effect of curing temperature on the development of hard structure of metakaolin-based geopolymer. Constr. Build. Mater. **24**, 1176–1183 (2010)
13. Basil, S.A., Tareq, S.A.: Effect of curing system on metakaolin based geopolymer concrete. J. Babylon Univ./Eng. Sci. **3**(24) (2016)
14. Provis, J.L., Duxson, P., Luckey, G.C., Separovic, F., Kriven, W.M., Van Daventer, J.S.J.: Modeling speciation in highly concentrated alkaline silicate solutions. Ind. Eng. Chem. Res. **44**, 8899–8908 (2005)
15. White, C.E., Provis, J.L., Llobet, A., Proffen, T., Van Daventer, J.S.J.: Evolution of local structure in geopolymer gels: An in situ neutron pair distribution function analysis. J. Am. Ceram. Soc. **94**(10), 3532–3539 (2011)
16. Fan, Z., Liying, Z., Ming, L., Chenzhong, M., Yen, N.L., Xiao, H.: Role of alkali cation in compressive strength of metakaolin based geopolymer. Ceram. Int. **43**, 3811–3817 (2017)
17. Bourlon, A.: Phys Physico-chimie et rhéologie de géopolymères frais pour la cimentation des puits pétroliers. Thèse de doctorat: Physique et Chimie des Matériaux, Paris: Université Pierre et Marie Curie (2011)
18. Duxson, P., Fernández-Jiménez, A., Provis, J.L., Luckey, G.C., Palomo, G.C., Van Daventer, J.S.J.: Geopolymer technology: the current of the art. J. Mater. Sci. **42**, 2917–2933
19. Pan, Z., Sanjayan, J.G., Rangan, B.V.: Fracture properties of geopolymer paste and concrete. Mag. Concr. Res. **63**(10), 763–771 (2011)
20. Rocha, T.S., Dias, D.P., França, F.C.F., Guerra, R.R.S., Marques, L.R.C.O.: Metakaolin-based geopolymer mortars with different alkaline activators (Na^+ and K^+). Constr. Build. Mater. **178**, 453–461 (2018)
21. Olivier, K.: Etude expérimentale et modélisation de l'auto-cicatrisation des matériaux cimentaires avec additions minerals. Thèse de doctorat: Sciences mécanique et énergétiques, matériaux et géoscienes, l'Université de Sherbrooke et de l'Université Paris-Saclay (2016)
22. Ma, Y., Ye, G.: The shrinkage of alkali activated fly ash. Cem. Concr. Res. **68**, 75–82 (2015)

The Effect of Curing Duration on Restrained Shrinkage Cracking of Concrete Containing Macro Polymeric Fibers

Alireza Bagheri, Armin Jamali, Mina Pourmir, and Hamed Zanganeh

Abstract The positive influence of longer duration of wet curing in improving the mechanical and durability properties of concrete is fairly well established. However, it is increasingly becoming apparent that longer duration of wet curing, unfavorably affects the drying shrinkage characteristics of concrete, resulting in both increased amount of shrinkage and also increased cracking potential under restrained conditions. Various types of fibers have been developed and used to enhance the durability of concrete structures by controlling the restrained shrinkage cracking. However, the influence of longer duration of wet curing on the age at cracking and crack characteristics of concrete containing macro polymeric fibers has not been previously determined. In this research, the effects of three different curing durations on strength characteristics and free and restrained drying shrinkage performance of mixes with and without macro polymeric fibers were studied. The results show that drying shrinkage and cracking potential of the control mixes increased with longer duration of wet curing, confirming the trend reported by recent literature. Interestingly the mixes containing fibers also showed the same trend and both the amount of free shrinkage and the restrained shrinkage potential increased with increased duration of wet curing. With regards to crack widths, the results show that with longer wet curing, the width of the single crack for the control mixes increased significantly while the cumulative width of the multiple cracks of the mixes with polymeric fibers increased only modestly.

Keywords Drying shrinkage · Curing duration · Polymeric fibers · Restrained ring test · Crack width

A. Bagheri (✉) · A. Jamali · M. Pourmir · H. Zanganeh
K. N. Toosi University of Technology, 470 Mirdamad Ave. West, 19697 Tehran, Iran
e-mail: bagheri@kntu.ac.ir

F. Kanavaris et al. (eds.), *International RILEM Conference on Early-Age and Long-Term Cracking in RC Structures*, RILEM Bookseries 31,
https://doi.org/10.1007/978-3-030-72921-9_15

1 Introduction

Restrained drying shrinkage is a common cause of cracking in concrete, which increases the ingress of aggressive agents and results in increased deterioration and maintenance cost of concrete structures [1]. The evaluation of cracking potential and restrained shrinkage crack details, can therefore provide insight into the short and long-term durability of concrete [2]. Structures with a low volume-to-surface ratio such as pavements, bridge decks and slabs on the ground, are more prone to restrained shrinkage cracking because a larger surface of these structures is exposed to drying [3]. In order to reduce the amount of drying shrinkage and cracking potential of concrete, the use of shrinkage reducing admixtures (SRA) has been advocated. These admixtures modify the properties of pore solution such as surface tension and reduce the amount of shrinkage, resulting in a reduction in cracking potential as well as crack width [3–6].

The use of steel fibers and recently polymeric fibers is another method to reduce the restrained shrinkage crack width by bridging and transmitting tensile stress across the crack opening. The effect of micro and macro polymeric fibers in different types and dosages on restrained shrinkage cracking has been considered by many researchers [3, 7–9]. The reported results indicate that although the use of these fibers does not significantly postpone the cracking time at lower dosages, it can control the crack width properly. Recent research by Bagheri et al. [7] showed that the use of 0.5% macro polymeric fibers reduced restrained shrinkage crack width by up to 50% compared to the mix without fibers. Yousefieh et al. [9] and Passuello et al. [3] also reported similar results.

The effect of increased curing duration in improving the mechanical and durability performance of concrete and extending its service life of concrete is well established [10, 11]. However its influence on the restrained shrinkage cracking potential and crack width is not clear, although it is increasingly becoming apparent that longer duration of wet curing, unfavorably affects the drying shrinkage characteristics of concrete. The results obtained by Beushausen and Bester [12] indicated that longer curing duration increased the crack resistance of the control mix, while the restrained shrinkage crack width was not directly correlated to the curing duration. However, the recent paper published by Bagheri et al. [6] showed that by longer curing duration, the shrinkage cracking potential and crack width of the control mix increased. The results obtained by Samouh et al. [13], Hajibabaee et al. [14, 15] also showed increased shrinkage due to longer curing duration whereas an opposite result was observed in other studies [16, 17].

Studies on the effect of curing duration on restrained shrinkage cracking have been limited to plain concrete and mixes containing SRA, and its effect on mixes containing fibers has not been previously investigated [6, 12, 13]. Therefore in this study, the effect of curing duration on the free shrinkage, shrinkage cracking potential and crack width of a mix containing macro polymeric fibers is investigated and compared to that of a control mix.

2 Materials and Methods

2.1 Materials

The cement used in this study was a type 2 Ordinary Portland Cement (OPC) produced by the Tehran cement factory. The chemical composition, physical and mechanical properties of the cement used, are summarized in Tables 1 and 2. Crushed gravel with a maximum nominal size of 12.5 mm and specific gravity (SSD) of 2.55 gr/cm^3 was used as coarse aggregates. A river sand with specific gravity (SSD) of 2.57 gr/cm^3 was also used as fine aggregates. The particle size distribution of aggregates is shown in Fig. 1. The single-strand macro polypropylene fibers having a length of 60 mm, a diameter of 0.5–1 mm and elastic modulus of 1.7 GPa, produced by EPC Barchip Company, were used in this study.

Table 1 Chemical composition of the cement (%)

SiO_2	Al_2O_3	Fe_2O_3	CaO	MgO	SO_3	Na_2O	K_2O	C_3S	C_2S	C_3A	C_4AF	Insoluble residue
21.3	4.8	3.55	62.8	3.9	2	0.18	0.51	50	22	6.72	10.8	0.6

Table 2 Physical and mechanical properties of the cement

Density (gr/cm^3)	3.15
Water of standard consistence (%)	21.5
Initial setting time (min)	130
Final setting time (min)	210
Compressive strength at 28 days (MPa)	37

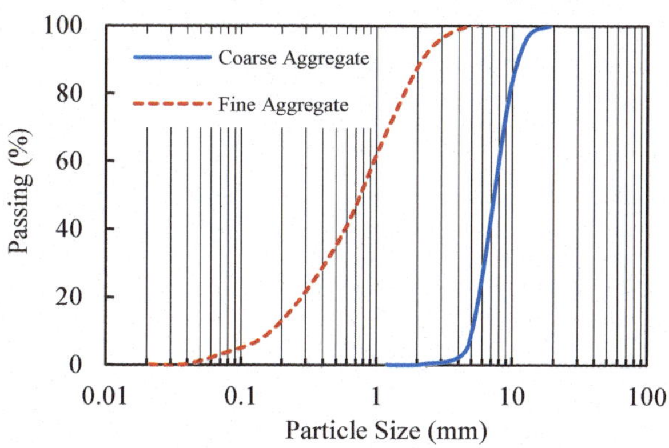

Fig. 1 Particle size distribution of the coarse and fine aggregates

2.2 Experimental Methodology

Mixed Design and Considered Curing Duration. The cement content of 420 kg/m^3 and the water-cement ratio of 0.5 were considered for all mixtures. The considered fiber dosage was 0.5% (by volume). Test specimens were cured in three different durations of wet curing involved 1, 3 and 7 days. The proportion of mixes used in this research is shown in Table 3.

Mechanical Properties. Three 10 cm cubic specimens were fabricated to determine the compressive strength of each mix at each testing age in accordance with EN 12390–3 [18]. Three 10 × 20 cm cylindrical specimens were also cast to measure the splitting tensile strength of each mix at each testing age according to ASTM C496 [19]. After considered initial curing including 24 h in the mold, covered with polyethylene sheet, and water tank (for specimens with more than 1 day curing duration), the test specimens were transferred to a cabinet with the relative humidity (RH) of 50 ± 4% and a temperature of 23°C ± 2 until the testing time. The compressive and splitting tensile strength tests were conducted at the ages of 1, 3, 7 and 28 days.

Free Shrinkage Test. The free shrinkage test was carried out on prismatic specimens with the dimension of 75 × 75 × 285 mm in accordance with ASTM C157 [20]. Three specimens were cast for each free shrinkage test. After casting, test specimens were in the molds, covered by polyethylene sheet, for one day at the RH of 95% and a temperature of 23°C. Thereafter, they were removed from the molds and the specimens with 1 day duration of curing were transferred to a drying cabinet with RH of 50 ± 4% and a temperature of 23°C ± 2, while the specimens with 3 and 7 days duration of curing were submerged in the water tank for 2 and 6 days respectively. After wet curing, they were also transferred to the mentioned drying cabinet. The length change of the specimens was measured by the comparator device with a precision of 0.002 mm. The length measurement started on the first day of specimens drying and continued up for a duration of 120 days.

Restrained Shrinkage Test. To evaluate the cracking potential of mixes, the restrained shrinkage test was conducted according to ASTM C1581 [21]. The details of the test equipment including strain gages, steel ring and base plate, an external mold and their dimension are shown in Fig. 2. Two ring specimens were cast for

Table 3 Concrete mixtures and curing times considered in the experimental program

Mixes code	C-1	C-3	C-7	F-1	F-3	F-7
Curing time (Day)	1	3	7	1	3	7
Cement (Kg/m^3)	420	420	420	420	420	420
Water (Kg/m^3)	210	210	210	210	210	210
Coarse (Kg/m^3)	733	733	733	733	733	733
Fine (Kg/m^3)	896	896	896	896	896	896
Fiber content (Vol %)	0	0	0	0.5	0.5	0.5
Slump (cm)	12	12	12	11	11	11

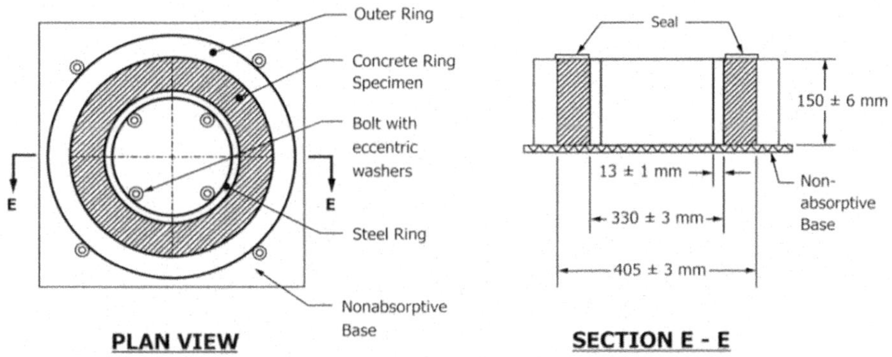

Fig. 2 Schematic view of restrained ring test equipment [21]

each restrained ring test. Before casting, the outer and top surface of the steel ring and base plate were coated with a release agent to reduce the friction between the concrete specimen and surface of the steel ring and base plate. After casting, the ring specimens were covered with wet sponge and polyethylene sheets to maintain RH at least 95%. After one day, the external molds were removed from the concrete rings and the top surface of the specimens with 1 day duration of curing was coated with paraffin to provide uniform circumferential drying conditions. They were then transferred to storage with RH of 50 ± 4% and a temperature of 23°C ± 2. The concrete specimens with 3 and 7 days curing duration were covered with wet sponges and were cured for 2 and 6 days respectively. After curing, their top surface was also coated with paraffin and transferred to the mentioned storage. All the specimens were immediately connected to the data logger after casting to record the steel ring strain induced by shrinkage of concrete ring specimens. According to ASTM C1581 [21], a sudden decrease in the recorded strain of the steel ring is considered as an indication of crack occurrence in concrete ring specimens. The *Age at Cracking* and the *Net Time to Cracking* was determined for each concrete ring specimens. The former is the elapsed time from casting to cracking while the latter is considered from the start of the drying to cracking.

Crack Width Measurement. After the crack occurrence, the restrained shrinkage crack width of the concrete ring specimens was monitored for six weeks. The crack width of the ten fixed points along the crack length was measured using a Crackscope with a precision of 0.05 mm and the average of which were reported as a restrained shrinkage crack width of each test specimen.

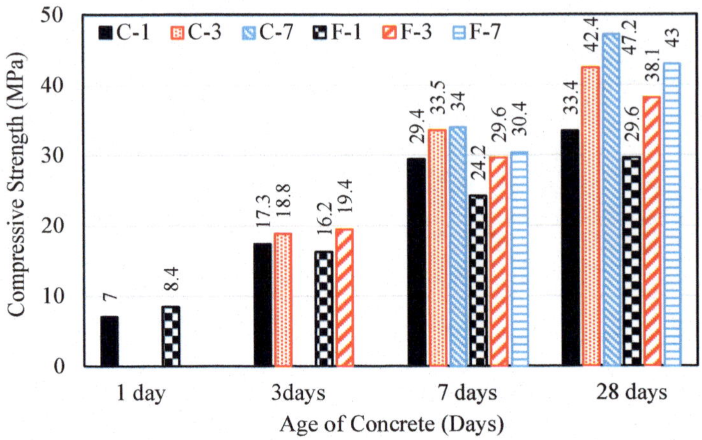

Fig. 3 The effect of curing duration on the compressive strength of the control mix and the mix containing 0.5% polymeric fibers

3 Results and Discussions

3.1 Compressive and Tensile Strength

As shown in Fig. 3, a similar increase was observed in the compressive strength of the mixes with and without polymeric fibers due to the longer duration of wet curing. The 28 days compressive strength of both mixes increased by 28% and 43% by increased curing duration from 1 day to 3 and 7 days respectively.

The effect of longer curing duration on tensile strength of the control mix and the mix with polymeric fibers are also depicted in Fig. 4. By increased the duration of curing from 1 day to 3 and 7 days, the 28 days tensile strength increased by 32% and 43% respectively for both control and polymeric fibers mixes. The hydration degree of the cement paste increases because of the longer curing duration, resulting in higher compressive and tensile strength of concrete [1].

The results also indicated that the use of 0.5% polymeric fibers reduced the 28 days compressive and tensile strength by 10% and 16% respectively. The lower elastic modulus of polymeric fibers compared to concrete and non-uniform distribution of fibers may result in a reduction in compressive and tensile strength of concrete [22].

3.2 Free Shrinkage

The results of the free drying shrinkage of the control mix and the mix with 0.5% polymeric fibers are shown in Fig. 5, plotted according to ASTM C596 [23] from the start of the drying. As can be seen, the control mix and the mix with polymeric fibers

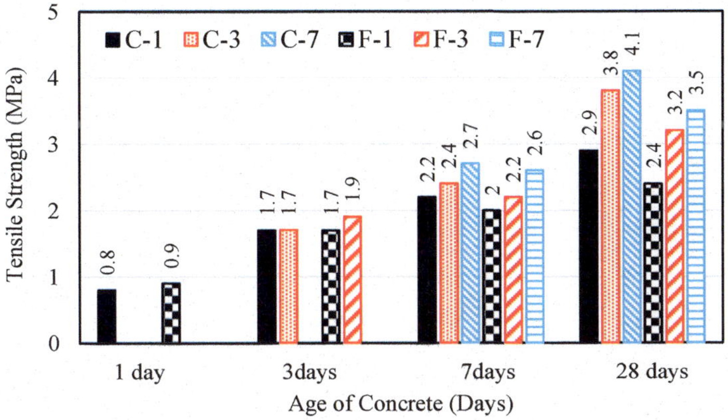

Fig. 4 The effect of curing duration on the tensile strength of the control mix and the mix containing 0.5% polymeric fibers

experienced a similar increase in drying shrinkage with increased curing duration. The obtained results revealed that by longer curing duration from 1 day to 3 and 7 days the drying shrinkage of both mixes increased by 3% and 11% respectively. Based on the previous study of the authors [6], the net effect of three main factors including *total pore volume*, *the proportion of finer porosity* and *internal restraint*, which are affected by curing duration, governs the amount of drying shrinkage of mixes. Longer curing duration increased the hydration degree of cement paste resulting in a reduction in its total pore volume as a result of which the amount of drying shrinkage reduced. On the other hand, the proportion of finer porosity increased with higher hydration and saturation degree of cement paste due to longer curing time. Finer porosity can increase the capillary tension resulting in higher drying shrinkage. Furthermore, the amount of non-hydrated cement particles, playing an internal restraint role, reduced due to the increased level of hydration. This results in increased drying shrinkage [1, 13, 14, 24]. Therefore, the net effect of the above mentioned three factors, appears to have resulted in increased drying shrinkage for both the control mix and the mix with polymeric fibers, due to longer curing duration.

The results also indicated that the use of 0.5% polymeric fibers has not significantly reduced the drying shrinkage compared to the control mix for the same curing duration. A similar result was also reported by Passuello et al. [3]. However, the results of Shen et al. [8] showed that the use of 0.6% polymeric fibers led to an 11.5% reduction in drying shrinkage. The authors claimed that the embossed surface of the macro polymeric fibers plays a restraint role in the cement paste, resulting in a reduction in the drying shrinkage [8].

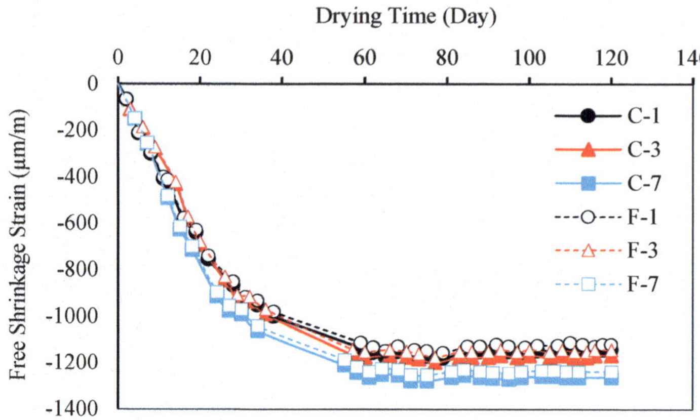

Fig. 5 The effect of different duration of curing on free drying shrinkage of the control mix, the mix containing 0.5% polymeric fibers

3.3 *Restrained Shrinkage*

The results of restrained drying shrinkage of the control mix and the mix with polymeric fibers are shown in Fig. 6. The obtained results revealed that the age at cracking of the control mix and the mix with polymeric fibers increased similarly by longer curing duration. By increasing the curing duration from 1 day to 3 and 7 days, the age at cracking for the control mix increased from 10 days to 11 and 13 days respectively and that for the mix with polymeric fibers also increased from 11 days to 12 and 14 days. However, the net time to cracking, considered as a criterion of cracking potential according to ASTM C1581 [21], showed an opposite trend for both mixes due to increased curing duration as illustrated in Fig. 7. By increasing the curing duration from 1 day to 3 and 7 days, the net time to cracking of the control mix reduced from 9 days to 8 and 6 days respectively and that for the mix with polymeric fibers also reduced from 10 days to 9 and 7 days indicating the increased cracking potential of both mixes.

With longer curing duration, the amount of drying shrinkage of mixes increased resulting in increased restrained shrinkage in the concrete ring specimens. Moreover, higher elastic modulus due to drying start at later age induces higher tensile stress in the concrete ring specimens. The effect of concrete relaxation, which reduces the tensile stresses over time, is less significant for later ages. It appears that the contribution of these factors outweighs that of the increased tensile strength of mixes with longer curing duration, which results in the increased cracking potential of the control mix and the mix with polymeric fibers [6].

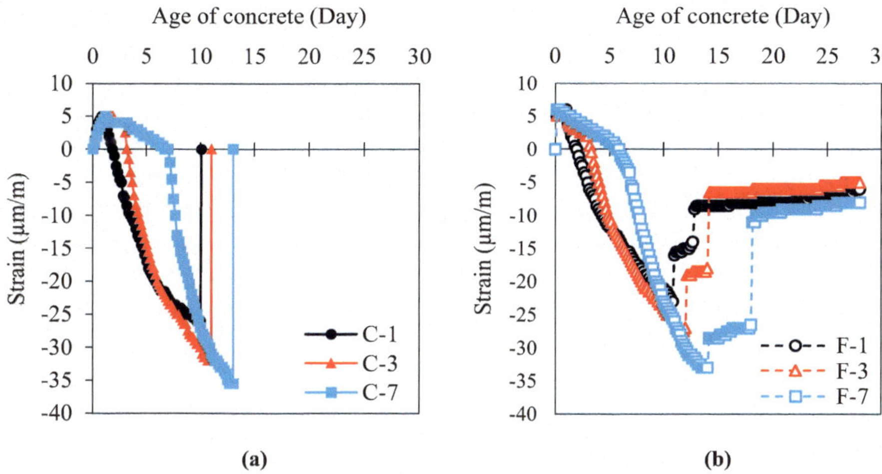

Fig. 6 Effect of the duration of curing on the cracking age of the control mix (**a**), and the mix containing 0.5% polymeric fibers (**b**)

Fig. 7 Effect the duration of curing on Age at cracking and Net time to cracking of the control mix and the mix containing 0.5% polymeric fibers

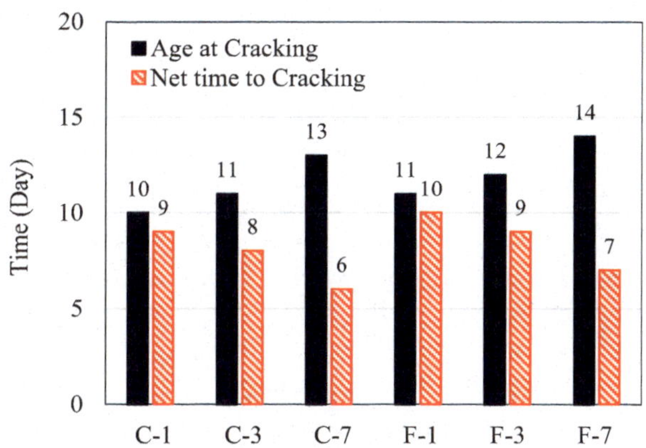

3.4 *Restrained Shrinkage Crack Width*

The results of restrained shrinkage crack width growth for 42 days after crack occurrence are shown in Fig. 8. As can be seen, the crack width almost remain unchanged after that period. The results indicate that crack width of the control mix increased significantly due to longer curing duration. By increasing the curing duration from 1 day to 3 and 7 days the crack width increased from 0.95 mm to 1.25 mm and 2.1 mm. Longer curing duration increased the shrinkage strain and reduced the net time to cracking of concrete ring specimens. This causes that larger part of induced tensile strains release as a single crack [6].

Unlike the control mix, some multi-hairline crack appeared in the mix with polymeric fibers. As expected, induced tensile stresses due to restrained shrinkage were transmitted by polymeric fibers across the crack opening after the crack occurrence resulting in reduced crack width in the mix with polymeric fibers up to 50% compared to the control mix as can be shown in Fig. 9.

The effect of curing duration on the crack width of the mix containing polymeric fibers was found negligible. As presented in Fig. 8, by increasing the curing duration from 1 day to 3 and 7 days, the restrained shrinkage crack width increased modestly from 0.5 mm to 0.6mm and 0.7 mm. Although a larger part of tensile stresses was released after cracking, polymeric fibers controlled the crack width growth by transmitting them across the crack opening [7].

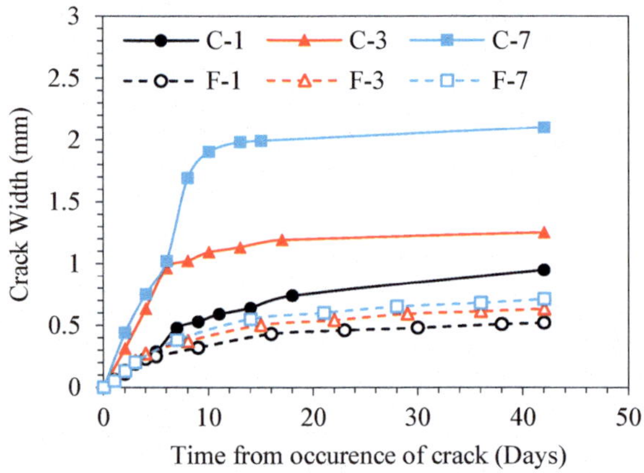

Fig. 8 Effect of curing time on crack width growth of the control mix and the mix containing 0.5% polymeric fibers

Fig. 9 Effect of 7 days curing duration on the crack width of the control mix **a** and the mix with polymeric fibers **b** after 42 days from the crack occurrence

4 Conclusion

- The amount of free drying shrinkage of the control mix and the mix with polymeric fibers increased by longer curing duration similarly. By increasing the curing duration from 1 day to 3 and 7 days, free drying shrinkage increased 3% and 11% respectively.
- By increasing the curing duration from 1 day to 3 and 7 days, the net time to cracking reduced by 10% and 30% respectively for both the control mix and the mix with polymeric fibers resulting in the increased cracking potential of both mixes.
- Longer curing duration increased the crack width of the control mix significantly. By increasing the curing duration from 1 day to 7days, the crack width of the control mix increased from 1mm to 2 mm.
- The crack width of the mix with polymeric fibers also increased with longer curing duration but not to the same extent as the control mix. As expected, the use of fibers reduced crack width significantly compared to the control mix for similar durations of curing.

References

1. Mindess, S., Young, J.F., Darwin, D.: Concrete, 2nd-ed. Prentice-Hall, Englewood Cliffs, NJ (2003)
2. Akhavan, A., Shafaatian, S.M.H., Rajabipour, F.: Quantifying the effects of crack width, tortuosity, and roughness on water permeability of cracked mortars. Cem. Concr. Res. **42**, 313–320 (2012)
3. Passuello, A., Moriconi, G., Shah, S.P.: Cracking behavior of concrete with shrinkage reducing admixtures and PVA fibers. Cement Concr. Compos. **31**(10), 699–704 (2009)
4. Sant, G., Eberhardt, A., Bentz, D., Weiss, J.: The influence of shrinkage-reducing admixtures on moisture absorption in cementitious materials at early ages. J. Mater. Civ. Eng. **22**(3), 277–286 (2010)
5. Lura, P., Pease, B., Mazzotta, G.B., Rajabipour, F., Weiss, J.: Influence of shrinkage-reducing admixtures on the development of plastic shrinkage cracks. ACI Mater. J. **104**(2), 187–194 (2007)
6. Bagheri, A., Jamali, A., Pourmir, M., Zanganeh, H.: The influence of curing time on restrained shrinkage cracking of concrete with shrinkage reducing admixture. Adv. Civ. Eng. Mater. **8**(1), 596–610 (2019)
7. Bagheri, A., Jamali, A., Gorgani-Firoozjah, M., Zanganeh, H.: Comparison of the performance of macro-polymeric fibers and steel fibers in controlling drying shrinkage cracks of concrete. Int. J. Struct. Civ. Eng. Res. **7**(4), 302–308 (2018)
8. Shen, D., Wena, C., Kang, J., Shi, H., Xu, Z.: Early-age stress relaxation and cracking potential of high-strength concrete reinforced with barchip fiber. Constr. Build. Mater. **258**, 119538 (2020)
9. Yousefieh, N., Joshaghani, A., Hajibandeh, E., Shekarchi, M.: Influence of fibers on drying shrinkage in restrained concrete. Constr. Build. Mater. **148**, 833–845 (2017)
10. Guide to external curing of concrete, ACI 308R-16, (2016)

11. Khanzadeh-Moradllo1, M., Meshkini, M.H., Eslamdoost, E., Sadati, S., Shekarchi, M.: Effect of wet curing duration on long-term performance of concrete in tidal zone of marine environment. Int. J. Concr. Struct. Mater. **9**(4), 487–498 (2015)
12. Beushausen, H., Bester, N.: The influence of curing on restrained shrinkage cracking of bonded concrete overlays. Cem. Concr. Res. **87**, 87–96 (2016)
13. Samouh, H., Rozière, E., Wisniewski, W., Loukili, A.: Consequences of longer sealed curing on drying shrinkage, cracking and carbonation of concrete. Cem. Concr. Res. **95**, 117–131 (2017)
14. Hajibabaee, A., Grasley, Z., Ley, M.T.: Mechanisms of dimensional instability caused by differential drying in wet cured cement paste. Cem. Concr. Res. **79**, 151–158 (2016)
15. Hajibabaee, A., Ley, M.T.: The impact of wet curing on curling in concrete caused by drying shrinkage. Mater. Struct. **49**(5), 1629–1639 (2016)
16. Deboodt, T., Fu, T., Ideker, J.H.: Evaluation of FLWA and SRAs on autogenous deformation and long-term drying shrinkage of high performance concrete. Constr. Build. Mater. **119**, 53–60 (2016)
17. Guide for Modeling and Calculating Shrinkage and Creep in Hardened Concrete, ACI 209.2R-08, Farmington Hills, MI: American Concrete Institute (2008)
18. Testing Hardened Concrete: Part3: Compressive Strength of Test Specimens, BS EN 12390–3. British Standards Institute, London (2019)
19. Standard Test Method for Splitting Tensile Strength of Cylindrical Concrete Specimens, ASTM C496/C496M, West Conshohocken, PA: ASTM International, approved October 1 (2017)
20. Standard Test Method for Length Change of Hardened Hydraulic-Cement Mortar and Concrete, ASTM C157/C157M, West Conshohocken, PA: ASTM International, approved August 15 (2017)
21. Standard Test Method for Determining Age at Cracking and Induced Tensile Stress Characteristics of Mortar and Concrete under Restrained Shrinkage, ASTM C1581/C1581M, West Conshohocken, PA: ASTM International, approved July 1 (2018)
22. Guide for specifying, proportioning and production of fiber reinforced concrete, ACI 544-3, pp. 4–8 (2008). ISBN 978-0-87031-311-0
23. Standard Test Method for Drying Shrinkage of Mortar Containing Hydraulic Cement, ASTM C596, West Conshohocken, PA: ASTM International, approved December 1 (2018)
24. Hobbs, D.W.: Influence of specimen geometry upon weight change and shrinkage of air-dried concrete specimens. Mag. Concr. Res. **29**(99), 70–80 (1977)

Modelling and Simulation

Crack Control of Upstream Polder Face Using Calibrated Thermo-mechanical Simulations

Vít Šmilauer, Petr Havlásek, Pavel Reiterman, and Petr Huňka

Abstract Concrete crack mitigation in hydro-engineering projects belongs to important tasks during design and operational stages. The paper focuses on a design of reinforced upstream concrete face of a polder. The face is made of cast concrete blocks, 15 m long, 2.5 m tall and up to 2.5 m thick. First, the paper summarizes design of two concretes with low hydration heat and their strength evolution, basic creep, calorimetry data, and freeze-thaw resistance. Heat release was further validated on a small insulated concrete cube. Second, a thermo-mechanical model has been assembled, taking into account heat of hydration, ageing basic creep, autogenous shrinkage, evolution of tensile strength. A multi-directional fixed crack model captures crack initiation, crack opening is controlled by exponential softening and dissipation of correct amount of energy is guaranteed by crack-band approach. Summer casting scenario is presented, showing crack evolution due to different reinforcement ratios and concrete mix design. The results show that minimizing cement amount in concrete is a must and crack width can remain under required 0.20 mm with the reinforcement ratio of 0.4–0.6%, depending on mix design.

Keywords Crack control · Thermo-mechanical simulations · Hydration heat · Basic creep · Mix design

1 Introduction

Thermal cracking of massive concrete structures has drawn research interest over a century; Hoover dam presents a notable example from 1930s. New challenges appeared during the development of massive concrete dams, massive foundations, nuclear power plants and other massive structures [1]. Massive structures should aim

V. Šmilauer (✉) · P. Havlásek · P. Reiterman
Faculty of Civil Engineering, Czech Technical University in Prague, Thákurova 7, 166 29 Prague 6, Czech Republic
e-mail: vit.smilauer@fsv.cvut.cz

P. Huňka
STACHEMA CZ, S.R.O., Hasičská 1, Zibohlavy, 280 02 Kolín, Czech Republic

F. Kanavaris et al. (eds.), *International RILEM Conference on Early-Age and Long-Term Cracking in RC Structures*, RILEM Bookseries 31,
https://doi.org/10.1007/978-3-030-72921-9_16

at holding a holistic view and balance between material and structural design in terms of mix design, crack development and durability. However, smaller structures tend to fall into specialization, fragmentation, or sub-contractor chains, loosing holistic view of the design.

Various requirements for durable concrete can outbalance material design at the expense of structural design. For example, upstream face of dam concrete falls to XF3 class according to EN 206+A1 and requires at least 320 kg/m^3 of cement, w/c $\leq$ 0.50 and strength class C30/37. Such requirements may work for non-massive structures, but in massive ones they cause high hydration heat, high temperature rise, high risk of cracking and high reinforcement ratio—all against lessons taught from historical massive structures. Using suitable mixes for massive structures have to follow non-traditional paths such as Equivalent concrete performance concept (ECPC) or Equivalent performance of combinations concept (EPCC) according to EN 206-1, proving to meet several criteria such as freeze-thaw resistance.

The Orlík dam on Vltava river, the largest concrete gravity dam in the Czech Republic built 1954–1961, contains almost 1 mil. m^3 of concrete. The interior concrete was designed from CEM II/B-S 32.5 (N, R) 130 kg/m^3, fly ash 50 kg/m^3, water 97.2 kg/m^3 and five aggregate fractions up to 100 mm. The characteristic 28-day concrete strength of 10.1 MPa increased after 50 years to 38.7 MPa [2]. The surface concrete increased cement content to 200 kg/m^3, keeping fly-ash at 50 kg/m^3. After 60 years in operation, surface concrete is still in good shape, proving good durability in freeze-thaw actions.

The weakest link of medium-strength concrete is usually cement paste. Reducing the amount of cement to technologically acceptable values has a positive effect on lower autogenous shrinkage, lower drying shrinkage, lower heat of hydration and long-term durability of in-situ tests [3]. In terms of the formation and propagation of cracks, a higher water/cement ratio exhibits higher creep and ductility [4, 5]. The low kinetics of hydration creates crack-resistant binders [3]. Other requirements may include inhibition of the alkali-silica reaction, for example Portugal requires at least 30% fly ash in a binder for the construction of water works [6].

2 Kutřín Polder

Kutřín polder focuses on flood protection, able to retain $Q_{100} = 3.65$ mil. m^3 while flooding 68 ha area. The polder is designed as a concrete gravity dam with the length of 136 m, the crown width of 5.6 m and a maximum height from the foundation joint of 24.9 m, see Fig. 1. In the middle part of the dam, there is a migration passage of 3.8 $\times$ 3.1 m with a closing shaft, two lower outlets DN 1200 and a monolithic tower object.

The upstream face is composed of waterproof L-shaped concrete (5 250 m^3) with the thickness of 0.8–2.8 m, see ocher color in Fig. 1. The dam is horizontally divided into 15 m long blocks. The blue color indicates rolled concrete (11 050 m^3), which

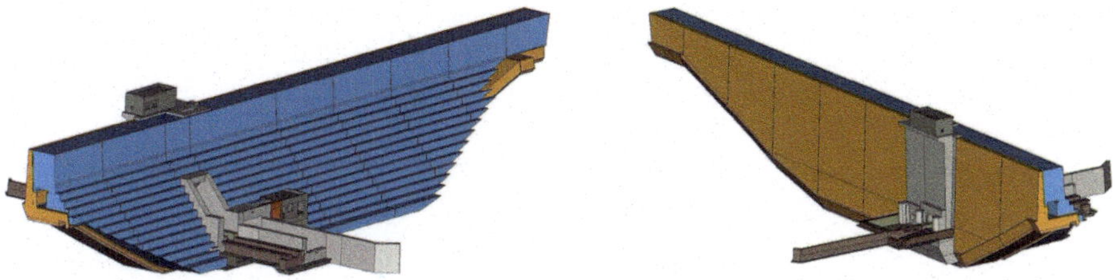

Fig. 1 Views of the downstream and upstream faces of the polder. Ocher color represents upstream face, blue color stands for roller-compacted concrete

Table 1 Mix design of two concretes

Component/concrete	C22-FA	C63-FA
CEM II/B-S 32.5 R Prachovice	270 kg/m^3	230 kg/m^3
Fly ash class F	90 kg/m^3	80 kg/m^3
Fine aggregates 0/4	860 kg/m^3	910 kg/m^3
Coarse aggregates	870 kg/m^3	870 kg/m^3
Plasticizer	3.50 kg/m^3	3.05 kg/m^3
Air-entraining agent	0.40 kg/m^3	0.40 kg/m^3
Water	170 kg/m^3	170 kg/m^3
Water/cement	0.63	0.74
Water/binder	0.47	0.55

ensures sufficient mass and stability of the dam. Reinforced concrete prefabricated elements will be installed on the rolled concrete, creating a downstream face.

2.1 Mix Design

Two mix designs were used for upstream face, different at maximum size aggregates of 22 and 63 mm. The binder consisted of a slowly hydrating slag-blended Portland cement and fly ash class F. Cement dosage was kept as low as possible to minimize hydration heat (Table 1).

2.2 Properties of Concrete

Table 2 summarizes basic properties for both concretes. Compressive strength was tested according to EN 12390-3 on 150 and 200 mm cubes. Secant elastic modulus followed ISO 1920-10 on cylinders ø150 mm × 300 mm. Freeze-thaw resistance

Table 2 Properties of concretes

Test/concrete	C22-FA	C63-FA
Consistency—slump in mm	S4-200	S4-190
Air entrainment	4.1%	5.7%
Compressive strength, 2 d	12.0 ± 0.2 MPa	9.3 ± 0.3 MPa
Compressive strength, 7 d	19.4 ± 0.7 MPa	15.7 ± 0.4 MPa
Compressive strength, 28 d	31.5 ± 0.6 MPa	26.8 ± 0.7 MPa
Compressive strength, 90 d	43.5 ± 0.3 MPa	36.8 ± 0.9 MPa
Splitting tensile strength, 28 d	3.2 ± 0.2 MPa	2.9 ± 0.1 MPa
Splitting tensile strength, 90 d	4.1 ± 0.1 MPa	3.6 ± 0.1 MPa
Strength class, 28 d	C20/25	C16/20
Strength class, 90 d	C25/30	C20/25
Secant elastic modulus, 28 d	25.0 ± 1.5 GPa	26.4 ± 3.0 GPa
Secant elastic modulus, 28 d	27.5 ± 0.9 GPa	27.6 ± 0.9 GPa
Relative flexural strength, 100 cyc, 28 d	0.72 ± 0.04	0.92 ± 0.04
Relative flexural strength, 100 cyc, 28 d	0.92 ± 0.04	0.90 ± 0.04
Water penetration, 90 d	44.8 mm	46.0 mm

used beams 100 × 100 × 400 mm for C22-FA and 150 × 150 × 700 mm for C63-FA according to ČSN 73 1322, loaded with 100 cycles. The prisms were cycled in 4 + 2 h regime between −20 and +15 °C sweep and relative residual flexural strength was determined on three samples. Both concretes reached the value above 0.90 at 90 days, which is much higher than required threshold of 0.75.

Water penetration was tested according to EN 12390-8 and reached the penetration depth bellow 50 mm, qualifying well both concrete for dams.

2.3 Basic Creep

Basic creep on both concretes was measured by vibrating wire strain gauges embedded in samples; they were cast into a KG-200 plastic pipe with the inner diameter of 190 mm and the height of 300 mm, see Fig. 2. A loading force of 100 kN yielded a uniform compressive stress of 3.53 MPa at the age of 14 days.

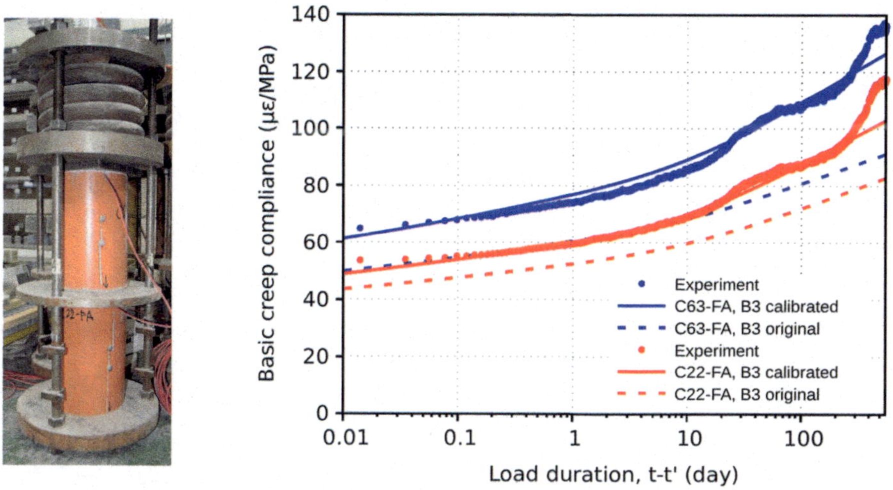

Fig. 2 Basic creep setup, compliance function for $t' = 14$ days and a fit using B3 model

Compliance functions for basic creep were compared with the blind prediction of model B3, where the fly ash is considered as an inert filler; the relative error of the creep function was up to 35%. The samples generally show a slightly higher creep compared to the blind prediction, which is mainly due to the blended binder. After about 6 months, it can be seen that there is also a slight drying of the samples and secondary shrinkage. To refine further calculations, the B3 model was calibrated by increasing the parameters q_2 and q_4 to a maximum of 1.8 times their original values, see Fig. 2.

3 Heat of Hydration

3.1 Isothermal Calorimetry

Kinetics and quantity of released heat are key parameters for massive concrete structures. Isothermal measurement of hydration heat took place on an 8-channel TamAir calorimeter at the Czech Technical University in Prague. Four potentially suitable binders with low heat of hydration were tested. The water/binder ratio was constant 0.45 and the ampule temperature was maintained at 20 °C. Heat flow integration started from 1 h after the mixing, so a small initial heat (typically 10–20 J/g) is neglected.

The proposed binder corresponds to a yellow curve in Fig. 3; the reaction of fly ash is not apparent until 7 days. For comparison, CEM III/B 32.5N-LH/SR cement is shown, which makes a binder for several massive concrete structures. Both mix

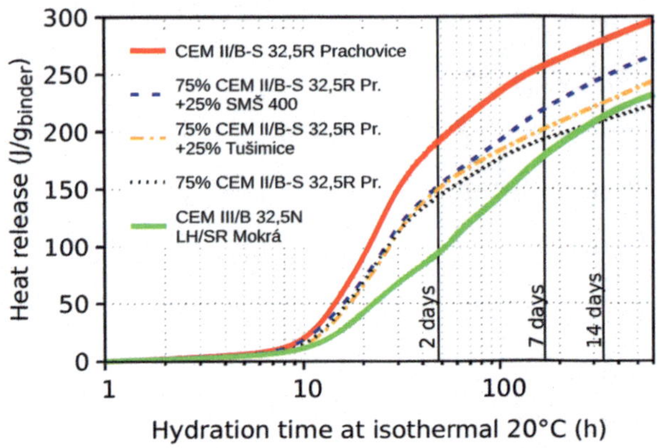

Fig. 3 Heat release of four binders. Yellow curve belongs to studied binder

designs are comparable with CEM III/B solution in terms of heat release at 7 days; at 20 °C, 1 m^3 of C63-FA releases 66.7 MJ and C22-FA yields 78.3 MJ, including small initial heat within 1 h.

3.2 Semi-adiabatic Validation on Cubes

In order to verify the results from the calorimeter, two concrete cubes with an edge of 0.3 m were made. Expanded polystyrene 100 mm thick was inserted into the wooden molds. Prior to casting, a three-dimensional diagonal made of a fiberglass rod with a Pt1000 temperature sensor was placed inside. The temperature was measured for 8 days including the ambient air temperature at interval of 5 min.

A four-parametric hydration model was calibrated from calorimetry data and the open software OOFEM [7] was used to calculate the non-stationary heat conduction problem. The polystyrene layer within the formwork was considered to have a heat transfer coefficient of 0.45 W/m^2/K. The initial concrete temperature was 21 °C and an ambient constant temperature 26 °C (Fig. 4).

3.3 Adiabatic Temperature Increase

The adiabatic temperature rise expresses the maximum temperature rise in a perfectly insulated sample, this situation often occurs in the core of massive elements. Table 3 gives these values for an initial temperature of 20 °C. The infinite time temperature should be taken as a guide, as it was obtained by extrapolating the measured data and with the estimated potential heat of hydration of the binder. Here it can be seen that the amount of cement 270 and 230 kg/m^3 causes a difference of ~7 °C after 100 h.

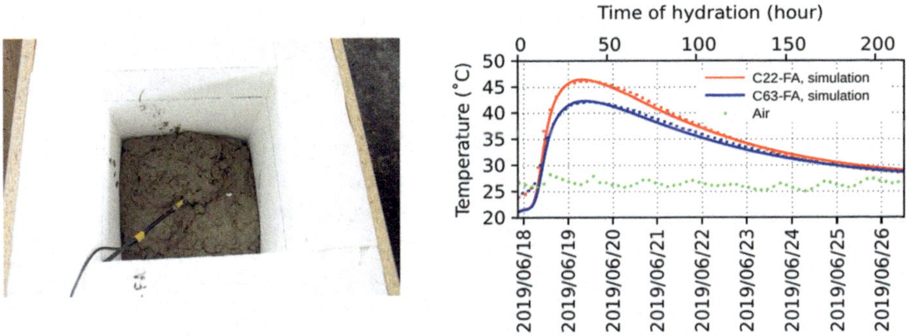

Fig. 4 Semi-adiabatic validation of thermal model on concrete cubes

Table 3 Adiabatic concrete temperature (initial temperature 20 °C)

Temperature increase/concrete	C22-FA (°C)	C63-FA (°C)
ΔT_{adiab} (50 h)	33.5	27.2
ΔT_{adiab} (100 h)	39.3	32.4
ΔT_{adiab} (200 h)	42.0	35.2
ΔT_{adiab} (∞)	42.8	36.7

4 H-beam Tests

The test of crack development in weakly reinforced concrete beams is intended to mimic block casting under restrained cooling. The original design followed real temperature control, however, this would be too challenging in handling fresh concrete and logistics. Instead, tensile stress was induced hydraulically using external struts. The middle part of the beam has a square cross-section of 0.3×0.3 m and a length of 1.32 m, reinforced longitudinally with 4ø14 B500 and stirrups ø8, see Fig. 5.

Fig. 5 H-beam reinforcement and hydraulic loading

The area of longitudinal reinforcement is 616 mm^2 and the degree of reinforcement is based on 616/300^2 = 0.68%. The mutual displacement of the beam heads at a distance of 1320 mm is measured by a gauge with a rod placed in a circular guard in the center of the cross-section.

At the age of 21 days, the beams were loaded using hydraulic cylinders. In both concretes, the formation of cracks was noticed when the stress in the concrete reached about 1.2 MPa and the cracks were visible in the middle part. The loading was stopped at visible cracks when the stress reached 1.2 MPa for C22-FA and 1.45 MPa for C63-FA and the position of the heads was fixed. The development of the tensile force in the reinforced concrete beam was monitored using a pair of string strain gauges placed on steel struts. Cracks 0.10–0.15 mm wide were formed in both beams, running through the entire middle cross-section, see Fig. 5 for C63-FA. After 77 days, there was a small relaxation of the stress in the reinforced concrete up to 7% and due to the drying of the concrete there was a further increase in the width of the cracks by a maximum of 0.10 mm. The experiment confirmed the expected behavior of concretes under long-term tensile stress compared to numerical models, at the same time giving a more accurate tensile strength on 1.32 long specimen.

5 Simulation of Summer Casting

A thermo-mechanical, weakly coupled model in the OOFEM software [7] was used for the simulation of block casting of the upstream face. The same four-parameter affinity model was used to calculate the heat conduction problem and was calibrated for isothermal calorimetry data. The geometry of the computational model followed summer casting of a 2.5 × 2.5 × 15 m block on the top of the same old block. The symmetry of the problem made it possible to model only a quarter of both blocks, yielding dimensions of 1.25 × 5.0 × 7.5 m, see Fig. 6. In total of 1020 finite elements of 0.25 × 0.42 × 0.44 m were used in the thermal part.

The initial temperature of the old and new blocks was chosen 25 °C as the characteristic summer values, the air temperature kept constant at 25 °C. Front face contained polyurethane matrices for surface texture, therefore a heat transfer coefficient of 0.733 W/m^2/K was assigned. After their removal, the coefficient increased to 1.765 W/m^2/K. The upper surface of the block is considered uninsulated with a heat transfer coefficient of 15 W/m^2/K.

The maximum temperature reached in the block core was 63 °C at 83 h for C22-FA and 57 °C at 83 h for C63-FA, see Fig. 6. Such an increase corresponds well with adiabatic temperature ΔT_{adiab} (100 h), i.e. C22-FA 38–39.3 °C and C63-FA 32–32.4 °C. This documents the almost adiabatic conditions in the block core and emphasizes the key importance of the type and amount of binder. Figure 6 shows evolution of maximum temperatures, complemented with theoretical scenario of cooling; two cooling pipes embedded in a block with 20 °C water and deactivated after 72 h.

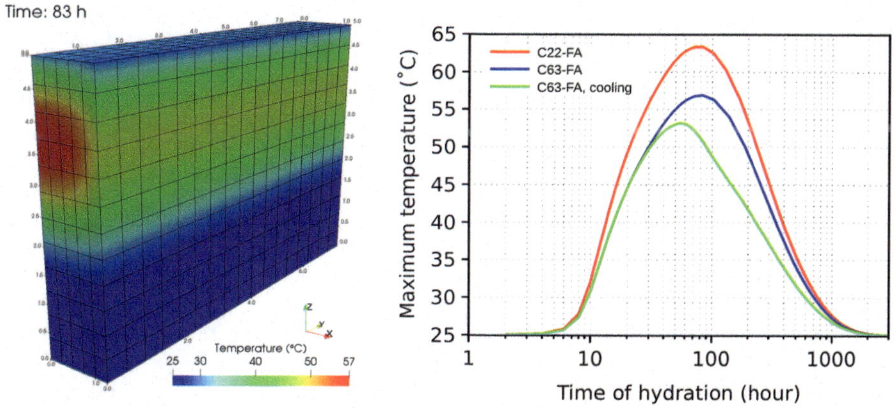

Fig. 6 Two blocks during summer casting of C63-FA at maximum and temperature evolution

The mechanical model combines a linear viscoelastic material model based on a modified MPS theory with a softening smeared cracking model in tension (*Concrete-FcmViscoelastic* model in OOFEM [7]). The models are linked in series such that the total deformation is obtained as a sum of the two contributions while the stresses remain the same. The model is formulated incrementally. The new stress vector $\boldsymbol{\sigma}_{\mathbf{i+1}}$ is computed from the preceding equilibrated stress vector $\boldsymbol{\sigma}_{\mathbf{i}}$ and from the product of the viscoelastic stiffness matrix $\mathbf{D}_{\mathbf{ve}}$ and the increment of the reduced strain vector.

$$\boldsymbol{\sigma}_{\mathbf{i+1}} = \boldsymbol{\sigma}_{\mathbf{i}} + \mathbf{D}_{\mathbf{ve}}\left(\triangle\boldsymbol{\varepsilon} - \triangle\varepsilon'' - \triangle\boldsymbol{\varepsilon}_{\mathbf{sh}} - \triangle\boldsymbol{\varepsilon}_{\mathbf{T}} - \triangle\boldsymbol{\varepsilon}_{\mathbf{crack}}\right) \tag{1}$$

where $\triangle\boldsymbol{\varepsilon}$ is the total strain increment, $\triangle\boldsymbol{\varepsilon}''$ strain increment due to creep, $\triangle\boldsymbol{\varepsilon}_{\mathbf{sh}}$ is the shrinkage strain increment, $\triangle\boldsymbol{\varepsilon}_{\mathbf{T}}$ is the thermal dilation increment, and $\triangle\boldsymbol{\varepsilon}_{\mathbf{crack}}$ is the cracking strain increment. The last term represents a smeared cracking strain, where the tensile cracks can form in mutually perpendicular directions and after their initiation the orientation remains fixed [8]. Model objectivity is achieved using the crack-band approach. Softening behavior was approximated with exponential softening law with fracture energy of 150 J/m^2.

The model takes into account the evolution of concrete tensile strength. In order to take into account probability of crack formation, we adopted approach described in [(8.30), 1] using two Gaussian distributions

$$P_f = \Phi\left(-\frac{0.75 f_{ctm,sp} - \sigma_t}{\sqrt{\left(0.75 f_{ctm,sp} \cdot V_R\right)^2 + (\sigma_t \cdot V_S)^2}}\right) \tag{2}$$

where $f_{ctm,sp}$ represents splitting tensile strength, σ_t is the principal tensile stress in concrete, V_R is the coefficient of variation for strength, V_S is the coefficient of

Table 4 Computed maximum crack widths inside (on surface) for longitudinal reinforcement ratios

Reinforcement ratio/concrete (%)	C22-FA (mm)	C63-FA (mm)
0.00 (unreinforced)	0.84 (0.70)	0.70 (0.60)
0.30	0.34 (0.25)	0.26 (0.20)
0.40	0.27 (0.20)	**0.19** (0.13)
0.50	0.22 (0.15)	0.19 (0.12)
0.60	**0.18** (0.11)	0.16 (0.10)

variation for stress and Φ returns the value of the distribution function of the standard normal distribution. By default, $V_R = V_S = 0.10$ is considered, however, we conservatively increased $V_R = 0.30$ due to a small number of tensile tests. Probability of crack formation 2–3% results to limit design stresses of 1 MPa for C22-FA and 0.8 MPa for C63-FA.

The mechanical model is activated at 12 h from the beginning of hydration and uses the calibrated basic creep compliance functions of B3 model with additional compliance due to transitional thermal creep. The mechanical problem contains 3000 finite elements of size 0.25 × 0.25 × 0.25 m and both tasks run for about 40 min. The thermo-mechanical model allows quantification of crack width with regards to reinforcement ratio. For simplicity reasons, the reinforcement was uniformly distributed in all orthogonal directions among all corresponding nodes, the stress of the reinforcement reaches values up to 40 MPa in any simulation.

The maximum width of cracks depends on reinforcement ratio, see Table 4. It is in the range of 0.10–0.84 mm where unreinforced concrete leads to strong crack localization. The crack widths inside the block are generally 35–65% larger than on the surface. The limit criterion of crack width for watertight structures is 0.20 mm, as long as concrete staining is of no concern and the cracks can heal [5]. This results in a minimum reinforcement ratio of 0.60% for C22-FA and 0.40% for C63-FA.

Figure 7 shows the state of cracks at 120 days for C22-FA and C63-FA, when the block has already cooled and no further crack increase occurs. In addition to transverse vertical cracks, horizontal cracks also form at the block interface, as the upper part tends to lift the edges during the cooling. The width of the surface cracks will increase slightly due to drying, which is not considered in the simulation. If water cooling takes place, reinforcement ratio of 0.2 and 0.1% would suffice to keep the crack width under 0.20 mm.

6 Conclusions

The paper presents more holistic approach for upstream face construction of a small dam, combining mix design with structural scale. Cement in two concretes was kept as low as 230–270 kg/m^3 to minimize hydration heat. Those concretes do not meet requirements specified in EN 206+A1 for XF3 and other national regulations,

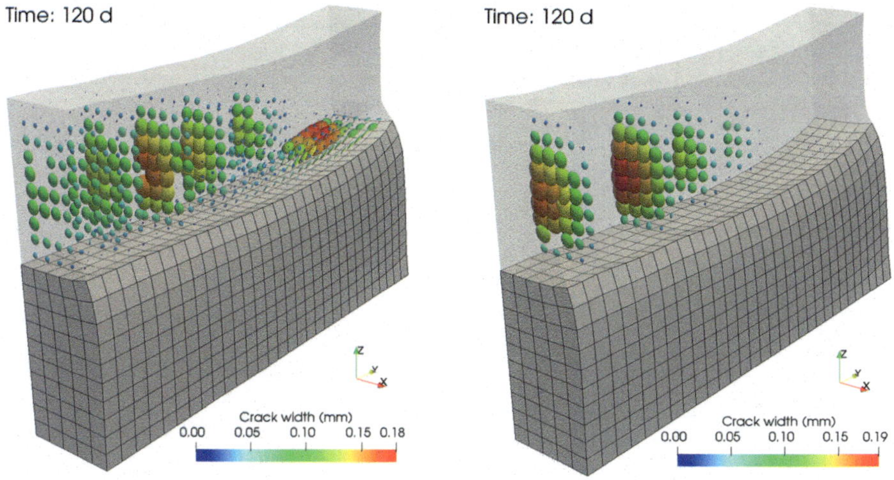

Fig. 7 Cracks for C22-FA at reinforcement ratio of 0.60% (left) and C63-FA at 0.40% (right)

requiring at least 320 kg/m^3 of cement. However, freeze-thaw tests showed excellent resistance and this could be used in equivalent concepts defined in EN 206+A1 and use them in real constructions.

Simulations showed that longitudinal reinforcement ratio needs to be kept in the range of 0.40–0.60% during summer casting, if crack width should remain under 0.20 mm. Combination of water cooling, spring and autumn casting and properly timed construction procedures would allow to use even unreinforced concretes, as done in several projects in history.

Acknowledgement We gratefully acknowledge financial support from the Czech Science Foundation (GA ČR) under projects 21-03118S and 19-20666S. We appreciate further support from Elbe River Basin, state enterprise (Povodí Labe s.p.) and Prefa PRO, a.s. conrete plant.

References

1. Fairbairn, E.M.R., Azenha, M., et al.: Thermal cracking of massive concrete structures. State of Art Report of the RILEM TC 254-CMS (2018)
2. Zobal, O., Kopecký, L., Padevět, P., Šmilauer, V., Bittnar, Z.: Analysis of concrete from the body of the Orlík dam after fifty years. Beton TKS **2**, 19–25 (2014)
3. Burrows, R.: The visible and invisible cracking of concrete. American Concrete Institute (1998)
4. RILEM Technical Committee TC-242-MDC (chaired by Z.P. Bažant): Model B4 for creep, drying shrinkage and autogenous shrinkage of normal and high-strength concretes with multi-decade applicability. Mater. Struct. **48**, 753–770 (2015)
5. Walraven, J., et al.: Fib model code for concrete structures 2010. In: International Federation for Structural Concrete (fib) (2013)
6. Camelo, A., Gaspar, V., da Silva, P.F: The use of fly ash in EDP hydro projects. In: Proceedings of EuroCoalAsh (2014)

7. Patzák, B.: OOFEM—an object-oriented simulation tool for advanced modeling of materials and structures. Acta Polytech. **52**(6), 59–66 (2012). https://www.oofem.org
8. de Borst, R., et al.: Discrete vs smeared crack models for concrete fracture: bridging the gap. Int. J. Numer. Anal. Meth. Geomech. **28**, 583–607 (2004)

CSI (Cement Science Investigation): Using Machine Learning to Guess the OPC Pastes Composition from the Elastic Response

Tulio Honorio, Sofiane Ait-Hamadouche, and Amélie Fau

Abstract Precise prediction of the elastic response is crucial to model cracking at early and late ages of cement-based materials and structures. Here, we use Machine Learning (ML) techniques to predict the elastic properties of Ordinary Portland Cement (OPC) pastes. A database with 365 observations is built on experimental studies from in the literature. We show that micromechanics-based estimations may provide missing data in databases to be interrogated by ML, increasing the accuracy of predictions. Finally, we explore the formulation space of OPC pastes using Monte Carlo computations, which enables guessing which are the compositions that can be associated with a given elastic response. Applications of our results include the development of tailored formulations for a target elastic response and also in the forensics of existing cement pastes.

Keywords Elastic constants · Machine learning · Data science · Composition-property correlations

1 Introduction

Modeling cracking in cement-based materials and structures requires a precise description prediction of the elastic response of the material. One of the main approaches to assess cracking risk in concrete is based on elastic analysis of the (e.g. crack risk index in JCI recommendation [1]), which of course require an accurate prediction of the elastic properties, especially at early-age when properties significantly change due to the progress of hydration process. The correct prediction of the evolution of mechanical properties is also a fundamental input in more sophisticated non-linear analysis of cracking of cement-based materials.

T. Honorio (✉) · S. Ait-Hamadouche · A. Fau
LMT - Laboratoire de Mécanique et Technologie, Université Paris-Saclay, ENS Paris-Saclay, CNRS, 91190 Gif-sur-Yvette, France
e-mail: tulio.honorio-faria-de-faria@ens-paris-saclay.fr

F. Kanavaris et al. (eds.), *International RILEM Conference on Early-Age and Long-Term Cracking in RC Structures*, RILEM Bookseries 31,
https://doi.org/10.1007/978-3-030-72921-9_17

Establishing composition-structure-property correlations is the central paradigm in the optimization of the formulation and understanding the behavior of cement-based materials. Most of the concrete research community has focused on an increasingly costly and laborious trial-and-error experimental exploration of the formulation space, while much less significant attention has been devoted to establishing composition-structure-property correlations via fundamental component data and validated modeling approaches. The latter is a strategy that has been recognized with the potential to "revolutionize" cement and concrete research [1].

In this work, we first use Machine Learning (ML) techniques to predict the elastic properties of OPC pastes. Understanding cracking process in cement pastes is a first milestone to understand cracking at more complex system such as mortar and concrete in which the heterogeneities due to interfacial transition zone (ITZ) may play a role. Here, we show that micromechanics-based estimations can be useful to provide missing data in databases to be interrogated by ML. Finally, we explore the formulation space of OPC pastes using Monte Carlo computations. This strategy allows guessing what are the compositions that can be associated with a given elastic response, which can be useful for application requiring a target elastic response and also in the forensics of existing cement pastes.

2 Data Science and Machine Learning Methods

2.1 Database Construction

A database with 365 observations is collected from experimental data from the literature [2–8]. The input and output, as well as statistical parameters associated, are displayed in Table 1. Both dynamic and quasi-static data are used to build the database. Dynamic and quasi-static methods are known to yield different values of elastic moduli (e.g. [9]). This fact adds to the uncertainties already present in experimental assessment of the elastic properties. To check the trustfulness of the data collected, we computed the theoretical bounds of Voigt-Reuss and Hashin-Shtrikman. All data collected regarding Young modulus respect the theoretical bounds.

2.2 Machine Learning Methods

We consider here two ML methods:

- **Artificial Neural Network (ANN)**: with this method a neural network is constituted of stacked layers. Each layer is related to simple computation. The information is then processed layer by layer starting at the input layer until the output

Table 1 Statistical parameters of the training dataset on OPC cement pastes. Datasize of 365 observations. *DOH* refers to degree of hydration, w/c to the (mass) water-cement ratio, and the mass fraction of the four main clinker minerals (mC_3S, mC_2S, mC_3A and mC_4AF) and gypsum (*mGyp.*) are reported in %

Data	Variable	Min	Max	Mean	St. dev	Ex. Kur.[a]
Input	*DOH* [−]	0.03	1.0	0.46	0.29	−0.68
	w/c [−]	0.25	0.80	0.44	0.10	−0.20
	mC_3S [%]	24.5	100	60.2	13.8	2.06
	mC_2S [%]	0	61.3	16.6	15.2	2.43
	mC_3A [%]	0	12.7	8.13	3.36	−0.70
	mC_4AF [%]	0	12.7	5.8	4.2	−1.53
	mGyp. [%]	0	6.8	2.9	2.9	−1.7
Output	*E* [GPa]	0.22	37.2	11.2	7.8	0.32

[a]normalized exceed kurtosis

layer. The neural network is trained to minimize a loss function on the training set. A gradient descent method is used to perform this minimization.

- **Gaussian Process (GP)**: with this method predictions are made using Bayesian inference on the Gaussian process conditioned to the training data (see, for example [10]). The assumption behind the method is that the prediction functions can be described by a Gaussian process, which is defined by its kernel. The training phase with GP consists of estimating the parameters of this kernel.

The algorithms are built in Mathematica®.

3 Results and Discussion

3.1 Validation

To validate our ML approach against the training dataset, we deploy the K-fold method [11]. This method consists of dividing the training dataset in K folds in which elements are randomly sampled. ANN or GP predictor is then trained on K-1 folds and, next, it is used to predict the values in the remaining folds. This step is repeated K times so that all folds have been used for validation. Figure 1 shows one of the stages of validation. The accuracy of the predictions is computed as the square root of the average variance obtained from the validation on all K-folds for the elastic constants. We obtain an average deviation of the Young modulus of 0.61 GPa for ANN and of 1.19 GPa for GP with $K = 5$ (which is a usual value adopted for validation).

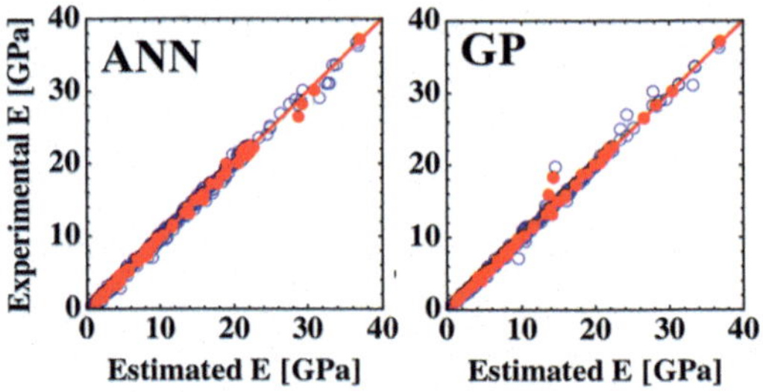

Fig. 1 An illustration of the K-fold (with K = 5) validation method: experimental Young modulus plotted against the predicted Young modulus at one stage (out of 5) of validation for ANN and GP. The empty symbols are the values associated with the (K-1) folds (with 292 observations in total) used for training, and full symbols depict the 73 elements in the remaining fold used in the validation

3.2 Predictions on Testing Dataset

We define a testing dataset composed of experimental data from the literature, which are not included in the training dataset. Prediction using ANN and GP of the Young modulus of OPC pastes are gathered in Table 2 and compared against their experimental counterpart.

Table 2 Comparison of predictions with ANN and GP using the testing dataset (not comprised in the training dataset) against the experimental data provided in the literature by various authors. The mass fraction of clinker mineral and gypsum as provided by the authors as well as the w/c and degree of hydration were used as input in the predictions

Data	w/c	Exp	ANN	GP
Constatinides et al. [12]	0.50	22.8	20.7	14.0
Tamtsia et al. [13]	0.35	20.70	14.2	22.6
	0.35	20.80	17.6	22.6
	0.35	21.50	18.5	22.7
	0.5	7.90	4.3	10.1
	0.5	10.40	7.4	12.4
	0.5	11.60	7.7	12.5
Savija et al. [14]	0.3	16.68	27.7	16.4
	0.4	12.79	22.9	14.1
	0.5	9.09	18.0	14.9
Haecker et al. [3]	0.6	1.7	6.9	10.6

3.3 Using Micromechanics to Provide Missing Data

We use Micromechanics, i.e. another technique to establish composition-property correlations in cement-based materials [15], to provide data that may be missing in our training dataset. In the sequel, we verify if the extended database performs better than the original database in the predictions.

For the micromechanics-based estimations, we adopt the representation of the microstructure of cement paste combining self-consistent (SC) scheme to estimations at the gel scale, and Mori-Tanaka scheme to estimations at the cement paste scale, as shown in Fig. 2a. Similar representations were successfully used in previous work to estimate mechanical [16, 17], thermal [18], and electromagnetic properties [19, 20] of cement pastes.

The accuracy of micromechanics-based estimations is illustrated in Fig. 2b, in which the data associated with late ages (DOH > 0.7) are identified. The average deviations in the Young modulus estimated by micromechanics are 6.9 GPa when all data is considered and 2.8 GPa when only late age data is taken into account.

Among the variety of possibility to guide the plan of experiments [20], the strategy deployed here to identify the domain in the 7-dimension space of input variables is to take a derivative of the ordered components of each input vector. The higher the

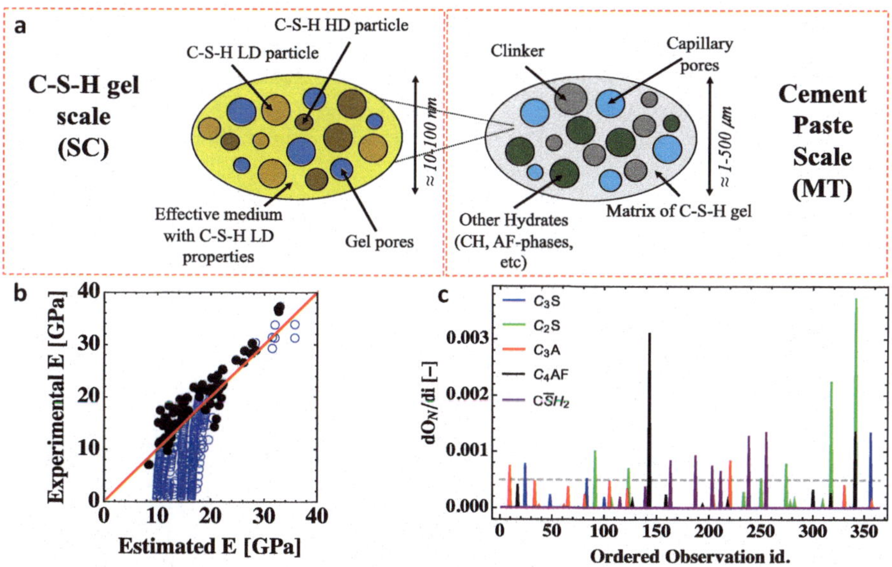

Fig. 2 **a** Adopted representation of the microstructure of cement paste in the micromechanics estimations: combinations of self-consistent scheme estimate at the gel scale, and Mori-Tanaka estimate at the cement paste scale. **b** Comparison of experimental and micromechanics predicted Young modulus taking into account the training database (black points corresponds to late ages (DOH > 0.7). **c** Identification of the main domains with missing data regarding OPC composition using the derivative approach of the ordered input

peak in the derivative, the most significant is the domain with missing data. Figure 2c shows the derivative for the mass fraction of OPC constituents. With this approach the points related to the most significant domain with missing data were identified as follows:

- DOH [−]: 0.70, 0.95
- w/c [−]: 0.28, 0.37, 0.72
- mC_3S [%]: 33.0, 52.4, 85.2
- mC_2S [%]: 5.40, 11.2, 16.7, 20.7, 31.7, 49.9
- mC_2S [%]: 1.10, 9.95
- mC_2S [%]: 4.40 11.2
- *Gyp.* [%]: 1.05, 2.00, 2.90, 3.65, 4.68, 6.09

where an arbitrary cut-off of 0.0001 for the w/c and degree of hydration, and 0.0005 for the clinker minerals and gypsum mass fractions was adopted.

In Fig. 3, we compare the average deviations of predictions in the testing dataset using the original (l_{orig}) and extended ($l_{Ex,6}$ and $l_{Ex,24}$) training datasets. The extended dataset is composed of a (i) $l_{Ex,6}$ with the minimum number of input points (6) comprising all the points above related to the most significative missing data domains; and (ii) $l_{Ex,24}$ with each point (24 in total) being associated with one of the points above related to the most significant missing data domains and the other input components are chosen also among these points related to the most significative missing data domains.

We observe that for both ANN and GP, using the extended dataset yields smaller average deviations than using the original dataset. This is an interesting result showing that micromechanics can be used to generate missing data in dataset for training ML methods.

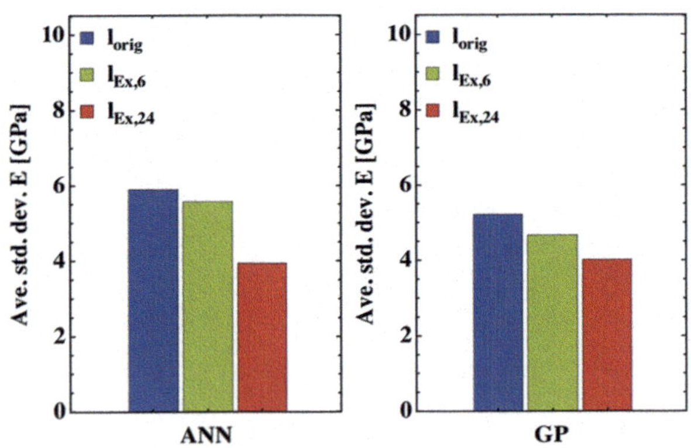

Fig. 3 Average deviations of predictions in the testing dataset using the original (l_{orig}) and extended ($l_{Ex,6}$ and $l_{Ex,24}$) training datasets

3.4 Towards the Estimation of the Composition from a Target Elastic Modulus

To estimate the composition from the elastic response, i.e. do the reverse path in the prediction done so far in this study, we adopt a Monte Carlo (MC) approach. We sample the given input following a uniform distribution with the limits defined by the minimum and maximum values in Table 1. We set the DOH at 1 to avoid the uncertainties related to hydration kinetics and early-age effects. Since we use as input the mass fraction of clinker minerals and gypsum, as a consistency condition, the sum of all masses should be 100%, which means that not all mass fractions can vary independently in the MC trials, one of the phases mass fractions (we adopted C_3S) must be computed from the consistency condition. A random sample with 10000 MC trials is computed for each independent input variable.

The probability distribution functions of each input from MC trials yielding predictions that comply with the target 16.68 ± 2 GPa (which corresponds to one of the experimental data from ref. [14]) are shown in Fig. 4. Yellow and blue histograms correspond to ANN and GP predictions, respectively. We also indicate the composition of the system according to the data provided in Ref. [14]. The reverse predictions

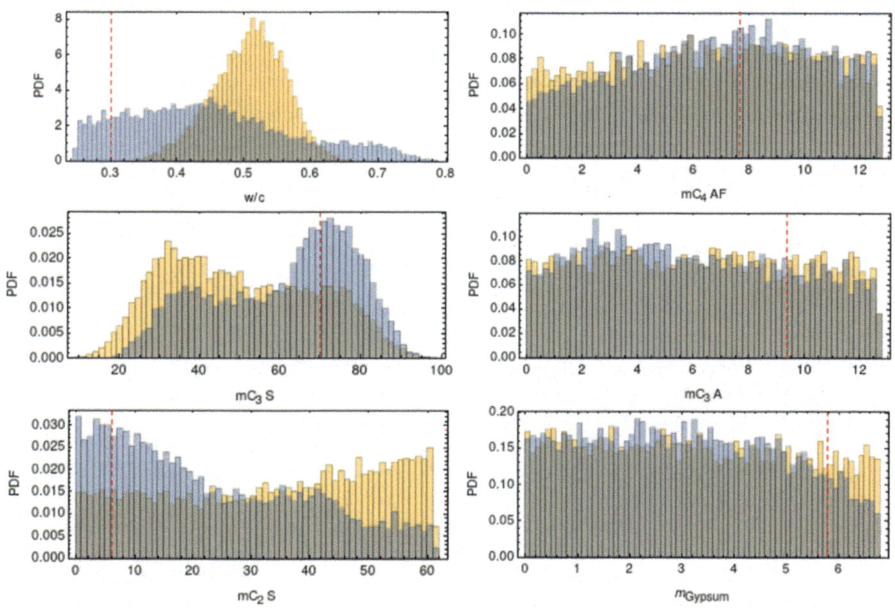

Fig. 4 Probability distribution function of each input from MC trials yielding predictions that comply with the target 16.68 ± 2 GPa (which corresponds to one of the experimental data from Ref. [14]). Yellow and blue histograms correspond to ANN and GP predictions, respectively. The red dashed line indicates the composition of the system according to data from Ref. [14]

using GP yield distributions of input variables containing the experimental composition data. The distribution obtained with ANN failed in containing the experimental w/c. For both ML methods, the distribution obtained can be rather broad, which may indicate that the reverse problem might be ill-posed.

4 Conclusion

In this work, Machine Learning methods were deployed to establish correlations between the composition and the Young modulus of OPC pastes. Both ANN and GP prediction performed here are not computer-intensive once the predictor function is trained, especially when one compares to the increasingly fastidious numerical homogenization approaches. This aspect indicates a clear advantage of ML methods, in particular in an application in which a larger exploration of the compositional and mixing design space is desired. ML prediction yields adequate precision with the training dataset proposed. Even in a relatively small dataset, ANN and GP have proven to be reliable and robust in the prediction of elastic constants of cement pastes from their composition. Micromechanics can be an ally of ML approaches by providing data missing in the database, increasing the accuracy of ML predictions. This result adds to the accumulating evidence showing that micromechanics-based approaches are a powerful tool to correlate composition and property in cement-based materials based on a few fundamental component data set and assumptions.

The strategy outlined in this work to explore the space of formulation design of cement pastes and correlating their composition to their effective Young modulus can be extended to other properties of cement-based materials.

References

1. JCI: Japan Concrete Institute Guidelines for Control of Cracking of Mass Concrete (2008)
2. Maruyama, I., Igarashi, G.: Cement reaction and resultant physical properties of cement paste. J. Adv. Concr. Technol. **12**, 200–213 (2014). https://doi.org/10.3151/jact.12.200
3. Haecker, C.-J., Garboczi, E.J., Bullard, J.W., et al.: Modeling the linear elastic properties of Portland cement paste. Cem. Concr. Res. **35**, 1948–1960 (2005). https://doi.org/10.1016/j.cemconres.2005.05.001
4. Helmuth, R.A., Turk, D.H.: Elastic moduli of hardened Portland cement and tricalcium silicate pastes: effect of porosity. REt EIVED 135 (1966)
5. Boumiz, A., Vernet, C., Tenoudji, F.C.: Mechanical properties of cement pastes and mortars at early ages: Evolution with time and degree of hydration. Adv. Cem. Based Mater. **3**, 94–106 (1996). https://doi.org/10.1016/S1065-7355(96)90042-5
6. Chamrova, R.: Modelling and measurement of elastic properties of hydrating cement paste. PhD Thesis, EPFL (2010)
7. Wang, X., Subramaniam, K.V.: Ultrasonic monitoring of capillary porosity and elastic properties in hydrating cement paste. Cement Concr. Compos. **33**, 389–401 (2011). https://doi.org/10.1016/j.cemconcomp.2010.11.001

8. Sun, Z., Garboczi, E.J., Shah, S.P.: Modeling the elastic properties of concrete composites: experiment, differential effective medium theory, and numerical simulation. Cement Concr. Compos. **29**, 22–38 (2007). https://doi.org/10.1016/j.cemconcomp.2006.07.020
9. Benboudjema, F., Carette, J., Delsaute, B., et al.: Mechanical properties. In: Fairbairn, E.M.R., Azenha, M. (eds.) Thermal Cracking of Massive Concrete Structures: State of the Art Report of the RILEM Technical Committee 254-CMS, pp. 69–114. Springer International Publishing, Cham (2019)
10. Williams, C.K.I., Rasmussen, C.E.: Gaussian Processes for Regression, vol. 7 (1996)
11. Bengio, Y., Grandvalet, Y.: No unbiased estimator of the variance of K-fold cross-validation. J. Mach. Learn. Res. **5**, 1089–1105 (2004)
12. Constantinides, G., Ulm, F.-J.: The effect of two types of C-S-H on the elasticity of cement-based materials: results from nanoindentation and micromechanical modeling. Cem. Concr. Res. **34**, 67–80 (2004). https://doi.org/10.1016/S0008-8846(03)00230-8
13. Tamtsia, B.T., Beaudoin, J.J., Marchand, J.: The early age short-term creep of hardening cement paste: load-induced hydration effects. Cement Concr. Compos. **26**, 481–489 (2004). https://doi.org/10.1016/S0958-9465(03)00079-9
14. Šavija, B., Zhang, H., Schlangen, E.: Micromechanical testing and modelling of blast furnace slag cement pastes. Constr. Build. Mater. **239**, 117841 (2020). https://doi.org/10.1016/j.conbuildmat.2019.117841
15. Wyrzykowski, M., Sanahuja, J., Charpin, L., et al.: Numerical benchmark campaign of COST Action TU1404—microstructural modelling. RILEM Tech. Lett. **2**, 99–107. https://doi.org/10.21809/rilemtechlett.2017.44
16. Honorio, T., Brochard, L., Bary, B.: Statistical variability of mechanical fields in thermo-poro-elasticity: Multiscale analytical estimations applied to cement-based materials at early-age. Cem. Concr. Res. **110**, 24–41 (2018). https://doi.org/10.1016/j.cemconres.2018.05.010
17. Honorio, T., Bary, B., Benboudjema, F.: Multiscale estimation of ageing viscoelastic properties of cement-based materials: a combined analytical and numerical approach to estimate the behaviour at early age. Cem. Concr. Res. **85**, 137–155 (2016). https://doi.org/10.1016/j.cemconres.2016.03.010
18. Honorio, T., Bary, B., Benboudjema, F.: Thermal properties of cement-based materials: multi-scale estimations at early-age. Cem. Concr. Compos. **87**, 205–219 (2018). https://doi.org/10.1016/j.cemconcomp.2018.01.003
19. Honorio, T., Carasek, H., Cascudo, O.: Electrical properties of cement-based materials: multi-scale modeling and quantification of the variability. Constr. Build. Mater. **245**, 118461 (2020). https://doi.org/10.1016/j.conbuildmat.2020.118461
20. Honorio, T., Bore, T., Benboudjema, F., et al.: Dielectric properties of the pore solution in cement-based materials. J. Mol. Liq. **302**, 112548 (2020). https://doi.org/10.1016/j.molliq.2020.112548

Development of Expansive Agent Model in a Multi-scale Thermodynamic Framework Based on Hydration and Microstructure Formation

Kolneath Pen, Tetsuya Ishida, Igarashi Go, Yuya Takahashi, and Shinya Ito

Abstract This study aims to develop an analytical tool to predict the expansion of Hauyne and free-lime based expansive additives in RC structures. Hydration of species was based on the multi-component heat of hydration model which incorporates referential heat rate at a certain temperature and thermal activities following Arrhenius' law of chemical reaction. All other time-dependent properties of concrete such as elastic modulus, temperature, pore pressure, creep, moisture status, total porosity of interlayer, gel and capillary pores are computed internally based on the micromodels of materials inside the system. Coupling this material information with poromechanics, volumetric change generated by cement hydration and shrinkage is systematically included in the modelling of concrete mechanics, which deals with macroscopic structural responses based on the space-averaged constitutive laws on the fixed four-way cracked concrete model. Finally, the models are then verified with experimental results that could prove the models' capability in predicting the amount of expansion under various replacement ratios.

Keywords Expansive additive · Multi-scale modelling · Concrete expansion

1 Introduction

Due to the nature of hydration and morphological characteristics of cement and aggregate matrix, concrete structures will undergo shrinkage. Especially concrete structures with high surface-to-volume ratios experience the highest effect of shrinkage stress. Then, once the tensile strain caused by the shrinkage stress exceeds the cracking strain, cracks will occur. Consequently, deleterious ions such as chlorides and sulfates could migrate into concrete and cause immense degradation, structurally

K. Pen (✉) · T. Ishida · I. Go · Y. Takahashi
Department of Civil Engineering, The University of Tokyo, Tokyo 113-8656, Japan
e-mail: kolneath@concrete.t.u-tokyo.ac.jp

S. Ito
Denka Company Limited, Niigata 949-0393, Japan

F. Kanavaris et al. (eds.), *International RILEM Conference on Early-Age and Long-Term Cracking in RC Structures*, RILEM Bookseries 31,
https://doi.org/10.1007/978-3-030-72921-9_18

and aesthetically. This phenomenon is inevitable and has remained a major challenge in the field of civil engineering.

Up until recent times, engineers have been tackling this issue from designing, constructing and material aspects. For example, in the case of designing, structural elements with different shapes have been used. In the case of constructing, sequential constructing has been considered for bridge decks to mitigate the effect of moment distribution on the top surface. From material viewpoints, they have successfully mitigated the effect of shrinkage through the use of shrinkage reducing admixtures, light-weight aggregates and expansive cements or expansive additives.

Due to the unstable nature of expansive additive, its practice in the civil engineering field is quite limited as the intended advantage of shrinkage compensating could not be attained if improper usage was conducted. Thus, the authors' objective is to extend a multi-scale thermodynamic analytical platform, coded as DuCOM-COM3, by incorporating an expansive additive model based on crystallization pressure theory. DuCOM [1] covers the thermodynamic multi-chemo-physical modelling part whereby numerous thermodynamic models, such as multicomponent hydration, micropore structure formation and moisture equilibrium/transport. Its counterpart, COM3 [2], is a nonlinear structural 3D finite-element analysis that implements constitutive laws of uncracked/cracked and hardening/aging/matured concrete. Time-dependent properties of concrete such as elastic modulus, temperature, pore pressure, creep, moisture status, total porosity of interlayer, gel and capillary pores are calculated consequently after inputting the basic boundary conditions for a structure such as mix proportions, geometry and environmental conditions based on the micromodels of materials inside the DuCOM system. The calculation sequence starts from the multicomponent heat of hydration model whereby the degree of hydration is determined through the amount of heat release by each phase in the system based on modified Arrhenius' equations. In the current model, expansive additive phases such as f-Lime (f-CaO), Hauyne ($C_4A_3\$$) and Anhydrite ($C\$$) could not be considered. Thus, the preliminary step is to determine the calculation of hydration degrees of these phases through the multicomponent heat of hydration model. Then, the corresponding degree of hydration and mass of phases information retrieved from the heat of hydration model would in turn be utilized to determine the volumetric balance and moisture status in the microstructure of cement and expansive additive system. Then, based on the solidification theory, time-dependent deformations such as creep and shrinkage would be superimposed to represent all the stresses existing in the cement paste matrix. In addition, macroscopic expansive pressure in the matrix from local crystallization pressure of expansive hydrates is also introduced. Finally, models of concrete mechanics will deal with macroscopic structural responses based on the space-averaged constitutive laws on the fixed four-way cracked concrete model [2].

2 Experimental Programme

In order to realize this expansive additive model, experimental investigations into the phenomena that lead to the expansion in the first place shall be done. As already mentioned previously, hydration, volume fraction of the expansive crystals, and local crystallization pressure are the precursors in the volumetric expansion of expansive additive system. Therefore, the experimental studies, which will be presented in the following sections comprise of hydration studies and unrestrained expansion of expansive additive cement paste with these parent phases: f-Lime (f-CaO), Hauyne ($C_4A_3\$$) and Anhydrite (C$). The experimental results would be presented together with modelling framework accordingly.

2.1 *Materials*

The chemical and mineral composition of various raw materials utilized in this study are listed in Tables 1 and 2. Three types of expansive additives with different amount of f-Lime, Hauyne and Anhydrite were used, which would enable us to view the correlation between the amount of CH and ettringite crystals that are formed and expansion magnitude.

Table 1 Chemical composition in expansive additive and ordinary Portland cement used

Expansive additive	Density	Chemical composition (weight %)						
	g/cm^3	LOI	SiO2	Al_2O_3	Fe_2O_3	CaO	MgO	SO_3
A	3.10	3	1.2	5.0	1.2	69.5	0.9	19.5
B	3.10	3	1.5	9.0	0.6	51.8	1.4	29
C	3.16	1.2	4.2	1.1	1.2	74.0	0.5	16.5
OPC	3.16	0.6	21.8	4.5	2.9	64	1.8	2.26

Table 2 Mineral composition of expansive additive and ordinary Portland cement used

Expansive additive	Mineral phases (weight %)						
	C_3A	C_2S	C_3S	C_4AF	f-CaO	$C_4A_3\$$	C$
A	–	–	–	–	44.0	12.2	31.0
B	–	–	–	–	17.0	27.0	46.0
C	–	–	20.4	–	43.0	–	30.9
OPC	5.6	18.3	62.2	5.6	–	–	–

2.2 X-ray Diffraction and Thermogravimetric Experiment

Two types of expansive additives were used in the study: A (free-lime content is higher than Hauyne content) and B (free-lime content is lower than Hauyne content). The water-to-binder ratio was 0.4. To derive a simple reference heat rate 100% replacement ratio with ordinary Portland cement was used in this study. In order to minimize bleeding, the pastes were stored in cylinder bottles whereby the bottles were rotated horizontally for approximately 6 h or more until the pastes set.

At these designated ages (0.0625, 0.125, 0.1875, 0.25, 0.3125, 0.375, 0.5, 1, 3, 7 and 14 days), the sample was crushed into pieces of below 1 cm and gently ground into powder by hand. The solvent exchange method by isopropanol was used to stop the hydration. After this hydration stoppage process, all samples were dried in a desiccator with silica gels for 1 week. The specimens are all stored in vacuum condition to avoid carbonation and environmental control room at 20 °C. After drying, all samples would be sieved through 100 μm and used for XRD and TG analyses. Internal standard method was used to determine the amount of amorphous inside the powders. Ten percent corundum (Al_2O_3), relative to the weight of cement, was blended with the sieved hydrated powder. The dry mixed powder was then placed in the sample holder for X-ray diffraction analysis via backloading approach. For Rietveld analysis, TOPAS software (Bruker) was utilized.

TGA for the calcium hydroxide and loss on ignition (LOI) was conducted at a heating rate of 2 °C/min up to 1000 °C under a nitrogen atmosphere. The weight percentage of calcium hydroxide was calculated between the temperature of 415 and 515 °C.

2.3 Isothermal Conduction Calorimetry

Cement paste samples were prepared at water-to-binder ratio of 0.40. 0, 10, 20, 50 and 100% replacement ratio of OPC were used for the corresponding expansive additives. Expansive additives A and B from Tables 1 and 2 were used at their respective replacement ratio. Three samples were conducted for each replacement ratio of the respective expansive additives for repeatability.

An isothermal conduction calorimeter was used to monitor the evolution of heat of hydration in A + OPC and B + OPC system. An isothermal calorimeter with four sample ampoules was used to determine heat evolution. The paste samples were mixed manually in the ampules for a period of 2 min before inserting them in the respective channels. It shall be noted that the data for the first 30 min was not considered for analysis to avoid the heat associated with mixing and placing the externally prepared samples, and to allow the samples to stabilize at the set temperature of 20 °C. The samples were monitored continuously for 72 h.

2.4 *Unrestrained Expansion of Expansive Cement Paste*

Cement paste samples were prepared at water-to-binder ratio of 0.40. 0, 3, 6 and 10% replacement ratio of OPC were used for B and C expansive additives. Three samples were prepared for each batch to check the consistency of the data.

The pastes were cast at room temperature of 20 ± 20 °C. To minimize bleeding, each paste was gently re-scooped at 30 min, 1 h, 2 h, 3 h and 4 h again inside the 4 × 4 × 16 cm^3 steel mold. After that, the mold gage, Tokyo Sokki PMFL-50T, was embedded in the center of the paste. Every mold gage was tested to check whether it functions normally or not before placing them inside the pastes. Then, each specimen was sealed with plastic wrap for sealed curing in an environmental control chamber with a temperature of 20 ± 20 °C. Then, after around 9 h, each specimen was gently removed from the mold, carefully wrapped with a plastic sheet and placed on the racks without any load on the top. The measurement of the strain starts around 30 min after the mold gage has been placed inside the paste.

3 Hydration Model for Expansive Additives

3.1 *Overview of Hydration Heat Model for Expansive Additives*

Referring to the model proposed by Suzuki et al. [3], which employed Arrhenius' equation to describe the hydration activity, two main parameters are required for each mineral phase: the reference heat rate and thermal activity. The hydration heat of each component is expressed as:

$$H_i = \beta_i H_{i,T_0}(Q_i)\exp\left[-\frac{E_i}{R}\left(\frac{1}{T} - \frac{1}{T_0}\right)\right] \quad (1)$$

where E_i is the activation energy of component *i*, *R* is the gas constant, $H_{i,To}(Q_i)$ is the reference heat rate of component *i* at constant temperature of T_0 equal to 20 °C as a function of accumulated heat Q_i and β_i is a parameter expressing reduction in precipitation space which subsequently reduces the heat rate. The term $-E_i/R$ reflects the thermal activity. To date, the mineral phases that are covered by the model are alite (C_3S), belite (C_2S), aluminate phase (C_3A), ferrite phase (C_4AF) and gypsum (C$H2). In the case of expansive additives, the main mineral phases which are hauyne (C4A3$), anhydrite (C$) and free-lime (f-CaO) are not present in the model. Thus, primarily, reference heat rate and thermal activity of the three aforementioned phases in expansive additives shall be determined.

3.2 Thermal Activity of Expansive Additive Phases

The thermal activity, $-E_i/R$, could simply be obtained once the activation energy of the expansive additive phases is known from literature. Hence, they could be determined as: $-E_{hauyne}/R = -27780$ WK/kg,$-E_{anhydrite}/R = -7541$ WK/kg and $-E_{free\text{-}lime}/R = -7000$ WK/kg.

3.3 Reference Heat Rate Determination

The accumulated heat of cement system is calculated via the equation:

$$Q_c = \int H_c dt = \int \sum \rho_i H_i dt \tag{2}$$

where Q_c is the accumulated heat of whole cement and ρ_i is the mass fraction of the individual mineral phases. The steps to determine the reference heat rate of the expansive additive phases are listed in the steps below. As an example, Free-Lime hydration in the case of 100% replacement ratio of expansive additive A was used.

STEP 1: Through a combined QXRD/TG method, $\alpha_{f\text{-CaO}}$ would be determined.

STEP 2: Using the experimental $\alpha_{f\text{-}CaO}$, $Q(t)$ (the heat liberated at time t) and $H(t)$ (the heat rate of the mineral phase) could be determined based on the equations:

$$\alpha(t)_{xp,i} = \frac{Q(t)_{xp,i}}{Q_{xp,i}^{\max}} \tag{3}$$

$$Q(t)_{xp,i} = \alpha(t)_{xp,i} \times Q_{xp,i}^{\max} \tag{4}$$

$$H(t)_{xp,i} = \frac{dQ(t)_{xp,i}}{dt} \tag{5}$$

To derive this heat rate, t that would be considered in QXRD/TG method consists of 1d, 3d and 7d. Therefore, at each t, corresponding $H(t)_{f\text{-}CaO}$ could be determined and empirically set at heat rate of each expansive additive mineral phase (Figs. 1 and 2).

Note that $Q_{f\text{-}CaO}{}^{\max}$ is the theoretical standard enthalpy of formation of free-lime (1166 J/g). For $Q_{hauyne}{}^{\max}$ and $Q_{anhydrite}{}^{\max}$, they are 555 J/g and 145 J/g, respectively. In Fig. 3, the accumulated heat was able to obtained according to Eqs. (3), (4) and (5). Then, based on the slope from Fig. 3a, the heat rate in Fig. 3b is consequently determined. However, it does not seem logical to assume such a step function for hydration process as hydration tends to stagnate due to denser micropore structure at time goes by. Thus, a linear heat rate as shown in Fig. 3c was adopted by keeping the same accumulated heat (hatched area under the curve of Fig. 3b, c. After that,

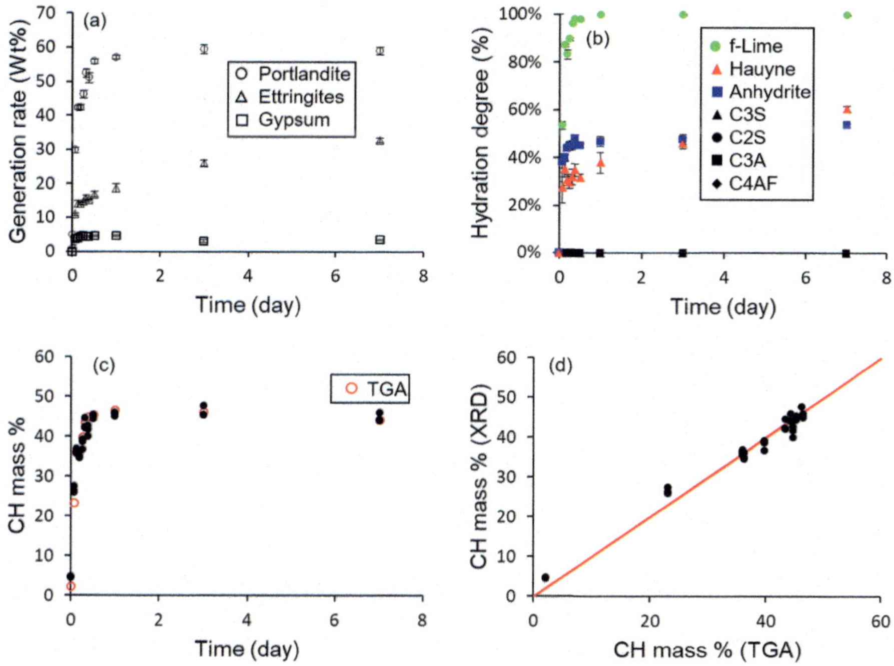

Fig. 1 XRD-TG experimental results for expansive additive A with 100% replacement ratio

plotting the heat rate against the accumulated heat, we would be able to obtain the linear heat rate against accumulated heat curve for free-lime based on the experiment conducted (Fig. 3d). As for the model, a bi-linear curve was adopted as shown in Fig. 3e after consideration of the fact that hydration in the very early stage when unreacted phases are immediately in contact with water should be much more rapid than that of later stage. Thus, the slope of the heat rate in the first 10% of accumulated heat was assumed as shown in Fig. 3e (Fig. 4).

STEP 3: Lastly, verification with isothermal calorimetric results at different replacement ratios would be shown.

3.4 *Verification of Analytical Results Based on Isothermal Calorimetry Tests*

Cement paste samples were prepared at water-to-binder ratio of 0.40. 0, 10, 20, 50 and 100% replacement ratio of OPC were used for the corresponding expansive additives. Expansive additive A and B from Tables 1 and 2 were used at their respective replacement ratio. Three samples were conducted for each replacement ratio of the respective expansive additives for repeatability (Fig. 5).

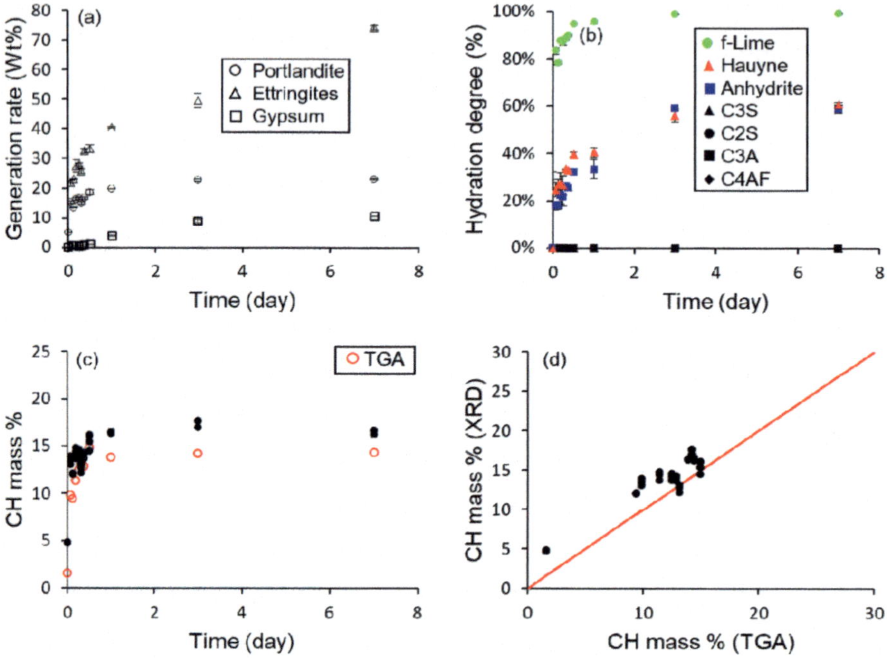

Fig. 2 XRD-TG experimental results for expansive additive B with 100% replacement ratio

4 Microstructure Model for Expansive Additives

A system of ordinary Portland cement and expansive additives comprises various hydrates. It is crucial that the information of the fraction of expansive hydrates, namely ettringites and calcium hydroxides, could be obtained. Thus, to keep the extension to the model in a simple manner, an assumption was made that only the ettringites from expansive additives could cause expansion and would be superimposed to the current system based on the stoichiometric balance. Similarly, only CH crystals from f-CaO could cause expansion and these expansive CH crystals would be superimposed to the current system in the same fashion as that of ettringites. However, not all of the expansive hydrates would be able to cause expansion. In the early stage of hydration, the dispersed particles most likely have an abundant amount of free room for precipitation to occur without introducing the built-up of stresses. As expansive pressure is a form of compressive stress between the adjacent particles, it is logical to assume that the expansion only arises once the particles around the vicinity can exert stress on each other through the form of contact.

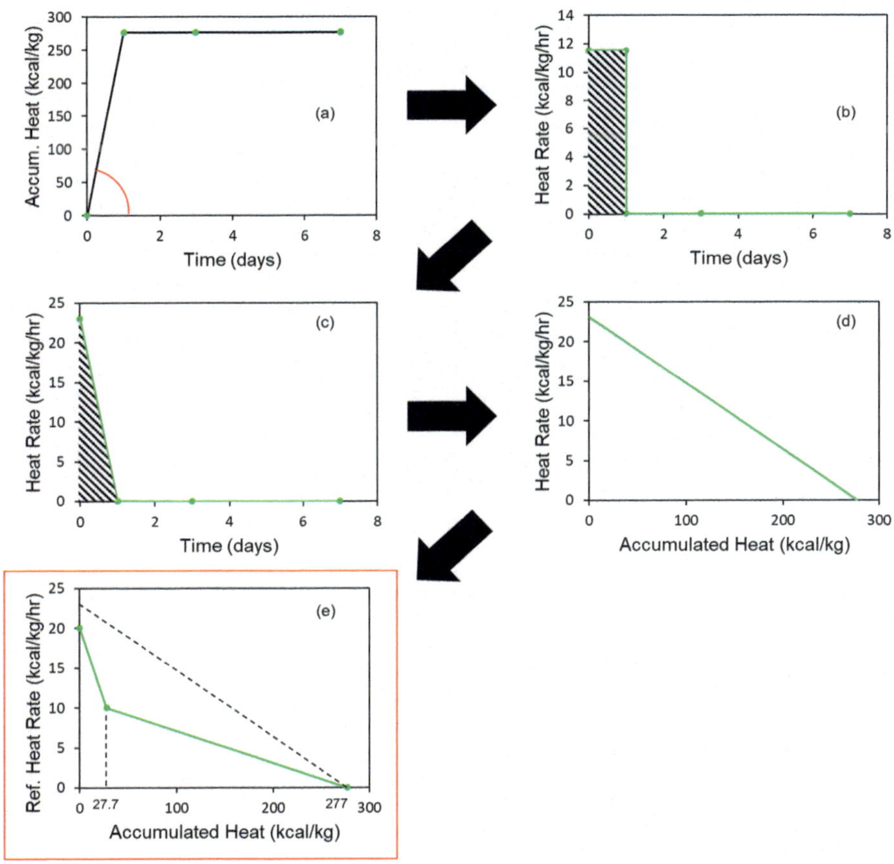

Fig. 3 Process of determining reference heat rate for expansive additive phases

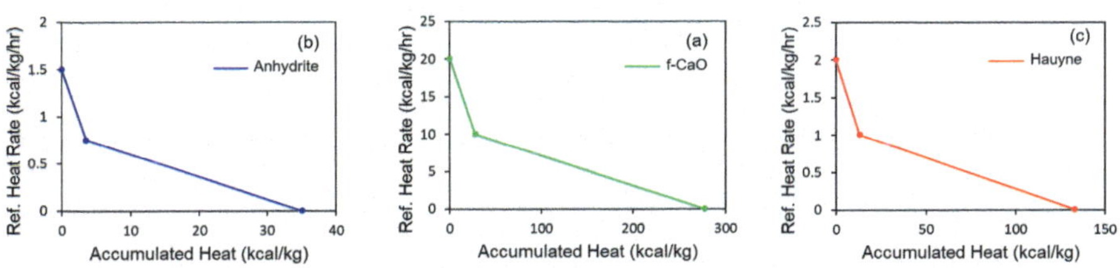

Fig. 4 Reference heat rate of expansive additive phases installed in heat of hydration model

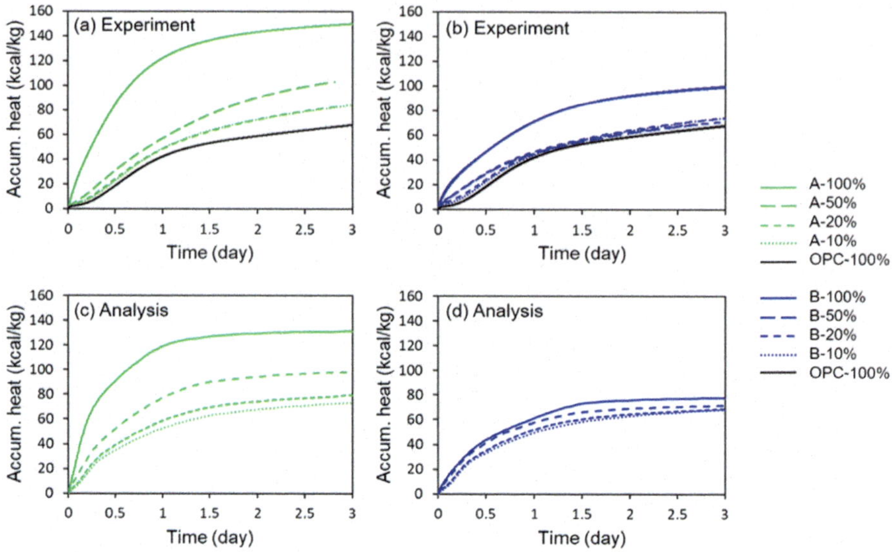

Fig. 5 Verification of isothermal calorimetric results versus DuCOM analytical results

5 Expansion Model for Expansive Additives

5.1 Local Crystallization Pressure

As supersaturation is one of the major factors which control the expansion, from a thermodynamic point of view, Correns [4] proposed the following equation to calculate the local crystallization pressure as:

$$p_c = \frac{RT}{v} ln\left(\frac{K}{K_{sp}}\right) \tag{6}$$

where R is the gas constant (8.314 J/K/mol), T is the absolute temperature, v is the molar volume of the crystals ($v_{Aft} = 735$ cm^3/mol and $v_{CH} = 33$ cm^3/mol), K is the ionic activity and K_{sp} is the solubility products of the ions that will form the crystal hydrates in the pore solution. Alternately, $\ln K/K_{sp}$ could also be referred as Saturation Index or Saturation Factor.

5.2 Determination of Saturation Index

As in the aforementioned Eq. (6), the most influenced variable or factor in causing the local crystallization pressure is $\ln K/K_{sp}$ as the other parameters are more or less

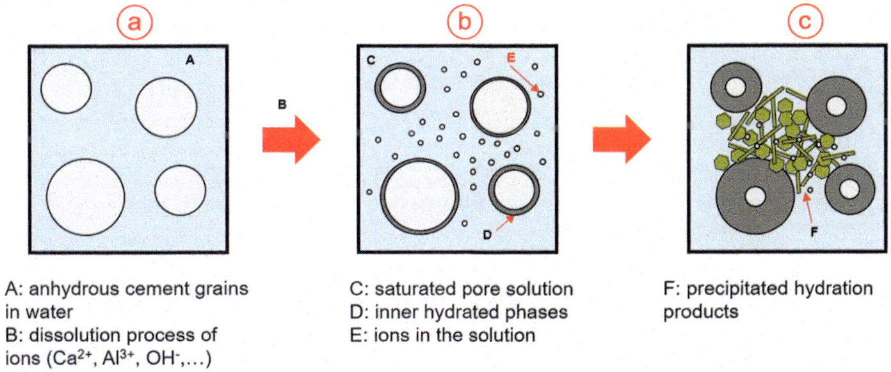

Fig. 6 Dissolution and precipitation process of cement clinker and expansive additive phases

constant. Typically, Saturation Index could be determined from the concentration of the ions in the pore solutions that would make up the crystal products. Refer to Fig. 6a, the dissolution process would give rise in the ionic concentration of relevant ions whereby these ions would determine whether the pore solution has reached the supersaturated state or not. Then, once the supersaturated state is achieved, precipitation of corresponding products based on the interaction between the ions in the pore solutions would occur. One limitation lies here in the case where the multi-component heat of hydration model is used. The calculation from this model allows us to obtain the hydration products immediately based on the amount of heat that is release by each clinker and expansive additive phase, without the need to consider the equilibrium between the solid phases and aqueous solution. In other words, phenomenon in Fig. 6b was disregarded. Therefore, obtaining the Saturation Index directly from DuCOM via pore solution is practically impossible as of now. However, a rather simple treatment could be proposed to imitate the Saturation Index through a function of unreacted parent phases instead of ionic concentration. It may be plausible if it is somehow boldly assumed that the unhydrated particles are like particles which are ready to dissolve in the pore solution and ready to precipitate to form hydration products. Chaunsali [5] determined the Saturation Index of ettringite in a Calcium Sulfo Aluminate and ordinary Portland cement system through pore solution extraction. Concurrently, the aforementioned author also conducted X-Ray diffraction to determine the mass of the phases in the system. Based on the result of Chaunsali [5], it shows that by 1 day of hydration process, the amount of Hauyne reacted was approximately 40% whereby there is no trace of Hauyne by the age of 7 days. Hence, a fictitious Saturation Index for ettringite based on unhydrated Hauyne is proposed through the equation shown below:

$$SI_{ett} = \left\{5 - 2.5 \cdot exp\left(-10 \cdot \frac{Q_{hauyne}}{Q_{max,hauyne}}\right)\right\} \cdot \rho_{hauyne} \tag{7}$$

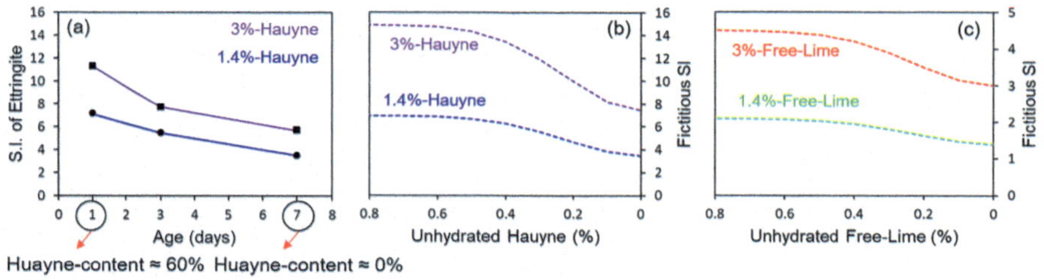

Fig. 7 **a** Saturation index of ettringites based on different contents of Hauyne (from Chaunsali [5]). **b** Fictitious Saturation Index of Ettringites. **c** Fictitious Saturation Index of Calcium Hydroxides

where Q_{hauyne} is the amount of heat released by hauyne hydration, $Q_{max;hauyne}$ is the maximum amount of heat that could be released from hauyne (complete hydration) and ρhauyne is the mass fraction of hauyne in the cement used.

As for free-lime, based on a study by Gartner et al. [6] on Saturation Index for CH in Portland cement paste, with free-lime content around 1%, the saturation index was approximately 1. Thus, the following equation is assumed for the Saturation Index of CH in a similar approach as that of ettringites (Fig. 7).

$$SI_{ch} = \left\{1.5 - 0.5 \cdot exp\left(-10 \cdot \frac{Q_{free-lime}}{Q_{max,free-lime}}\right)\right\} \cdot \rho_{free-lime} \tag{8}$$

5.3 *Upscaling of Local Crystallization Pressure to Macroscopic Stress*

Based on poromechanics, under the effect of pore pressure and heat, Coussy [7] introduced a relationship between the mean stress and strain in a linear isotropic body as:

$$\sigma_{exp} = b(S_c p_c + S_L p_L) + 3\alpha K \Delta T \tag{9}$$

where σ is the mean stress, K is the bulk modulus and α is the thermal expansion coefficient, b is the Biot coefficient, ΔT is the difference in temperature, S_c and S_L are the saturation level of the crystal and liquid, respectively and p_c and p_L are the local crystallization pressure of the crystal and pressure of liquid, respectively. As DuCOM have already considered the effect of liquid pressure and thermal dilation, Eq. 5 reduces to:

$$\sigma_{exp} = bS_c p_c \tag{10}$$

Tentatively, the biot coefficient is assumed to be 0.61. S_c could be obtained as a fraction of the expansive hydrates in the total volume of cement paste over the capillary porosity. Then, in coherent with shrinkage model, the linear isotropic expansive stress is added to the system by superimposing the macroscopic crystallization stress with the current model as:

$$\sigma_{cp} = \dot{\sigma}_{cp} + \sigma_s + \sigma_{exp} \tag{11}$$

where σ_{cp} is the total stress on cement paste matrix, $\dot{\sigma}_{cp}$ is the volumetric creep stress induced by solidification of clusters, σ_s is the shrinkage stress and σ_{exp} is the macroscopic expansive stress. Thus, this model has the capability in assessing the mechanical behavior of expansive cement paste, mortar and concrete from its expansion in early hydration period to its shrinkage in the later stage of hydration (Fig. 8).

The concrete specimens were cured in water for 7 days before exposing to an environmental condition with relative humidity of 60% at 20°C (Fig. 9). N, BB, H and X represent ordinary Portland cement, Blast-furnace slag cement (40% Slag), High early-strength cement and expansive additive A, respectively. It could be seen that the current model could capture the behavior of both expansion and shrinkage

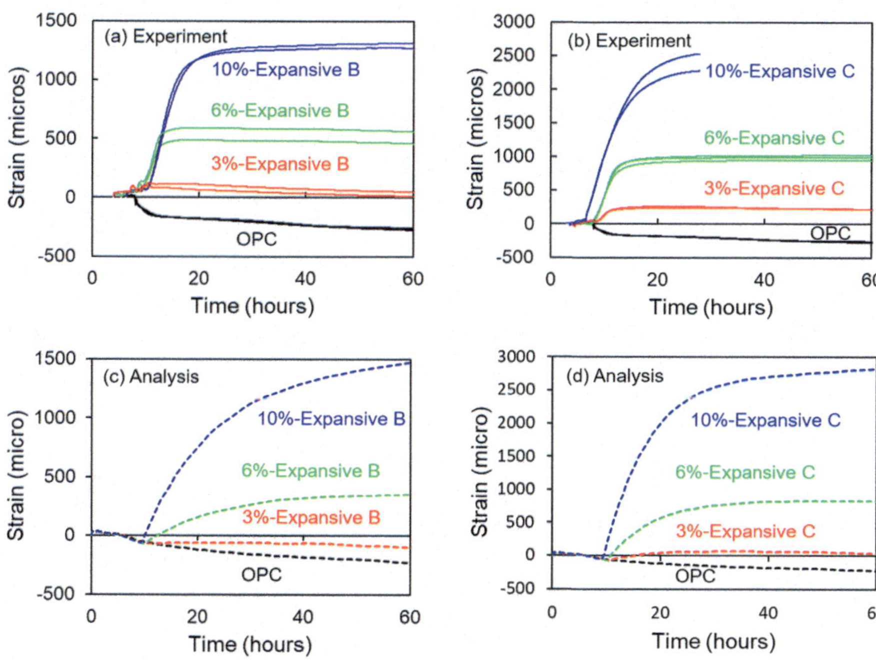

Fig. 8 Analytical and experimental results on unrestrained $4 \times 4 \times 16\,\text{cm}^3$ cement paste of ordinary Portland cement and expansive additive systems

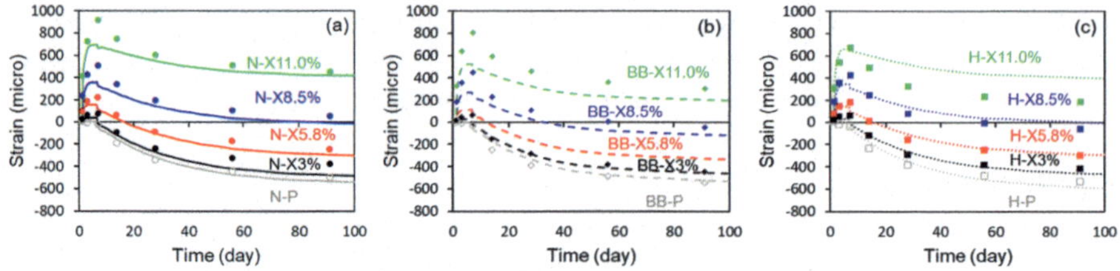

Fig. 9 Analytical and experimental results on expansive additive A-OPC concrete under JIS-A-6202 (10 × 10 × 40 cm^3 with 0.97% rebar ratio under 60% RH after 7-day water curing at 20 °C)

of concrete at various mix design quite considerably at typical replacement ratio for shrinkage compensating effect whereby, for higher replacement ratio in the case of ordinary Portland cement and Blast-furnace slag cement, the discrepancy might be linked to the decrease in modulus of elasticity of the cement matrix.

6 Conclusion

Based on multi-component heat of hydration model, hydration of Hauyne, Anhydrite and f-Lime were added to the system to compute the reaction degree of those phases. Then, on the basis of calculated degree of hydration, water consumption and mass balance, the volumetric change of the expansive hydrates were added into the microstructure of the current system. After particle contacts have been made, the expansive hydrates produced afterwards would contribute in volumetric expansion based on macroscopic crystallization stress upscaling from the local crystallization pressure. The model has proved to work for both cement paste and concrete under recommended practice (6% or 20 kg/m^3 of replacement ratio of expansive additives). Further studies are being conducted regarding the effect of temperature, low water-to-binder ratio, bi-axial and tri-axial restraint and different amount of restraining force.

Acknowledgement This study was financially supported by Denka Company Limited. The authors would like to extend their greatest gratitude for the ongoing support.

References

1. Maekawa, K., Ishida, T., Kishi, T.: Multi-scale Modeling of Structural Concrete. Crc Press (2008)
2. Maekawa, K., Okamura, H., Pimanmas, A..: Non-linear Mechanics of Reinforced Concrete. CRC Press (2003)

3. Suzuki, Y., Tsuji, Y., Maekawa, K., Okamura, H.: Quantification of Hydration-Heat Generation Process of Cement in Concrete. JSCE16 (1990)
4. Correns, C.W.: Growth and dissolution of crystals under linear pressure. Discuss. Faraday Soc. **5**, 267–271 (1949)
5. Chaunsali, P.: Early-age hydration and volume change of calcium sulfoaluminate cement-based binders. Ph.D. thesis, University of Illinois at urbana-Champaign (2015)
6. Garner, E., Tang, F.J., Weiss, S.J.: Saturation factors for calcium hydroxide and calcium sulfates in fresh Portland cement pastes. J. Am. Cera. Soc. **68**(12), 667–673 (1985)
7. Coussy, O.: Poromechanics. Wiley (2004)

Double Wall Containment Building Leak-Tightness Prediction: Strategy and Application

Laurent Charpin, Charles Toulemonde, Jean-Luc Adia, Florian Escoffier, Sylvie Michel-Ponnelle, Guillaume Boulant, Benoît Masson, and Julien Niepceron

Abstract EDF operates a fleet of 56 nuclear reactors. For 24 of these reactors, the concrete containment building is a double wall structure. The inner wall is prestressed and has no metallic liner. Every ten years, the leak-tightness of the inner wall is verified by performing a pressure test. The leakage rate has to remain below a given threshold. As time passes, the leakage rate is getting closer to this threshold, so important coating programs are underway to keep the leakage rate within the regulatory limits. Therefore, it is important for EDF to be able to forecast the evolution of leakage, which depends on the concrete saturation and the level of prestress. The prestress decrease is related to creep and shrinkage of concrete and tendons relaxation (this last factor is considered negligible at ambient temperature). To tackle this issue, EDF has launched an important research program around the VERCORS mock-up, which is a containment building at scale 1/3 and has used this research program to improve its ability to predict the evolution of leak-tightness with time. In this paper, the different tools of the digital twin are presented, as well as an example of the use of the models originally developed on the VERCORS mock-up to help the choice of coating strategy for the third pressure test of an industrial containment building.

Keywords Concrete containment building · Leak-tightness · Digital twin

L. Charpin · C. Toulemonde · J.-L. Adia (✉)
EDF Lab Les Renardières, Ecuelles, France
e-mail: jean-luc.adia@edf.fr

F. Escoffier · S. Michel-Ponnelle · G. Boulant
EDF Lab Saclay, Palaiseau, France

B. Masson · J. Niepceron
EDF DT, Lyon, France

F. Kanavaris et al. (eds.), *International RILEM Conference on Early-Age and Long-Term Cracking in RC Structures*, RILEM Bookseries 31,
https://doi.org/10.1007/978-3-030-72921-9_19

1 Introduction

EDF operates a large fleet of Nuclear Power Plants (NPPs). The leak-tightness of the Concrete Containment Buildings (CCBs) is checked and validated every 10 years during the Integrated Leakage Rate Test (ILRT) which success is mandatory to continue the plant operation.

24 of EDF French containment buildings are double-walled containments whose leak-tightness is primarily dependent on the inner prestressed reinforced CCB. The post-tensioning is calibrated in order to ensure that in case of an accident the concrete remains in compression. Tendons are grouted on these containments; therefore increasing post-tension is not an option. Therefore, EDF has acquired a long experience of repair solutions that are thoroughly tested and qualified. The in situ feedback is positive but their operational implementation remains a technical and organizational challenge that benefits from optimization. To maximize the safety of the containments, maintenance policies require a detailed knowledge about (i) the behavior of the concretes used in the containment buildings, in order to be able to predict (ii) the future evolution of their delayed strains and (iii) their subsequent effects on leak-tightness.

Many research programs have been conducted to understand and characterize the delayed strain effect, but in order to bridge remaining knowledge gaps towards leak prediction and refurbishment, an important research program was started. This program is called VERCORS and is presented in the following section.

This paper presents the strategy for leakage prediction which has been built thanks to the VERCORS program, and an application to Belleville 1 containment building.

2 Strategy for the Containment Buildings Leakage Prediction

2.1 The VERCORS Program

VERCORS (Vérification Réaliste du Confinement des Réacteurs/realistic assessment of reactors containment) is a 1/3 scale mock-up (Fig. 1) of a double wall CCB that has been built at EDF Lab Les Renardières research facility, near Paris, France. An overview of the project as well as details and lesson learned from the first benchmark dedicated to modelling of the mock-up in its early age phase can be found in [1]. The second VERCORS benchmark was the object of a restitution workshop during TINCE conference in Saclay, France in 2018 and a special session at SMIRT conference in 2019. An overview paper has been submitted [2].

Since drying is a diffusive phenomenon, the drying characteristic times proportional to the square of the characteristic drying length. Moreover, shrinkage is driven by drying, and creep is partly driven by drying (only the basic creep component is not accelerated by drying). Therefore, in order to achieve faster ageing, and according to

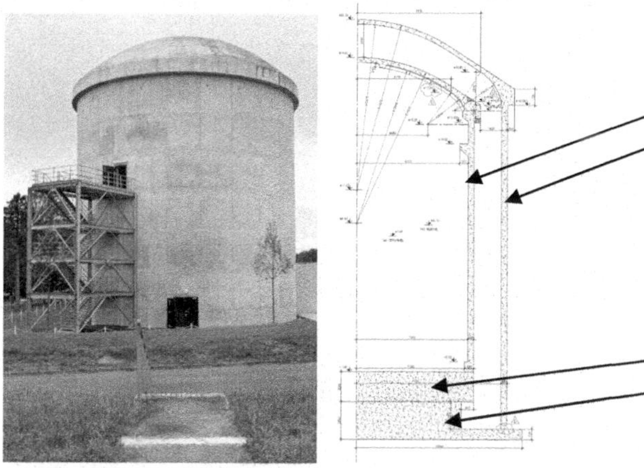

Fig. 1 Outer containment wall of the VERCORS mock-up and sectional schematic view

the assumption that drying will be the main phenomenon causing delayed deformation, VERCORS has been built with the expectation of an accelerated ageing factor on deformations. The 1/3 scale was thus chosen as the best compromise between scalability of construction and acceleration of ageing. All the construction elements are scaled to the best possible extent.

The mock-up construction started in 2013, inner containment raft foundation was concreted in July 2014, and the dome was finished in April 2015. Prestressing started after 28 days of curing and was completed in august 2015. The first ILRT (so-called "preoperational") was conducted on the 5th of November 2015. After the end of the construction (March 2016), two auxiliaries air-conditioning units were started to impose realistic relative humidity and temperature inside and outside the inner containment and simulate the drying effect of regular operation of a nuclear reactor. IRLTs are conducted on a regular basis (approximately every 13 months according to observed delayed strain acceleration).

VERCORS is highly instrumented: more than 300 vibrating wire strain sensors, 200 PT100 temperature sensors, 2 km of optic fibers, 20 TDR moisture sensors, 80 strain gages (on rebars) as well as classical pendula and invar wires were installed.

The VERCORS concrete has also been widely studied (and is still under study for higher temperature applications), with many French and European laboratories involved in the measurements of properties such as: Young's modulus, tensile and compressive strength, drying, creep and shrinkage, fracture energy, early-age properties, permeability. A more detailed information about VERCORS, its monitoring systems and experimental program, can be found in [3].

A third VERCORS benchmark will start in January 2021 for 1 year, giving an opportunity for participants to test their structural models and leakage prediction models against the VERCORS experimental results.

2.2 Leakage Prediction Strategy

The VERCORS experiment is part of a global strategy to build a tool for the optimization of the coatings that are installed on the containment buildings as a preparation to ILRTs. This strategy can be summarized as follows:

1. Validation of drying, shrinkage and creep models at moderate temperature by calibrating these models on laboratory data only (using results from the VERCORS program and MACENA PIA project) and comparing the results of the full-scale VERCORS staggered Thermo-Hygro-Mechanical (T-H-M) analysis to the monitoring data obtained on VERCORS.
2. Proposition of the global analytical leakage model using the results of the staggered T-H-M computation as input in an averaged way. The model is calibrated using the existing ILRT results on VERCORS and validated by predicting the next ILRT global leakage.
3. Proposition of a calibration procedure for the drying and mechanical models which can be used even in absence of laboratory data (which is the common case for the real containment buildings). Validation of this calibration procedure on CCBs by calibrating the models on the monitoring data except for the last 10 years and predicting the remaining 10 years.
4. Proposition of a local leakage model, using as inputs the local drying profiles and local stresses (or loss of prestress). Calibration using the existing measurements of local leaks and validation using the results of the next ILRT test when available.
5. Proposition of a post-processing procedure for the leakage model results to test the influence of different coating strategies on the global leakage of the next ILRT.

This strategy is detailed in [4]. The present paper is focused on recalling the tools used for the containment digital twin and the application to a French containment building, which corresponds to an application of steps 2, 3 and 5.

2.3 Digital Twin Factory

The prediction of leak-tightness of containment building has a very important strategic value for EDF: the repair works are very expensive; the cost of failing at meeting the regulatory leak-tightness requirement during an ILRT is enormous.

At the same time, looking at each containment building individually, the simulations are not performed often: the IRLT occur every 10 years for each containment building.

Therefore, it has been decided to build a "digital twin factory" for the containment buildings, in order to be able to:

- Capitalize and version all the developments performed to contribute to the simulations on the long-term
- Capitalize and version the input data for the studies
- Automatically chain the different steps of the study to avoid output/input errors, and facilitate running again the study for verification or parametric study
- Mutualize all that can be mutualized between the simulation of different containment buildings
- Test the individual functions and well as the studies made by chaining a number of function with a specific set of input data
- Facilitate the migration of different parts of the study to follow environment evolutions (change of Python interpreter, change of FE solver, ...)
- Facilitate experiment/simulation comparison to give access to easy understanding of the results
- For VERCORS, automatize visualization, post-processing of experimental data.

More details on the VECORS digital twin can be found in [3, 5].

3 Application to Belleville 1 Containment Building

In this section, the strategy presented in the introduction and applied to VERCORS mock-up in [4] is presented. It was chosen here to present an illustration on Belleville 1 containment building, which is one of the containment buildings of the EDF fleet which has already necessitated repair works and then, has been studied in detail.

A parallel on the life of VERCORS and Belleville 1 is shown in Table 1.

The temperature/humidity history on VERCORS is controlled. Is it quite close to Belleville 1 case. Taking into account the ageing factor of 9 affecting VERCORS (which was expected and has been verified), it is already older than Belleville 1 NPP!

A new feature of the leak-rate prediction on real containment buildings is the necessity to account for the effect of coatings, which are installed before ILRT in case of a risk that the regulatory criterion is exceeded. The leakage prediction on a real containment is a complex sequence of many analysis steps, as presented on Fig. 2.

The main steps are described in the following sections.

3.1 Model Calibration

The information available on real EDF containment buildings to calibrate drying and delayed strains constitutive laws are embedded strain measurements (by vibrating wire sensors). In this study, the information available on the Belleville 1 concrete are scarce: Young's modulus, porosity and thermal properties only. Therefore, a desorption isotherm form another containment building has been used, and the parameters

Table 1 Comparison between VERCORS and Belleville 1 characteristic dates

Event	Description	VERCORS		Belleville 1	
		Date	Operation time (y.)	Date	Operation time (y.)
Start of construction	Concreting of the raft	24/07/2014		26/05/1981	
End of construction	End of concreting of the dome	28/04/2015		13/07/1984	
Start of prestressing (excluding raft prestressing for Belleville 1)	Prestressing was performed in numerous phases, ensuring of VERCORS that concrete age was minimum 1 month before prestressing	06/05/2015		24/05/1984	
End of prestressing		12/08/2015		07/02/1985	
Containment kept wet until this date	To avoid early-age drying, VERCORS was kept wet until the end of prestressing	12/08/2015			
Pre-operational test	This is the first pressure test of the CCB which technically marks the start of the commercial use of the NPP	05/11/2015		07/06/1985	
Start of the reactor	Date of the beginning of operation (it took approximately 1 year to reach full operation power)			02/1988	0
VC1 test	Control visit (performed twice on Belleville 1)	27/01/2016		29/06/1989	

(continued)

Table 1 (continued)

Event	Description	VERCORS		Belleville 1	
		Date	Operation time (y.)	Date	Operation time (y.)
Start of the heating system	For VERCORS, the heating system could not be turned on before VC1, as it should have. The drying and by extension the "aging" process start at this point	01/03/2016	0		
VD1 test	First decennial test	14/03/2017	1	27/10/1998	10
VD1bis test	This first test was repeated on VERCORS to check whether results changed or not	21/03/2017	1		
VD2 test	Second decennial test	29/03/2018	2	26/08/2009	21
VD3 test	Third decennial test	19/03/2019	3	04/08/2019	31
VD4 test	Fourth decennial test (Unfortunately, the fourth "Decennial Visit" didn't happen because of the COVID-19 and the confinement obligation in France)	Cancelled			

for the drying law are also taken from previous studies, without the possibility to verify that drying is correctly simulated on saturation measurements, since this data is not available.

The validation of the drying model is indirect. We assume that if we manage to calibrate the mechanical parameters in a way that strains of the containment building are correctly reproduced, we are catching correctly the "ageing" kinetics of the containment. Of course, given the difficulties to calibrate drying laws even when

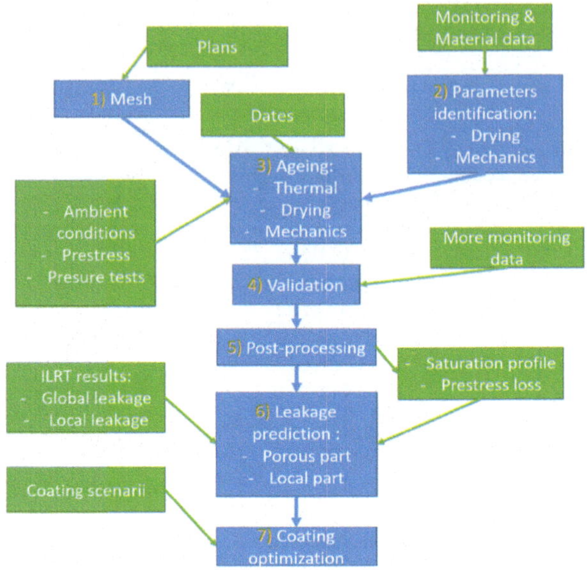

Fig. 2 Sequence of steps to predict the leak rate for the next IRLT of a containment building

laboratory data is available [6], we are aware that the simulation drying profiles might not be identical to the real profile.

The calibration computations are performed using a small model representative of the central part of the cylinder allowing for quick iterations (the tendons are explicitly represented, but their relaxation is neglected). The comparison of the prediction and measurement after calibration of the parameters is shown on Fig. 3. Note that the presence of coating on the inner side of the containment after VD1 (10 years of

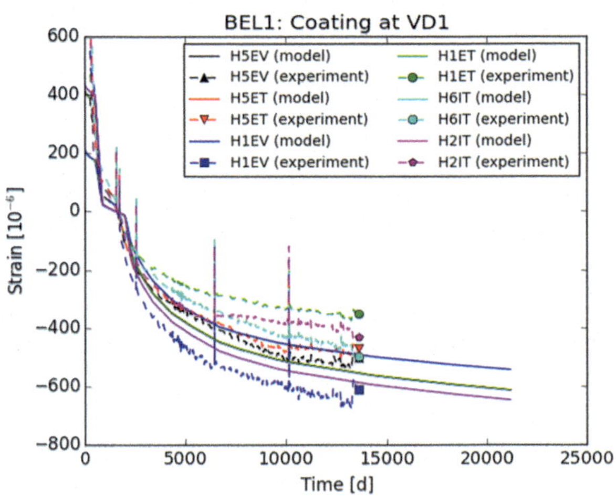

Fig. 3 Mechanical parameters calibration. Comparison on simulated strains to strains measured at mid-height of the cylinder part of Belleville 1 containment building. Strain reference taken right after the pre-operational pressure test. Each color corresponds to a specific sensor (E = outer face of the wall, I = inner face of the wall, T = tangential, V = vertical)

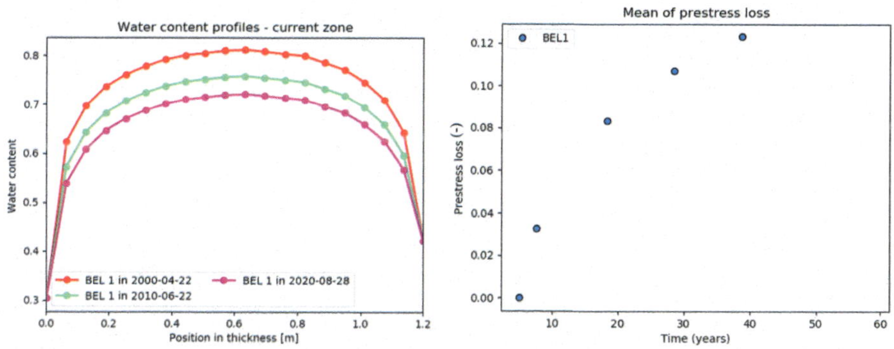

Fig. 4 Saturation profile and prestess lose in time used as input to the leakage model

operation) has been accounted for during the calibration computations. The uncertainty of measurements on the field is non-negligible in absolute value, but the trends measured after 30 years of operation are quite consistent between measurements and simulations, which makes it possible to use the model for the prediction of strains and prestress evolution.

3.2 Outputs from the Structural Computation Used for the Leak-Rate Prediction

In a next step, a full structural computation is performed. This computation includes staggered thermal, drying and mechanical analysis. In the mechanical analysis, shrinkage and creep of concrete are taken into account (similarly to the calibration computation, but on a full scale model). Strain at different locations are compared to measurements for a wider validation of the structural model (not shown here). Two outputs from the full computation are used as input to the leak-rate model:

- The saturation profile across the wall
- The average prestress loss computed on all the prestressing cables

These quantities are shown on Fig. 4.

3.3 Leak-Rate Prediction Without Adding New Coatings

The next step of the prediction is to predict the leakage that would occur during the next IRLT (here VD3, after 30 years of operation) if no additional coating were added.

As explained in [4], the leaks are categorized into 2 contributions:

- Local leak: air flux going through cracks, assumed dependent on the prestress loss. For this contribution, the local air flux is assumed proportional to the prestress loss (which has been shown to be quite accurate for VERCORS mock-up). The proportionality coefficient is calibrated using VD2 results. Experimentally this contribution is quantified by collecting the air from defects that are visible when soapy water is sprayed.
- Porous leak: air flux going through the concrete porosity, assumed dependent on the saturation profile. For this contribution, an unsaturated airflow simulation is performed through the wall taking into account the saturation profile simulated for VD2. The intrinsic permeability is calibrated against experimental results. The VD3 porous leak flux is predicted. Experimentally this contribution is determined by subtracting local leak to the total leak (which is computed by measuring pressure, temperature and relative humidity in the containment).

Using this methodology, it is predicted that without adding new coatings[1]:

- The porous leak would rise by approximatively 10%
- The local leak would rise by approximatively 15%.

3.4 Leak-Rate Prediction with New Coatings

The aim of this work is to evaluate different coating strategies of coating to help optimize the repair program. Here, only the coating configuration which has been decided and implemented before the VD3 test (November 2020) is considered. The results were not available when the work has been performed but are now available and can be compared to the predictions.

The Belleville 1 containment building had been already coated on its inner surface to prepare for VD1 and VD2. As a preparation for VD3, coatings are only added to the outer surface, as shown on Fig. 5.

The coatings are taken into account differently for porous or local leaks:

- For the porous leaks, since it is not known where they occur, they are reduced in proportion to the surface of coating added relative to the containment surface, applying an efficiency factor.
- For the local leaks, if a local leak is in a zone where new coating is added, it is also reduced of a portion using a different coating efficiency factor.

The coating efficiency factors were determined by inverse analysis using leakage measurement results on other containment buildings.

In substance, the prediction was first made with a crude value of the coating efficiency factor, and the coating scenario implemented was considered safe enough.

Then, the repair works were performed.

In a third phase, the prediction was improved by calibrating the efficiency factors on a larger number of test results.

[1] The numerical values are not given for confidentiality reasons.

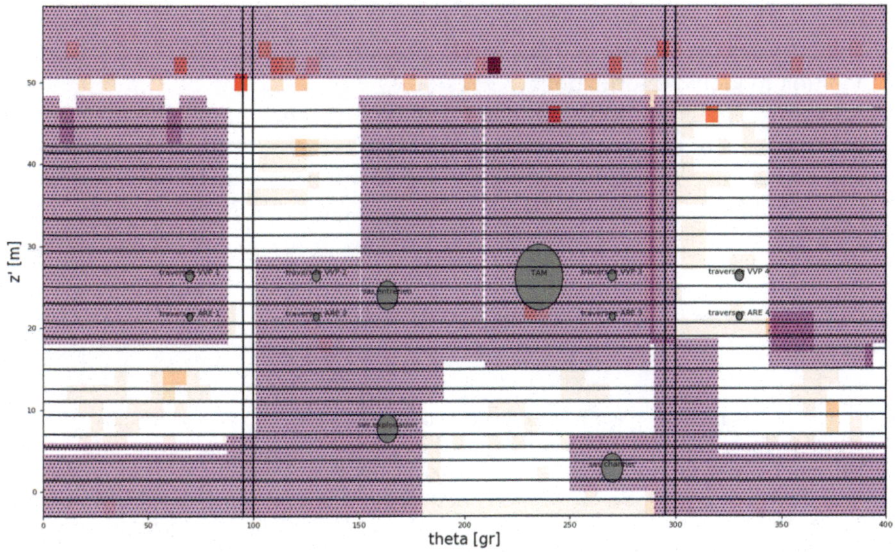

Fig. 5 Coatings added to the outer surface of Belleville 1 containment building as a preparation for VD3 (purple zones represent the coating placed on the outer surface of the containment between VD2 and VD3, orange and red areas show the localization of the main leaks during the VD3 pressure test, an information which was not available when the work was performed)

The final result obtained during the VD3 pressure test is low enough to meet the regulatory requirement, and very close to the second adjusted prediction.[2]

4 Conclusion and Perspectives

- The leakage rate prediction works have intensified thanks to the VERCORS program and a global leakage model validated on VERCORS (without coating) is available since 2018.
- This leakage model has been adapted to containment buildings to account for the presence of coatings, and has now been applied to three different containment buildings including Belleville 1.
- In the present paper, the VERCORS program, the leakage prediction methodology as well as an application to Belleville 1 are presented.
- The predictions obtained thanks to the model are of important help to optimize the repair programs.
- To improve theses approach and make them more reliable, different paths will be followed:
- The coating model must be refined;

[2]The numerical values are not given for confidentiality reasons.

- Local leakage prediction models are being validated on VERCORS mock-up. These models make use of local stress fields from the structural calculation in addition to global information.
- An uncertainty analysis on the model will be performed in order to quantify the safety margin of the leakage rate predictions.
- The number of containment buildings studied will be increased, so as to improve the calibration of the model. Two more will be studied in 2021.

References

1. Corbin, M., Garcia, M.: International benchmark VERCORS 2015—overview, synthesis and lessons learned. EDF, SEPTEN (2015)
2. Charpin, L., Niepceron, J., Corbin, M., Masson, B., Mathieu, J.-P., Haelewyn, J., Hamon, F., Ahs, M., Aparicio, S., Asali, M., Azenha, M., Bouhjiti, D., Calonius, K., Chu, M., Herman, N., Huang, X., Jimenez, S., Mazars, J., Nahas, G., Stepan, J., Thénint, T., Torrenti, J.-M.: Ageing and Air Leakage Assessment of a nuclear reactor containment mock-up: VERCORS 2nd. Submitted to Nuclear Engineering and Design (2021)
3. Mathieu, J.-P., Charpin, L., Sémété, P., Toulemonde, C., Boulant, G., Haelewyn, J., Hamon, F., Michel-Ponnelle, S., Hénault, J.-M., Taillade, F.: Temperature and Humidity-Driven Ageing of the VeRCoRs Mock-Up. Euro-C, Austria (2018)
4. Charpin, L., Haelewyn, J., Cherki El Idrissi, A., Niepceron, J., Masson, B., Toulemonde, C., Boulant, G., Mathieu, J.-P., Hamon, F., Michel-Ponnelle, S., Henault, J.-M., Taillade, F., Adia, J.-L., Escoffier, F.: Predicting Leakage of the VERCORS Mock-up and Concrete Containment Buildings: A Digital Twin Approach. Submitted to ICCS20–21, Prague (2021)
5. Haelewyn, J., Semete, P., Mathieu, J.-P., Escoffier, F., Michel-Ponnelle, S., Hamon, F., Buchoud, E.: A numerical clone for VeRCoRs mock-up. 23ème Congrès Français de Mécanique. Lille, France (2017)
6. Charpin, L., Courtois, A., Taillade, F., Martin, B., Masson, B., Haelewyn, J.: Calibration of Mensi/Granger Constitutive Law: Evidences of Ill-Posedness and Practical Application to VeRCoRs Concrete TINCE 2018. Paris-Saclay, France (2018)

In-situ X-CT Test on Mortar Micro-specimen Coupled with Mesoscale Numerical Simulations of Fracture

Aliaksandra Tsitova, Fabien Bernachy-Barbe, Benoît Bary, and François Hild

Abstract The mesoscale fracture modeling of concrete requires explicit descriptions of the mesostructure and the fracture behavior of its phases. To identify the fracture properties, an in situ three-point bending test on a notched beam was carried out in an X-ray scanner. The kinematic fields at different loadings and crack propagation steps were measured via Digital Volume Correlation. A realistic mesh was created for mechanical simulations to be run. The experimental displacement fields were used as kinematic boundary conditions of a region of interest around the crack. Fracture of the cement paste was modeled with a phase field method. The numerical cracking pattern is compared to the experimental observation.

Keywords Mesoscale · Crack · X-ray tomography · Digital volume correlation · Phase field model

1 Introduction

The study of creep/damage coupling requires the identification of damage models at the mesoscopic scale. The mesoscale fracture modelling of materials requires an explicit description of the mesostructure and the behaviour of its phases. In concrete, aggregates, cement paste and ITZ (Interfacial Transition Zones) need to be characterized with their intrinsic fracture properties. The calibration of these parameters is not trivial given the challenge of direct measurements of fracture energies of these three phases. In situ bending tests on concrete notched beams with direct 2D [1, 2]

A. Tsitova (✉) · B. Bary
Université Paris-Saclay, CEA, Service d'Étude du Comportement des Radionucléides, 91191 Gif-sur-Yvette, France
e-mail: aliaksandra.tsitova@cea.fr

A. Tsitova · F. Hild
LMT-Laboratoire de Mécanique et Technologie, Université Paris-Saclay, ENS Paris-Saclay, CNRS, 91190 Gif-sur-Yvette, France

F. Bernachy-Barbe
CEA, DES, IRESNE, DEC, Cadarache, 13108 Saint-Paul-lez-Durance, France

F. Kanavaris et al. (eds.), *International RILEM Conference on Early-Age and Long-Term Cracking in RC Structures*, RILEM Bookseries 31,
https://doi.org/10.1007/978-3-030-72921-9_20

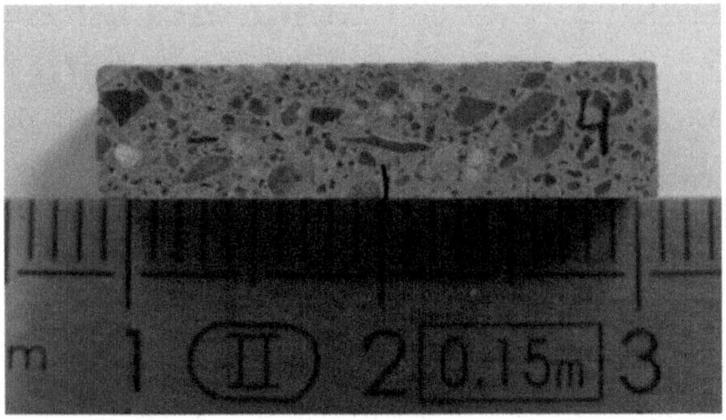

Fig. 1 Mortar microbeam

and 3D numerical simulations [3] allowed to study the influence of cement paste and ITZ fracture parameters on beam strength and crack path.

To identify the fracture properties by inverse analyses, an experimental approach that included microstructure characterization and in situ testing with controlled fracture propagation was designed. An in situ three-point bending test on a notched beam was carried out in an X-ray scanner. The test was performed on a small-scale ($5 \times 5 \times 20$ mm) mortar specimen (Fig. 1). The kinematic fields at different loadings and crack propagation steps were measured by processing the reconstructed images via regularized Digital Volume Correlation (DVC). A realistic mesh was created for mechanical simulations to be run based on the aggregate and external surface geometries obtained from tomographic scans. The experimental displacement fields were used as kinematic boundary conditions of a region of interest around the crack.

Numerical simulations were carried out using the Cast3m finite element code (www-cast3m.cea.fr) and the MFront code generator (www.tfel.sourceforge.net). Fracture in the cement paste was modelled with a phase field method. The full dataset consisting of realistic geometry and experimental boundary conditions allowed for full simulations of the performed test and for the calibration of parameter sets for the chosen damage models that could realistically reproduce the observed cracking pattern.

2 Specimen

A mortar batch was mixed using a CEM I cement at W/C = 0.525. A low sand-to-binder ratio was used (0.5). Sand was thresholded within the range 200 μm − 2 mm fraction. The reduction of sand volume fraction and elimination of fine fractions make the resulting microstructure easier and more suitable for efficient image processing as less geometric details are distorted or lost.

A thin mortar plate with dimensions $65 \times 35 \times 8$ mm was cast in a silicone mold and demolded after 1 day. By means of X-ray CT (Computed Tomography), the mortar plate was checked for absence of cracking and defects. After control, the plate was selected for further fabrication. First, the top surface was polished until a 5 mm thickness was obtained with simultaneous control of parallelism of the top and bottom faces. The thickness deviation was 0.1 mm that gives a $\angle 1.5‰$ inclination between the top and bottom surfaces. Second, the plate width was adjusted to 20 mm with a wiresaw. Third, $5 \times 5 \times 20$ mm beams were cut from the plate with a diamond disk cutter to ensure planar and parallel sides. Fourth, the notch was cut with a wiresaw equipped with a fine wire (0.1 mm) and low tension to avoid sample degradation. The notch width was approximately 0.15 mm and the height was 1.5 mm (Fig. 1). The specimens were stored at 100% air humidity. During preparations, the specimen surface was rewetted to prevent from drying cracking.

3 Microstructure Characterization

The central part of the beam was initially scanned with a 5 μm resolution, and a tube voltage of 180 kV using a GE phoenix v/tome/x m X-ray CT system. Each X-ray projection used for volume reconstruction is the mean of 5 acquired projections with 0.5 s acquisition time. The scan duration was 2.5 h. Image processing and microstructure segmentation were conducted in the Avizo® data visualization and analysis software. The sub-volume region of interest (ROI) containing the crack with dimensions $2.6 \times 5 \times 5$ mm was selected for further segmentation. First, a median filter $3 \times 3 \times 3$ pixel was applied to mitigate noise. Secondly, the segmentation of inclusions was carried out as follows. Two types of aggregates (limestone—$CaCO_3$, silica—SiO_2) are present in mortar with different morphologies and levels of X-ray attenuation that define their grey level representation on images (Fig. 2). Silica aggregates have similar gray levels as the cement paste matrix, but show less heterogeneity. The calculation of the image gradient norm gives a field with low values in homogeneous zones and high values in heterogeneous regions. This tool allows the homogeneous aggregates to be distinguished from the heterogeneous matrix. Limestone aggregates are brighter than the cement matrix and can be separated by simple thresholding. Last, the notch surface was segmented from the volume. Macroporosity was present in the specimen in a small amount and during the bending test did not affect the crack path. For these reasons, macropores were not segmented as a separate phase and were incorporated into the cement paste matrix.

Based on the segmentation results, the Avizo® tools allow surface triangular meshes to be generated and exported in .stl format. The surface meshes were reduced to the minimum possible size (300k faces for the inclusion surface mesh, 50k faces for the notch surface mesh) with simultaneous control of excessive deformation of triangles. The surface meshes were then processed with the tools from the freeware Salome® platform (https://www.salome-platform.org/) to obtain volume meshes (Fig. 3).

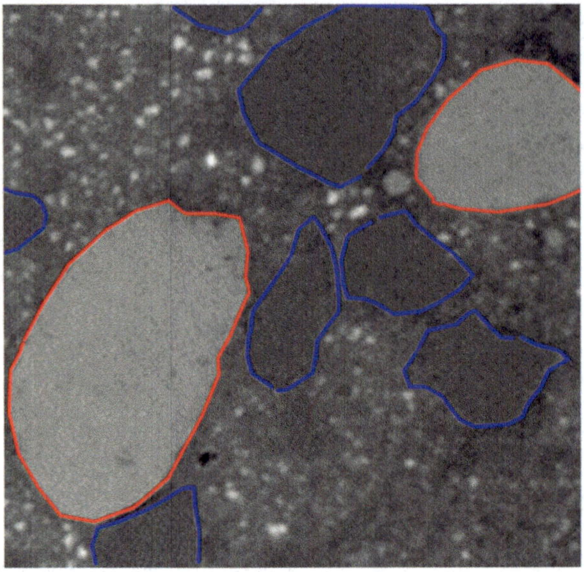

Fig. 2 Mortar morphology: red—limestone aggregates, blue—silica aggregates

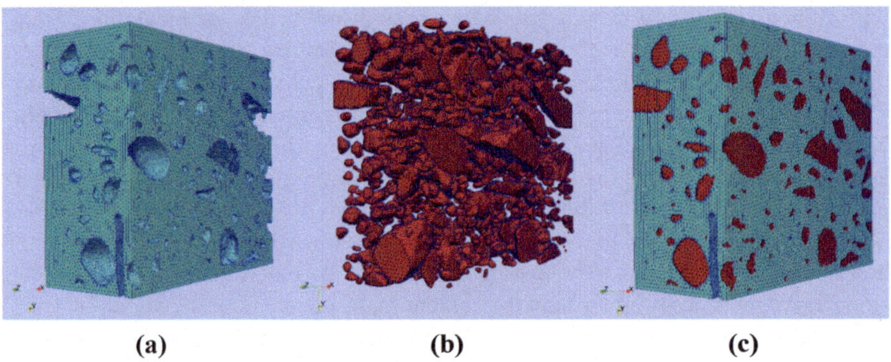

Fig. 3 Volume mesh generated with Salome® platform based on surface meshes. **a** Matrix, **b** aggregates, **c** complete mesh

4 In Situ Tests

In situ bending tests were performed with the Tension-Compression (TC) $\pm$ 1 kN capacity testing machine of LMT similar to that used in Refs. [4, 5]. The applied force was measured with the axial load sensor of the testing machine. The adapted set-up for indirect three point bending tests is presented in Fig. 4. The vertical motion of the bottom actuator with part A1 causes a horizontal motion of support parts B and C thanks to 45° contact surfaces. The shortening of the distance between B and

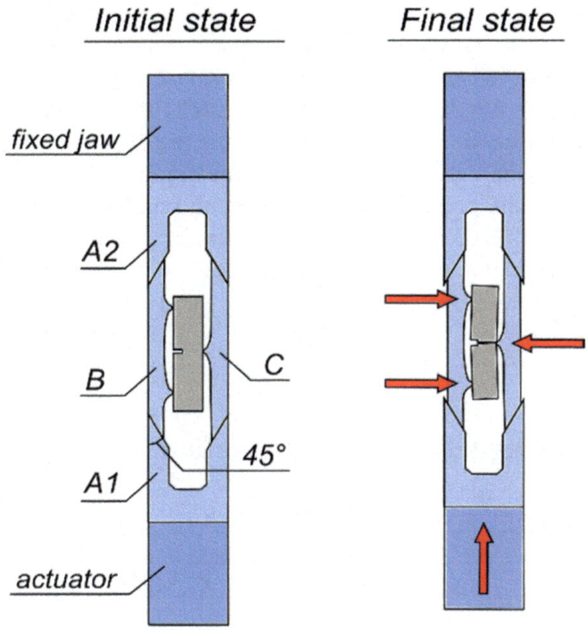

Fig. 4 Schematic view of the bending set-up

C results in point load application to the sample that is equivalent to three-point bending. The parts of the bending set-up were fabricated in aluminium alloy as it has a high stiffness and reasonable X-ray attenuation compared to mortar, which allows additional artifacts to be avoided. Adhesive tape was attached to the top and bottom faces of the beam for maintaining the sample and improving its stability during testing. Adhesive tape was also used for fastening the parts of the set-up to ensure better control over the sample positioning inside it, and for safe insertion of the set-up into the tube of TC machine.

The in situ test consists of several loading steps and CT scans performed at sustained and constant load. The scans were acquired in the same CT equipment at 200 kV. Each X-ray projection used for volume reconstruction is the mean of 3 acquired projections with 0.334 s acquisition time. The scan duration was 45 min with a resolution of 8.8 μm per voxel. First, two successive scans were performed on the undeformed specimen for uncertainty quantifications.

The experimental configuration along with the specimen geometry were not allowing for a fine control over crack initiation and propagation. Failure was brittle and followed by substantial force drop and instantaneous crack propagation on the major part of the ligament.

The second scan was carried out after fracture. Third, an increment of axial displacement of 20 μm was applied for further crack propagation and opening. After each loading, a 10 min dwell was applied before the next scan (i.e. force relaxation and short-term creep were stabilized).

The measured peak force in the presented test was 28 N. However, it is likely that the tension of applied on specimen adhesive tape and friction of aluminium parts could interfere in force measurements. For example, after fracture, the residual force does not reduce below 6 N. For modelling purposes, the peak force corrected with the residual force in the end of the test will be assumed as the indicative strength although with some irreducible uncertainty.

In the presented test, the loading point was not exactly aligned with the notch. The off-axis loading resulted in a bifurcation of the fractured surface into two main cracks (Fig. 6). In image processing, the ROI was chosen to encompass all fractured surfaces. The crack crossed the cement paste matrix, ITZ and some limestone sand particles.

5 Digital Volume Correlation

Finite Element (FE) regularized DVC was applied [6] on the scans acquired during in situ tests. DVC measures displacement fields in the ROI by registering deformed images g and the reference image f assuming grey level conservation

$$f(\mathbf{x}) = g(\mathbf{x} + \mathbf{u}(\mathbf{x})) \tag{1}$$

where $\mathbf{u}(\mathbf{x})$ is the displacement field, and $\mathbf{x}$ any pixel location within the ROI. The sought displacement field minimizes the gap to grey level conservation

$$\min_{u} \Phi_c^2 = \sum_{\text{ROI}} [g(\mathbf{x} + \mathbf{u}(\mathbf{x})) - f(\mathbf{x})]^2 \tag{2}$$

with

$$\mathbf{u}(\mathbf{x}) = \sum_{n} u_n \psi_n(\mathbf{x}) \tag{3}$$

where $\psi_n(\mathbf{x})$ are shape functions, and u_n the associated Degrees Of Freedom (DOF). In global DVC, the sought DOFs are the nodal displacements of a finite element discretization. In regularized DVC, elasticity is enforced at a local level and an additional penalty term based on mechanical equilibrium is introduced

$$\Phi_m^2 = \{\mathbf{u}\}^T [\mathbf{K}]^T [\mathbf{K}] \{\mathbf{u}\} \tag{4}$$

where $\{u\}$ is a column vector gathering all nodal displacements, and $[\mathbf{K}]$ the stiffness matrix of the regularized nodes. The grey level residuals are penalized by the equilibrium gap so that the weighted sum $\Phi_c^2 + w_m \Phi_m^2$ is to be minimized with respect be minimized with respect to $\{\mathbf{u}\}$. The weight w_m is proportional to a regularization length ℓ_{reg}^4. If ℓ_{reg} is greater than the element size, the mechanically inadmissible displacement fluctuations are filtered out over a spatial domain of size ℓ_{reg}. In the case of damage or plastic strains in the volume, inadmissible displacements are smeared over a domain that depends on ℓ_{reg} [6]. DVC was performed within the Correli 3.0 framework developed at LMT and implemented within the MATLAB™ software.

The central part of length 4 mm was cropped for DVC analyses. A homogeneous mesh was reconstructed. The external surface of the specimen was segmented in Avizo®, and a surface mesh with 10k faces was generated. Then, the surface mesh was used to construct the volume mesh with gmsh freeware (http://gmsh.info/) with four-noded tetrahedrons. The mesh dimensions were 537 × 472 × 547 voxels with 42.5k elements. The mesh size (i.e., cubic root of mean volume) was 15 voxels.

For the first iterations, the regularization length was set to a high value (120 voxels). Then, the mechanically regularized displacements fields were used as initial values for the next iterations where the regularization length was gradually decreased. The iteration process was continued until ℓ_{reg} was set to 40 voxels. Such iterative process allows most of the uncertainties to be filtered out in the first iterations and then iteratively converge to a good solution. In this study, too small ℓ_{reg} values (i.e. less than twice the element size) were observed to be ineffective for mechanical regularization.

The deformed mesh is presented in Fig. 5a and the normal strain field $\boldsymbol{\varepsilon}_{zz}$ in Fig. 5b. Strain concentration is well pronounced in the crack propagation zone. The grey level residual field represents the difference between the volume in the reference configuration and that in the deformed configuration corrected by the measured displacement field. The residual fields are reported in Fig. 6. The highest residual levels correspond to the fractured surface, which is continuous and well defined. The

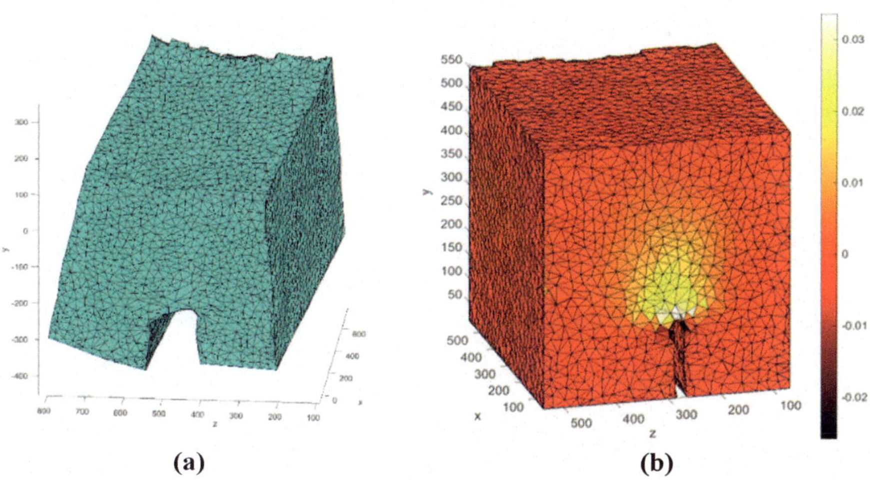

Fig. 5 **a** Deformed mesh (amplification × 18), **b** normal strain fields $\boldsymbol{\varepsilon}_{zz}$

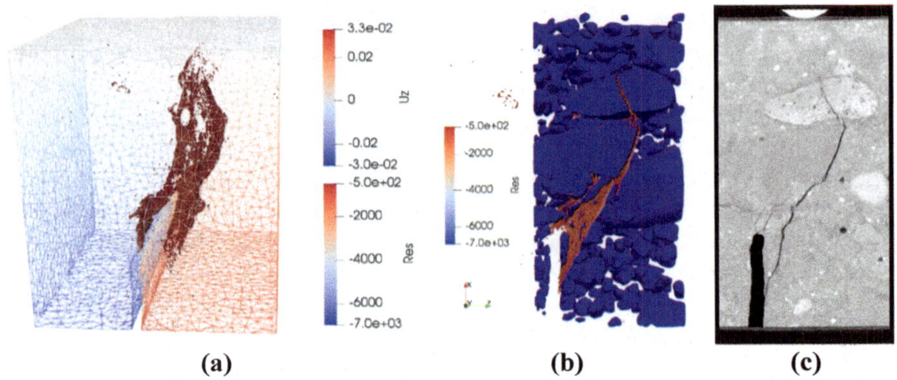

Fig. 6 **a** Residual field, **b** residual field superimposed on the microstructure, **c** original image of the crack

absence of high residuals in the intact zones proves a good overall scan quality, low influence of the short-term creep and efficient mechanical regularization of DVC. The residual field can be superimposed with the sample microstructure (Fig. 6b). It allows to track the crack propagation based on the spatial distribution of aggregates and to compare it directly with numerical damage fields.

6 Mechanical Simulations

6.1 Damage Model

The phase field approach has been proven to be an effective tool for crack modelling in brittle materials. In this section, a short review of the thermodynamic and energetic approach of Miehe et al. [7, 8] is presented. Sharp crack topologies are regularized by introducing a diffusive crack zone. The crack surface is described by a functional Γ dependent upon an auxiliary variable d (*crack phase field*) and a *characteristic length* l

$$\Gamma_l(d) = \int_{\Omega} \gamma_l(d) d\Omega \tag{5}$$

$$\gamma_l = \frac{d^2}{2l} + \frac{l}{2}|\nabla d|^2 \tag{6}$$

where γ_l is the crack surface density (with $d = 0$ characterizing the unbroken state and $d = 1$ the fully broken state of the material). To bring in continuous principles, the damage variable d is gradually varying from $x = 0$ corresponding to the cracked surface, and $x = l$, where the body is considered intact (Fig. 7).

Fig. 7 1D diffusive crack zone [8]

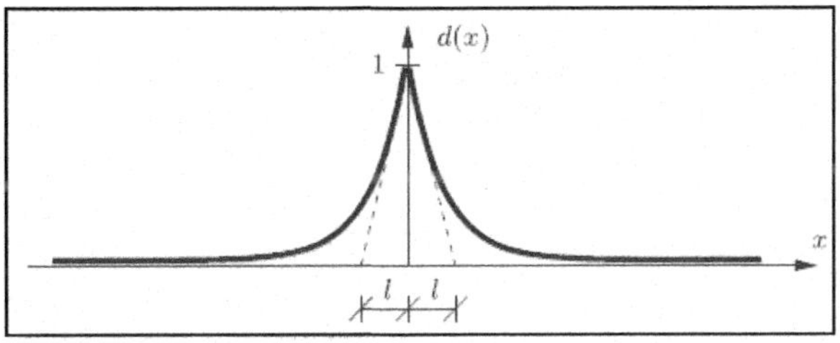

The minimization problem for the cracked surface $d = \arg\left\{\inf_{d} \Gamma^{l}(d)\right\}$ gives the crack phase field function

$$d(x) = e^{-|x|/l} \tag{7}$$

The total energy is equal to the sum of the bulk energy E_e and the energy required for crack propagation E_d. Based on Griffith's theory

$$E_d = \int_{\Omega} \psi_d d\Omega = \int_{\Omega} G_f \gamma_l(d) d\Omega \tag{8}$$

where G_f is the fracture energy. The damaged bulk energy is derived for the case of tensile induced failure with a split in positive and negative energy contributions

$$E_e(d, \varepsilon) = \int_{\Omega} \psi_e d\Omega \tag{9}$$

$$\psi_e(d, \varepsilon) = (g(d) + k)\psi_0^{+}(\varepsilon) + \psi_0^{-}(\varepsilon) \tag{10}$$

where $\psi_0^{+}(\varepsilon)$ is the "positive" extension energy, $\psi_0^{-}(\varepsilon)$ the "negative" contraction energy, $g(d) = (1-d)^2$ the *degradation function.* As $g(d)$ is specifically associated with the extension energy, it allows the stiffness recovery to be modelled in contraction due to crack closure. The variational derivative of the potential energy for irreversible process reads

$$\sigma : \dot{\varepsilon} = \dot{\psi} + \phi, \phi \geq 0 \tag{11}$$

where σ is the Cauchy stress tensor, $\dot{\varepsilon}$ the strain rate tensor, $\dot{\psi}$ the free Helmholtz power, and ϕ the dissipated power. The derivative of the free Helmholtz power is expressed as

$$\dot{\psi} = \dot{\psi}_e + \dot{\psi}_d = \frac{\partial \dot{\psi}}{\partial d} \cdot \dot{d} + \frac{\partial \dot{\psi}}{\partial \varepsilon} : \dot{\varepsilon} \tag{12}$$

The minimization of Eq. (11) gives a thermodynamically consistent phase field equation for the damage variable d

$$\left(\frac{G_f}{1} + 2\mathcal{H}_n\right)\mathrm{d} - G_f \mathrm{l}\Delta \mathrm{d} = 2\mathcal{H}_n \tag{13}$$

$$\mathcal{H}_n = \max\left\{\psi_{n-1}^+, \psi_n^+\right\} \tag{14}$$

The temporal changes of the phase field is driven by a *local history field* $\mathcal{H}$, which is equal to the maximum tensile strain obtained in the history of $\psi_0^+(\varepsilon)$.

6.2 Results

The volume mesh generated with Salome® consisted of 4.5 million elements. Performing nonlinear calculations on such fine meshes demands large computing resources. For the first trials, the size of the mesh was reduced from dimensions $2.6 \times 5 \times 5$–$2.6 \times 5 \times 1$ mm and from 4.5M elements to 800k. The experimental displacement fields measured via DVC were applied to the top and side faces. The displacement field measured directly after failure is linearly interpolated over 100 time steps. The final displacement field corresponds to time $\mathrm{t} = 1$.

In this study, all aggregates were considered to be linear elastic, although the limestone aggregates located on the crack path were observed to fail. The modulus of elasticity of the aggregates was estimated by nanoindentation method. The phase field model was applied to the cement paste matrix. The Young's modulus and Poisson's ratio for the cement paste were equal to experimentally measured values [9]. The cement paste matrix was assumed to be brittle at mesoscale. Therefore, the characteristic length l was taken to be small while satisfying the necessary condition to be at least twice the element size (i.e. $l = 0.05$ mm). Then, the fracture energy G_f was calibrated so as to retrieve the experimental peak force (Fig. 8). All parameters are gathered in Table 1.

Despite the abovementioned simplifications, the cracking pattern was correctly reproduced due to the representative microstructure and boundary conditions (Fig. 9). When only the matrix was subjected to damage, the choice of fracture parameters did not alter significantly the crack path. However, it may be changed by the introduction of interface elements whose behavior is different from that of the matrix. Crack bifurcation was not reproduced as the major part of the microstructure was omitted in this study.

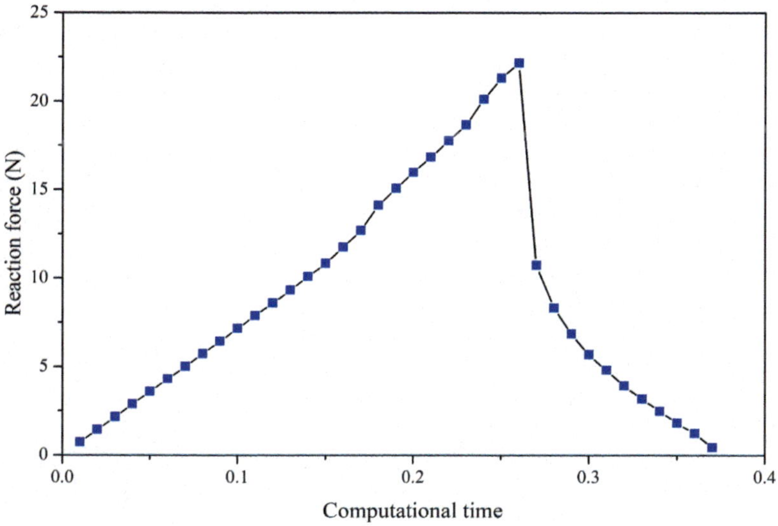

Fig. 8 Numerical reaction force history

Table 1 Models and input parameters for the mesh phases

Phase	Matrix	Limestone aggregate	Silica aggregate
Constitutive model	Phase field model	Linear elastic isotropic	
E, GPa	13.3	50.0	70.0
ν, –	0.25	0.2	0.2
$G_{f,m}$, N/m	4.0	–	–
l, μm	50	–	–

7 Conclusion

To calibrate the fracture properties by inverse analyses, an in situ three-point bending test on a notched beam was carried out in an X-ray scanner. The test was performed on a small-scale mortar specimen (5 × 5 × 20 mm). The kinematic fields were measured at different loadings and crack propagation steps by processing the reconstructed images with mechanically regularized DVC. A realistic mesh was created for mechanical simulations to be run based on the aggregate and external surface geometries obtained from tomographic scans. The experimental displacement fields were used as kinematic boundary conditions of a region of interest around the crack.

In further studies, mechanically regularized DVC will be adapted to heterogeneous materials. Realistic volume meshes adapted for DVC will be constructed. Mechanical regularization will take into account the difference in elastic properties of the different phases of mortar such as HCP matrix and aggregates. The measured kinematic fields

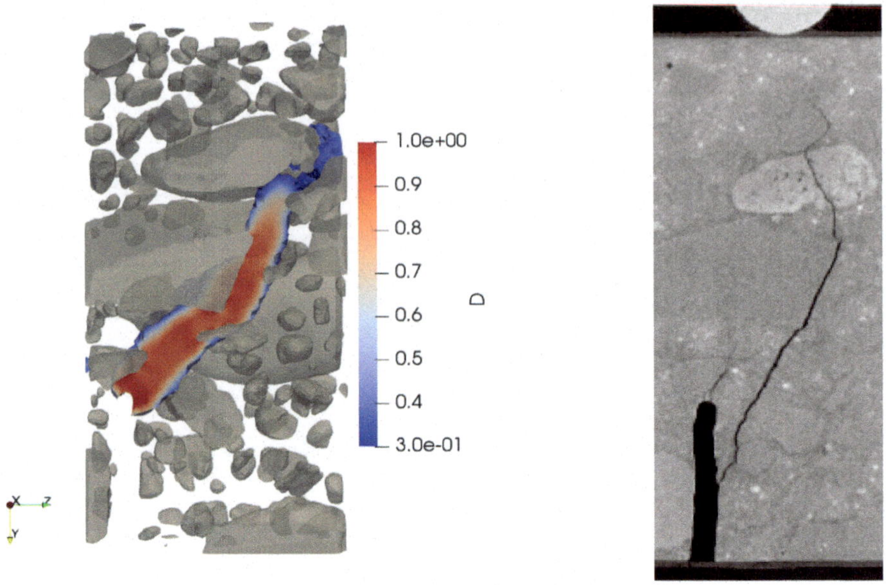

(a) (b)

Fig. 9 **a** Damage field, **b** original image of the crack

will be able to represent the discontinuity in strain fields and strain heterogeneities related to the underlying microstructure. The modified method will improve the trustworthiness of displacement fields to be checked with the residual fields.

Fracture in the cement paste was modelled with a phase field damage approach for brittle material. In further studies, debonding of ITZs will be accounted by a Cohesive Zone Model (CZM). The interface parameters will be calibrated relative to the bulk damage model. A damage model will be also applied to limestone aggregates to further improve the reproduction of the crack path. The identified parameters will have to be validated on larger ROIs and other bending tests.

Acknowledgements The authors gratefully acknowledge financial support from EDF and CEA.

References

1. Skarżyński, Ł., Tejchman, J.: Experimental investigations of fracture process in concrete by means of x-ray micro-computed tomography. Strain **52**(1), 26–45 (2016). https://doi.org/10.1111/str.12168
2. Trawiński, W., Bobiński, J., Tejchman, J.: Two-dimensional simulations of concrete fracture at aggregate level with cohesive elements based on X-ray μCT images. Eng. Fract. Mech. **168**, 204–226 (2016). https://doi.org/10.1016/j.engfracmech.2016.09.012

3. Trawiński, W., Tejchman, J., Bobiński, J.: A three-dimensional meso-scale modelling of concrete fracture, based on cohesive elements and X-ray μCT images. Eng. Fract. Mech. **189**, 27–50 (2018). https://doi.org/10.1016/j.engfracmech.2017.10.003
4. Buffiere, J.-Y., Maire, E., Adrien, J., Masse, J.-P., Boller, E.: In situ experiments with x ray tomography: an attractive tool for experimental mechanics. Exp. Mech. **50**(3), 289–305 (2010). https://doi.org/10.1007/s11340-010-9333-7
5. Buljac, A., et al.: Digital volume correlation: review of progress and challenges. Exp. Mech. **58**(5), 661–708 (2018). https://doi.org/10.1007/s11340-018-0390-7
6. Taillandier-Thomas, T., Roux, S., Morgeneyer, T.F., Hild, F.: Localized strain field measurement on laminography data with mechanical regularization. Nucl. Instrum. Methods Phys. Res., Sect. B **324**, 70–79 (2014). https://doi.org/10.1016/j.nimb.2013.09.033
7. Miehe, C., Hofacker, M., Welschinger, F.: A phase field model for rate-independent crack propagation: Robust algorithmic implementation based on operator splits. Comput. Methods Appl. Mech. Eng. **199**(45), 2765–2778 (2010). https://doi.org/10.1016/j.cma.2010.04.011
8. Miehe, C., Welschinger, F., Hofacker, M.: Thermodynamically consistent phase-field models of fracture: variational principles and multi-field FE implementations. Int. J. Numer. Meth. Engng. **83**(10), 1273–1311 (2010). https://doi.org/10.1002/nme.2861
9. Tsitova, A., Bernachy, F., Bary, B., Hild, F.: Experimental and numerical study of crack propagation with the phase field method: application to three-point bending test. In: *Proceedings of 10th international conference on fracture mechanics of concrete and concrete structures*, Bayonne, France, p. 7 (2019). https://doi.org/10.21012/fc10.233322

Numerical Analysis of Size Effects on the Stabilized Cracking Stage in RC Members

David E. M. Bouhjiti, Noam Demri, Miquel A. Huguet, Silvano Erlicher, and Julien Baroth

Abstract The control of random cracking in reinforced concrete members is of crucial interest to assess their serviceability and durability. From a regulatory point of view, such control within the design phase is ensured using a set of guidelines (definition of the reinforcement ratio, limitation of the crack width, limitation of the deflection, etc.). One can find corresponding formulae in the Eurocode2 (Design of concrete structures—Part 1-1: General rules and rules for buildings, [1]), Code Models (fib Model Code for concrete structures, [2]) or others. On the one hand, those regulatory formulations do not provide accurate results for all concrete types as the used experimental results for fitting show a large discrepancy. On the other hand, and mostly, they do not account for size effects observed on the tensile strength of concrete. This contribution aims at analyzing size effects on the cracks' distribution in axially reinforced concrete members subjected to pure tension. For the most part, the paper presents an alternative approach (to existing regulatory formulae) based on the use of spatially correlated Weibull random field coupled to an energetic-statistical size effect law. The main interest is geared towards the estimation of the mean and maximal spacing values once the stabilized cracking stage is reached. The method is validated based on the set of tests performed by Farra and Jacoud (Rapport des essais de tirants sous déformation imposée de courte durée. Projet: Influence du béton et de l'armature sur la fissuration des structures en béton. Publication N°140. IBAP. Ecole Polytechnique Fédérale de Lausanne, [3]). The obtained results show a logarithmic decrease of the mean and maximal spacing values with the effective size of the ties which should be considered for future regulatory formulations.

Keywords Random cracks' spacing · Reinforced concrete members · Axial tension · Spatial correlation · Size effects

D. E. M. Bouhjiti (✉) · N. Demri · M. A. Huguet · S. Erlicher
EGIS Industries, 4 Rue Dolorès Ibarruri, 93188 Montreuil Cedex, France
e-mail: david.bouhjiti@egis.fr

J. Baroth
Laboratoire Sols, Solides, Structures (3SR), Gières, France

F. Kanavaris et al. (eds.), *International RILEM Conference on Early-Age and Long-Term Cracking in RC Structures*, RILEM Bookseries 31,
https://doi.org/10.1007/978-3-030-72921-9_21

1 Brief State of the Art

In current engineering practice, the control of the cracking state is aimed at using a set of design guidelines (Eurocode 2 [1]; Model Code 1990; Model Code 2010 [2] and others) to limit the stress state, the crack width and the deflection level in reinforced concrete elements. It all comes eventually to defining an accurate reinforcement ratio ρ to keep a maximal spacing value $s_{r,\max}$ smaller than a given criterion for a given Limit State s_r^{LS}. So, the challenge consists of predicting for each concrete type what would be its associated maximal spacing representing a stabilized cracking state and allowing the quantification of the maximal crack opening based on the strain gap between steel and concrete.

The maximal spacing value is a random quantity; meaning that one gets a maximal value given a certain probability of exceedance or a confidence interval. The uniqueness of such value is only obtained when the probability threshold (on the high side of the mean value) is fixed. Accordingly, for an accurate measurement of the maximal spacing for a given concrete type, a large set of tests should be achieved to explore tail distributions. As this remains hard to achieve, usual practice consists of considering a limited number of experimental tests to quantify the mean spacing value $s_{r,m}$ and associate an amplification factor $\lambda = s_{r,\max} / s_{r,m}$ (from 1.5 in Model Code 1990 to 1.7 in Eurocode 2 and Model Code 2010) representing the tail distribution to the best of our knowledge. However, one can note that such empirical approaches—or at best semi-empirical—do not allow a reliable quantification of the risks associated with the control of cracking in structures. Indeed, they remain intrinsically dependent on the selected experimental design plan, which explains the observed discrepancies among existing regulatory codes [4].

To overcome the limitations related to the experimental characterization of tail distributions of cracks' spacing values, numerical analyses are usually of interest. In the literature, one can find several approaches to deal with cracks' spacing quantification. Early works on the topic involve the demonstration of the experimentally observed dependence of the mean spacing $s_{r,m}$ on the ratio φ/ρ of steel diameter φ to the effective reinforcement ratio ρ (bond theory in [5]) and its dependence on the cover distance c (no bond theory in [6]). In terms of recent advancement on the matter, [7] proposes a model based on a steel to concrete transfer length and demonstrates theoretically the linear relationship between the mean spacing value $s_{r,m}$ and the product of ratios $(\varphi/\rho)(R_t/\tau_{\max})$ with φ the bar diameter, ρ the reinforcement ratio, R_t the tensile strength and $\tau_{\max}$ the maximal steel-concrete bond. The same formulation is retained in Model Code 2010 [2]. Yet, the constants of the linear equation remain defined based on empirical fitting (best-fit equation based on statistical analysis of experimental data). Later on, [8] extended the model in [7] to a probabilistic framework using a normal and uncorrelated spatial random field associated with the tensile strength R_t over a discretized tie's length. Then, through several runs, authors compute iteratively the positions of cracks and define the probability density function of spacing values. Though it is an interesting approach, authors did not consider any size effects in their developments nor justify the nature of the used random field. Indeed, the tensile strength follows rather a Weibull distribution than a normal one, with different tail behaviors [9]. Also, [7, 8] neglected the spatial correlation of R_t which is a strong hypothesis given experimental evidence on the matter [10]. Such spatial correlation is directly linked to the material intrinsic heterogeneity (voids and aggregates spatial distribution).

Hence, the main purpose of this paper is to suggest a new theoretical and practical methodology allowing the accurate and physical prediction of cracks' spacing distribution in reinforced concrete structures. The foreseen improvements concern:

(a) The introduction of size effects in the physical formulation of the model.
(b) The introduction of the spatial distribution of the tensile strength in a realistic way (spatially correlated Weibull distribution).
(c) The probabilistic coupling using non-intrusive techniques.
(d) The validation of numerical results based on various concrete types and reinforcement ratios.

2 The New Simplified Model

The proposed model hereafter is based on the same basic principle as the one used in [7] using a representative reinforced concrete tie with a length l, a concrete section of A_c, a steel section of A_s (associated to a diameter φ) and a bond-slip law subjected to a normal and axial load N (Fig. 1).

2.1 Theoretical Aspects

Given the hypotheses of perfectly brittle concrete behavior, linear elastic behavior of steel prior to crack stabilization and a linear bond-slip law ($\tau_b(s) = a\,s$), one can easily solve the bond-slip differential equation. In terms of stress distribution in concrete between two cracks (at positions $-\frac{l}{2}$ and $\frac{l}{2}$ with the element's origin at mid-distance), this writes:

$$\sigma_c(x) = \frac{N \pi \varphi}{A_c\, A_s\, E_s}\left(1 - \frac{\cosh\left(\sqrt{a\chi}\, x\right)}{\cosh\left(\sqrt{a\chi}\, \frac{l}{2}\right)}\right) \tag{1}$$

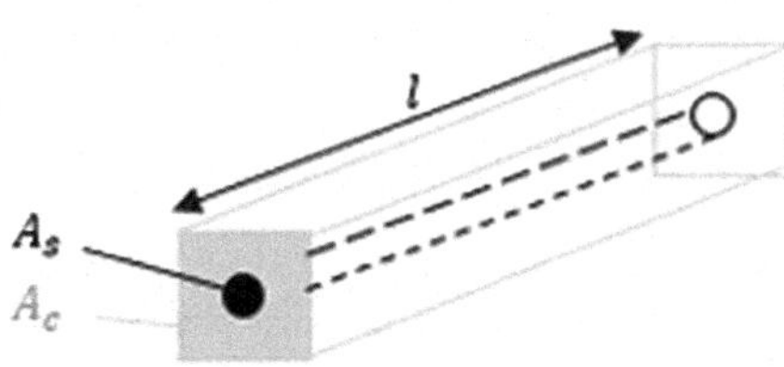

Fig. 1 Visualization of the tie element between two boundaries (c and s stand for concrete and steel materials respectively)

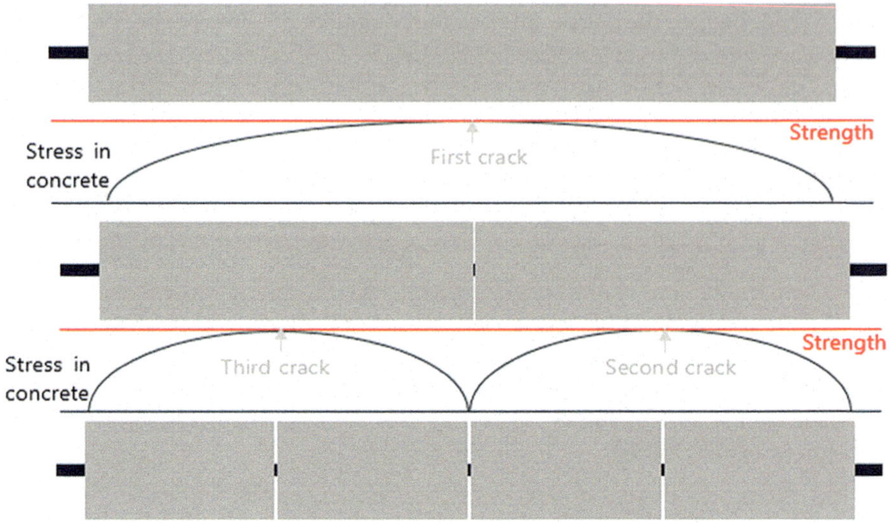

Fig. 2 Evolution of the cracking state within a deterministic framework

One obtains a new crack within the element length l when the stress level σ_c at a given position reaches the material tensile strength at the same point: $\sigma_c = R_t$. In the absence of spatial variation, the problem is solved iteratively as depicted in the following figure (Fig. 2).

To be more physically representative, two main improvements are introduced:

(a) Definition of the tensile strength in concrete using an adapted Size Effect Law. In this work, we are only interested in Type I size effect (denoted also energetic-statistical size effect in [11]) in the absence of deep notches or a large traction-free cracks in the structural volume (this is in line with the multi cracking behaviour observed in the case of 1D-tensionned ties). The size effect law generally involves an effective volume V at the structural element scale, a reference volume V_0 at which the measurement of the tensile strength is achieved $R_{t,0}$ and its associated Weibull modulus m.

$$R_t(V) = R_{t,0}\left(\frac{V}{V_0}\right)^{-\frac{1}{m}} \quad (2)$$

The effective volume is bounded on the upper side so as to a have a non-null and minimal tensile strength for high volumes and on the lower side so as to have a maximal nominal strength for small volumes. Such effective volume is estimated using the criterion defined in the WL^2 model in [12].

(b) Definition of a spatially correlated Weibull random field associated to the tensile strength using an exponential quadratic correlation function a given fluctuation length l_{flu} [13–15] (Fig. 3).

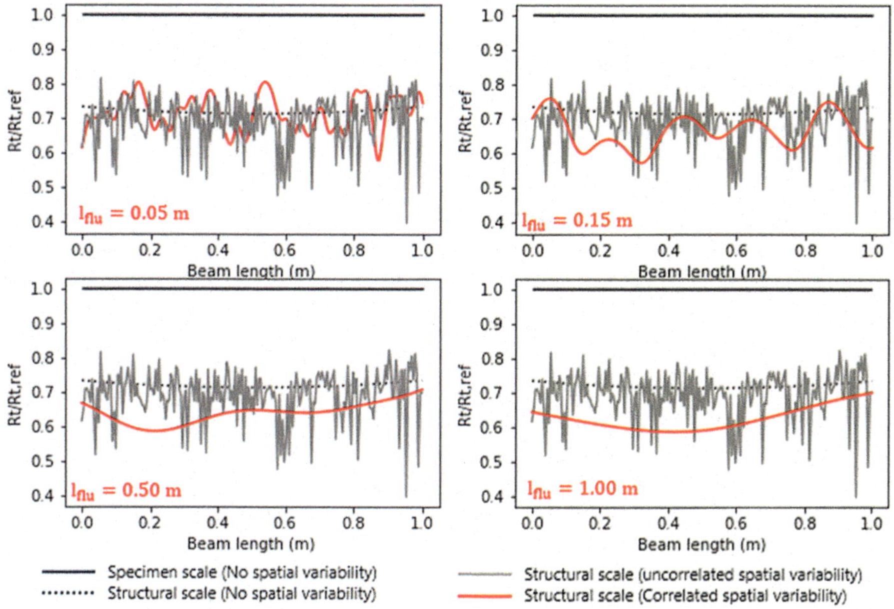

Fig. 3 Effect of spatial correlation on the Weibull random field realizations

2.2 Step-by-Step Instructions for a Calculation

- Define a representative tie section and the material properties.
- Discretize the tie into finite elements (the number of elements should be sufficient to describe objectively the spatial correlation and the number of cracks).
- Evaluate size effects on the tensile strength in the tie.
- Generate several realizations of the random field associated to the tensile strength.
- For each realization, use an iterative process to increase the axial load activating one crack at a time (where the stress in concrete reaches the tensile strength) until the stabilized stage is reached.
- Consider all cracks' position to post-process the probability density function and achieve reliability analysis if necessary. (one should note that for a number of simulations N_{simu}, one can only quantify tail probabilities higher than $p_f = 10^{2-\log_{10} N_{simu}}$).

3 Validation Based on the Experimental Work of Farra and Jaccoud [3]

In the experimental work of Farra and Jacoud [3], several reinforced concrete ties are subjected to tensile loads for cracks' spacing identification at the stabilized state.

The cross section of these ties is 10 cm × 10 cm. The length is 1.15 m. Three reinforcement ratios (0.79, 1.56, and 3.24%) are considered. Seven concrete types with compressive strength between 29.9 and 55.4 MPa are explored. The reference tensile strength values are measured using cylindrical specimen of 16 cm × 32 cm under direct tensile loads. Finally, for each specimen, three tests are performed. This is rather limited for an objective experimental probabilistic quantification. However, the observed results can still be considered for the model validation; especially in terms of physical tendencies and the quantitative analysis of the mean spacing values.

Numerical simulations are limited herein to $N_{simu} = 1000$ realizations of the Weibull random field for each tie.

- In terms of the mean spacing values the obtained results are summed up in the following table (Table 1).
  Predictive results are compared to the ones obtained using the Model code formula. Compared to experimental results, the Model Code shows a mean absolute error of 1.2 cm; whereas the proposed model leads to a mean absolute error of 2.8 cm. One can note that the fully predictive model shows a satisfactory estimation of the mean spacing compared to other explicit models fitted empirically to the data base. Nevertheless, for those empirical models, such accuracy is not guaranteed if other data are considered. It is important to underline that size effects are not that pronounced at the scale of Farra and Jacoud ties ($R_t(V_{eff})/R_{t,0} \approx 0.93$) which explains why empirical models using the tensile strength at the specimen scale still work accurately.
- As for the randomness of the spacing values for each tie, the Model Codes applied within a probabilistic framework underestimates the variation of the spacing values. In part, this is directly due to the lack of spatial variation in the Model Code formula. On the other hand, the newly proposed model shows higher variation that depends on the considered concrete type and the measured randomness of the tensile strength.
- Also, given the probabilistic framework of the newly suggested method, one can access to the cumulative distribution functions of the cracks' spacing values. Based on a selected probability of failure, one can also define a maximal spacing value not to be exceeded for the crack control within the design phase. In practice, one selects a $1 - \alpha$ quantile which represents the risk of having values at tail distribution. For the spacing values, the interest is geared towards the high range of values (as it increases the crack opening values). So, by giving the value of α, one defines a probability of not exceeding a maximal value corresponding to the $1 - \alpha$ quantile. One should note that the value of α varies depending on the considered application and the operational context (Fig. 4).

Finally, one can note that, compared to other methods, the suggested model offers a rigorous theoretical framework to (a) compute the crack spacing values in a fully predictive way based on measurements achieved at the specimen scale (empirical fitting is absent at the tie scale) (b) predict representative variation of the spacing values accounting for size effects and for the spatial randomness of the tensile strength

Table 1 Mean spacing values and their coefficients of variation: experimental versus numerical

Tie reference	Experimental	Models			
		New model (Fully predictive)		Model code (Empirical)	
	Mean value (mm)	Mean value (mm)	CoV (%)	Mean value (mm)	CoV (%)
N10-10	237	213	21	214	3
N10-14	173	146	19	174	2.6
N10-20	129	118	19	138	2.2
N20-10	222	238	22	221	3.6
N20-14	146	156	20	179	3.2
N20-20	141	118	18	142	2.7
N30-10	208	233	23	216	3.6
N30-14	180	150	20	175	3.1
N30-20	142	114	17	140	2.6
N40-10	193	239	22	215	2.4
N40-14	180	147	20	175	2.1
N40-20	142	110	16	139	1.8
N12-10	238	200	24	191	4.8
N12-14	165	129	21	158	4.1
N12-20	153	104	12	128	3.4
N22-10	192	197	25	187	5.2
N22-14	168	126	21	154	4.4
N22-20	131	102	11	126	3.6
N32-10	236	194	24	185	3.7
N32-14	152	120	20	153	3.1
N32-20	137	100	8	125	2.6
N42-10	208	213	25	191	4.3
N42-14	160	126	20	158	3.7
N42-20	130	101	9	128	3.0

(c) predict the full cumulative distribution function (or probability distribution function) to deduce the maximal spacing value not to be exceeded for a given probability of exceedance.

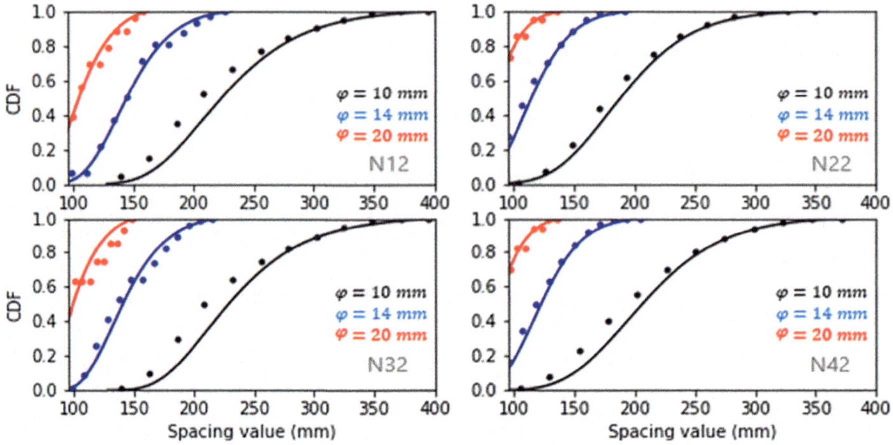

Fig. 4 Cumulative distribution functions (CDFs) of the spacing values

4 A New Explicit Regulatory Formulation for the Cracks' Spacing Estimation

In this part, it is proposed to investigate numerically the size effects on the computed mean and coefficient of variation of the spacing values. To do so, the same reinforcement ratios defined by Farra and Jacoud [3] are retained and extended to the following numerical design plan for the following tie sections: 0.1 x 0.1 (reference tie), 0.2 x 0.2, 0.5 x 0.5, 0.8 x 0.8, 1.2 x 1.2. One should note that the last sections are rarely encountered in standard engineering applications and are merely considered to explore the full size effect law (for small and large volumes). For the sake of conciseness, results' analyses are limited to the evolution of the mean spacing value and the coefficient of variation with the size of the tie. Ultimately, numerical results are fitted to an explicit formula (to facilitate its application in engineering applications). The foreseen function shape is inspired from the Model Code formulae [2] with the introduction of size effects parameters (based on the uses size effect law).

$$s_{rm} = A_1 * c + A_2\left(\frac{R_{t,0}}{\tau\mathrm{max}}\right)\left(\frac{V}{V_0}\right)^{-\frac{1}{m}}\left(\frac{\varphi}{\rho}\right) \tag{3}$$

with c the cover distance, $R_{t,0}$ the tensile strength measured at the reference volume V_0, τ_{max} the maximal bond stress.

Based on the achieved numerical calculations, the unknown parameters verify $A_1 = 1.36$, $A_2 = 0.65$ and $m = 0.2 + 1.2/CoV_{R_t}$ with $\mathrm{CoV}_{\mathrm{R_t}}$ the measured coefficient of variation of the tensile strength (Fig. 5).

This illustrates the logarithmic evolution (with a slope of −1/m) of the mean spacing value with the effective volume.

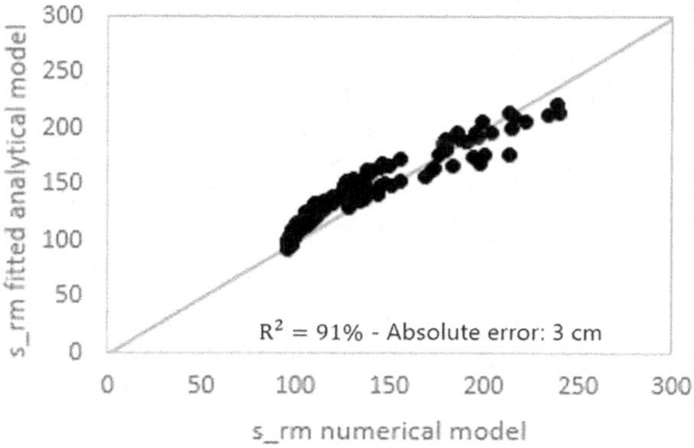

Fig. 5 Comparative analysis of numerical and fitted mean spacing values

As for the maximal spacing value, defined using the ratio $\lambda = s_{r,\max}/s_{r,m}$ in current regulatory codes, the proposed model allows the identification of the following relationship (associated to a Lognormal distribution of spacing values):

$$\lambda(\alpha) = \frac{\exp\left(\sqrt{2}\sigma k_\alpha + \mu\right)}{s_{rm}} \tag{4}$$

with $k_\alpha = 1.1e^{-1,92\alpha}$, $\mu = \ln\left(s_{rm}/\sqrt{1 + CoV_{sr}/s_{rm}}\right)$, $\sigma = \sqrt{\ln(1 + CoV_{sr}/s_{rm})}$ and $CoV_{sr} = 0.279CoV_{R_t} + 0.19$.

Eventually, as one defines a given $1 - \alpha$ quantile, one gets a corresponding ratio λ and a maximal spacing threshold for design purposes. As one can see, the values of λ are not constant and not forcibly equal to 1.5 and 1.7 as provided in Eurocode 2 [1] and Model Codes [2]; at least not for low values of α. In other words, the values in regulatory codes refer to high values of $20\% \leq \alpha \leq 30\%$ (and not 5% as usually used in engineering applications).

In fine, by using the set of equations here above, one can easily deduce the mean and maximal spacing values of a given tie characterized by its size, stress distribution (tensile vs. bending), ratio φ/ρ, variation of the tensile strength CoV_{R_t} and a given probability of exceedance or a given $1 - \alpha$ quantile.

5 Conclusion

The main aim of this research paper is to propose a probabilistic formulation of the crack's spacing values for axially reinforced members subjected to pure tensile loads. Such effort is of interest to enhance the predictive crack control in terms of mean and maximal crack spacing values and its associated crack openings. To reach the foreseen goal, the following theoretical advancements are achieved (a) considering an analytical solution for the bond-slip differential equation based on a linear bond-slip law (b) considering an energetic-statistical size effect law relating the tensile strength property to the effective structural volume (c) considering a spatially correlated Weibull random field associated to the tensile strength property. By the use of Monte Carlo Methods (which involves in this work the generation of several random fields), the stochastic model leads to the identification of a probability density function of the spacing values. From this distribution, one gets a mean estimate, a coefficient of variation and, given a certain probability threshold, the maximal crack spacing not to be exceeded.

Eventually, by exploring an extended numerical design plan, a set of equations is suggested for engineering applications. These expressions relate the mean spacing value to the ratio φ/ρ and the ratio $R_{t,0}/\tau_{\max}$ as already suggested in the Model Code formulae. In this work, it is recommended to include also a term proportional to the volume of the tie to the power (−1/m) to account for size effects on the tensile strength. As for the maximal spacing value, a theoretical approach is proposed based on the hypothesis of a Lognormal distribution of spacing values and of a coefficient of variation strongly correlated to the coefficient of variation of the tensile strength. Both of these hypotheses are verified in the present work by numerical means.

References

1. NF EN 1992-1-1 Eurocode 2—Design of concrete structures—Part 1-1: General rules and rules for buildings
2. Model Code. fib Model Code for concrete structures. 1990 & 2010
3. Farra, B., Jacoud J.-P.: Rapport des essais de tirants sous déformation imposée de courte durée. Projet: Influence du béton et de l'armature sur la fissuration des structures en béton. Publication N°140. IBAP. Ecole Polytechnique Fédérale de Lausanne (1993)
4. Lapi, M., Orlando, M., Spinelli, P.: A review of literature and code formulations for cracking in R/C members. Struct. Concr. **19**(5) (2018). https://doi.org/10.1002/suco.201700248
5. Watstein, D., Parsons, D.E.: Width and spacing of tensile cracks in axially reinforced concrete cylinders. Research **31** RP1545 (1943)
6. Broms, B.B.: Technique for Investigation of Internal Cracks in Reinforced Concrete Members, vol. 62. ACI (1965)
7. Beeby, A.W.: The influence of the parameter φ/ρeff on crack widths. Struct. Concr. **5**(2) (2004). https://doi.org/10.1680/stco.2004.5.2.71
8. Beeby, A.W., Scott, R.H.: Insights into the cracking and tension stiffening behaviour of reinforced concrete tension members revealed by computer modelling. Mag. Concr. Res. **56**(3) (2004). https://doi.org/10.1680/macr.2004.56.3.179
9. Bažant, Z.P., Le, J.-L.: Probabilistic Mechanics of Quasibrittle Structures. Strength, Lifetime, and Size Effect. Cambridge University Press (2017). https://doi.org/10.1017/9781316585146
10. Vořechovský, M.: Simulation of simply cross correlated random fields by series expansion methods. Struct. Saf. **30** (2008). https://doi.org/10.1016/j.strusafe.2007.05.002
11. Bažant, Z.P., Chen, E.P.: Scaling of structural failure. Appl. Mech. Rev. **50**(10) (1977). https://doi.org/10.1115/1.3101672
12. Sellier, A., Millard, A.: Weakest link and localisation WL2: a method to conciliate probabilistic and energetic scale effects in numerical models. **18** (2014). https://doi.org/10.1080/19648189.2014.906368
13. Baroth, J., Schoefs, F., Breysse, D.: Construction Reliability: Safety, Variability and Sustainability, pp. 21–37. Edition Lavoisier (2011)

14. Loève, M.: Probability Theory, 2nd edn. Princeton, NJ, Van Nostrand (1960)
15. Nataf, A.: Détermination des distributions dont les marges sont données. Comptes-rendus hebdomadaires des séances de l'académie des sciences, 42–43 (1962)

Numerical Estimation of Crack Openings in Concrete Structures Under Multi-physic Loadings

François Soleilhet, Farid Benboudjema, Xavier Jourdain, and Fabrice Gatuingt

Abstract Cement based material structures are designed to work in pair with steel reinforcement. The cracking of reinforced concrete is a normal phenomenon given the working principle of this material. But, for aesthetic or durability reasons, the openings of these cracks must be controlled. In the modelling, there is often an ambiguity between damage (micro-cracking at the scale of a REV) and crack opening at the scale of the material. The goal of this study is to estimate the crack opening in concrete structures under various multi-physic loadings. To do so, a method based on continuum mechanics is developed. This method post-processes the mechanical damage induced by the diverse loadings. To illustrate the procedure, a numerical application is run on an experimental reinforced shear wall made of ordinary concrete submitted to drying prior to loading. A hygro-mechanical model previously developed by the authors is firstly calibrated on experimental data and secondly used to compute the shear wall's damage induced by drying shrinkage. The resulting computed crack openings are compared to the experimental results. Interestingly, the post-processing technique managed to locate the areas where cracks are of major importance. However, the assumption of perfect adhesion between concrete and steel leads to a slight underestimation of the amplitude of the crack opening. In a first attempt to capture cracks, the proposed method based on continuum mechanics seems to be a good alternative to sophisticated methods.

Keywords Concrete · Damage · Crack opening · Multi-physic loadings

F. Soleilhet (✉)
EDF R&D, Les Renardières, 77250 Ecuelles, France
e-mail: francois.soleilhet@edf.fr

F. Soleilhet · F. Benboudjema · X. Jourdain · F. Gatuingt
CNRS, LMT – Laboratoire de Mécanique et Technologie, Université Paris-Saclay, ENS Paris-Saclay, 91190 Gif-sur-Yvette, France

F. Kanavaris et al. (eds.), *International RILEM Conference on Early-Age and Long-Term Cracking in RC Structures*, RILEM Bookseries 31,
https://doi.org/10.1007/978-3-030-72921-9_22

1 Introduction

Cementitious material are often used in pair with steel reinforcement. They are thus intended to crack. These cracks are of importance regarding durability issues. To control and maintain durability, design codes and standards give recommendations regarding crack opening.

Nowadays, the assessment of the mechanical state of structure is regularly obtained with numerical simulations. Different approaches with different levels of complexity are available to model concrete cracks [1]. To assess the mechanical state of a structure, damage modelling can be used. This method is affordable regarding calculation time and allows to model correctly the concrete failure process. However, the main drawback of this method is that it does not model explicitly the cracks and gives only information regarding diffuse micro-cracks with the damage index. To extract crack opening from damage model, post-processing technics were developed [2, 3]. Nevertheless, these methods were initially formulated for mechanical loadings only. As soon as additional strains (thermal, swelling, shrinkage) are involved, some deviations might come. The aim of this contribution is to provide a method that allows to estimate the crack opening in concrete structures under various multi-physic loadings.

The study is divided into two parts. First, the damage mechanical modelling is presented and the crack opening routine is introduced. Second, applications are run: a rod is submitted to thermal and hydric loadings then a shear wall is studied under drying.

2 Concrete Crack Modelling

2.1 Mechanical Modelling

The mechanical model used in this work is an isotropic damage model; the decrease of the material stiffness is described by the evolution of a variable D ranging from 0 to 1. In an isotropic case, this scalar variable is introduced in the behaviour law "Eq. (1)".

$$\sigma = (1 - D)\mathbb{C} : \varepsilon \tag{1}$$

with $\mathbb{C}$ the four order stiffness tensor undamaged. The evolution of the damage depends on a load threshold function: $f = \varepsilon_{eq} - \kappa(D)$, where κ the hardening-softening parameter is equal to the maximum of ε_{eq} or ε_{d0} and ε_{d0} the tensile damage threshold corresponding to f_t/E. The equivalent strain (ε_{eq}) is taken as proposed by Mazars [4] and the evolution of damage in tension is taken as:

$$D = 1 - \frac{\epsilon_{d0}}{\epsilon_{eq}} \exp\left(-B_t\left(\epsilon_{eq} - \epsilon_{d0}\right)\right) \tag{2}$$

ε_{d0} (-) the damage threshold in tension and B_t, (-) the model parameters controlling the post-pic phase evolution of the material behaviour. To avoid mesh dependency and non-uniqueness of the solution due to softening behaviour of concrete, an energetic regularization is used [5, 6]. It is based on the parameter B_t (-) computed as a function of the size of element h (m), the tensile strength f_t (Pa), the fracture energy G_f (J.m^{-2}) and the threshold ε_{d0} (-):

$$B_t = \frac{hf_t}{G_f - \frac{h\epsilon_{d0} f_t}{2}} \tag{3}$$

2.2 Crack Opening Extraction Model

To compute the crack opening, let us assume that for a given time step *i*, the stress tensor, the scalar damage variable were computed. The post-processing method is composed of the following steps:

1. The total strain is computed with regard to the stress and the damage variable:

$$\varepsilon = [(1 - D)\mathbb{C}]^{-1} : \sigma \tag{4}$$

2. The elastic strain is computed according to the stress and the initial stiffness:

$$\varepsilon_e = [\mathbb{C}]^{-1} : \sigma \tag{5}$$

3. The anelastic strain is obtained combining Eqs. (4) and (5):

$$\varepsilon_{an} = \varepsilon - \varepsilon_e \tag{6}$$

4. Finally the crack opening is computed as follow with h_e the element size:

$$u_c = h_e \varepsilon_{an}. \tag{7}$$

3 Practical Case Studies

3.1 Rod Under Multi-physic Loadings

In the first application a rod under thermic and hygric loadings is modelled. The rod's dimensions are a length L of 0.20 m, a height of b of 0.02 m and each element size (h) are equal to 0.02. The modelling is carried out in 2D plane stress and is composed of 10 linear quadrangular elements. Regarding the boundaries conditions two cases scenario have to be distinguished. The first one models an imperfect rod submitted to free thermal expansion and drying shrinkage (Fig. 1a) and the second one simulates the same imperfect rod submitted to restrain thermal expansion and drying shrinkage (Fig. 1b). The model's parameters associated to the damage mechanical model are presented in the Table 1. In order to localize the damage and the related cracking, the sixth element is randomly introduced as a weak element with a strength value reduced by 50%. Thus, there is not regard according to regularization and B_t is adapted manually to the element size.

To model the thermal strains and the drying shrinkage a fictitious evolution is designed. The final strain related to drying shrinkage is chosen to be equal to 600 $\mu\varepsilon$ and the thermal strains are designed to present a succession of levels (Fig. 2).

To validate the computed crack opening, the analytical value is calculated according to "Eq. (8)".

$$u_c = u_x - \frac{F_x}{E \times b} L - \sum_n \varepsilon_n \times L \tag{8}$$

with u_c (m) the analytical crack opening, u_x (m) the total axial displacement, F_x (N) the axial strength, E (Pa) the Young's modulus, b (m) the element height, L (m) the

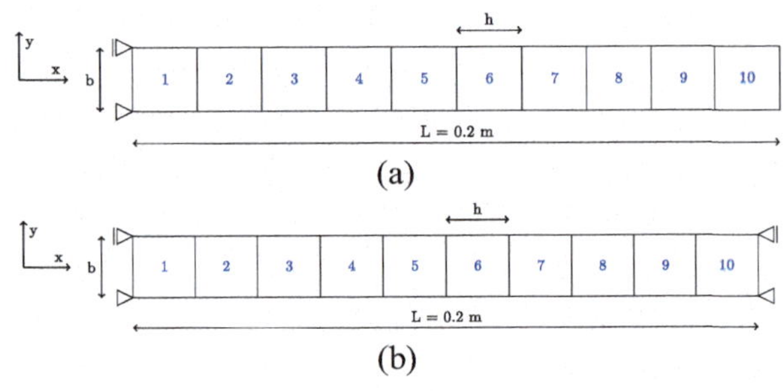

Fig. 1 Two dimension rods with one imperfect element: Two cases scenario **a** free and **b** restrained

Table 1 Set of mechanical parameters of Mazars original model [4]

f_t (MPa)	E (GPa)	ν (-)	G_f (Jm^{-2})	Ac (-)	Bc (-)	At (-)	Bt (-)
3.0	30	0.2	300	1.4	1900	-20	0.01

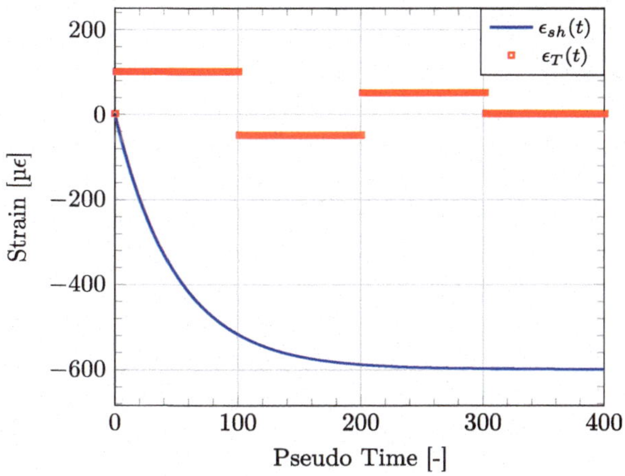

Fig. 2 Designed imposed multi-physics strains

rod length and ε_n (-) the various additional strains. This analytical crack is computed according to numerical values.

The crack opening displacements computed in both scenario are plotted (Fig. 3). In the first modelling, the crack opening is expected to be equal to zero due to the free boundaries conditions. The theoretical value and the value computed with the proposed method are similar (no cracks) (Fig. 3a). However, the method published by Matallah et al. [2] found deviations from the analytical solution. The resulting cracks values are negative, which is not physical. Regarding the second modelling with restrained boundaries conditions (Fig. 3b), it is found that the proposed method meets the analytical value. As damage occurs, crack openings develop. There is a first stage where the rod is elastic then drying shrinkage and thermal expansion induced cracks. The final crack opening value is equal to 115.24 μɛ and 153 μɛ for the proposed method and the Matallah et al. [2] method respectively.

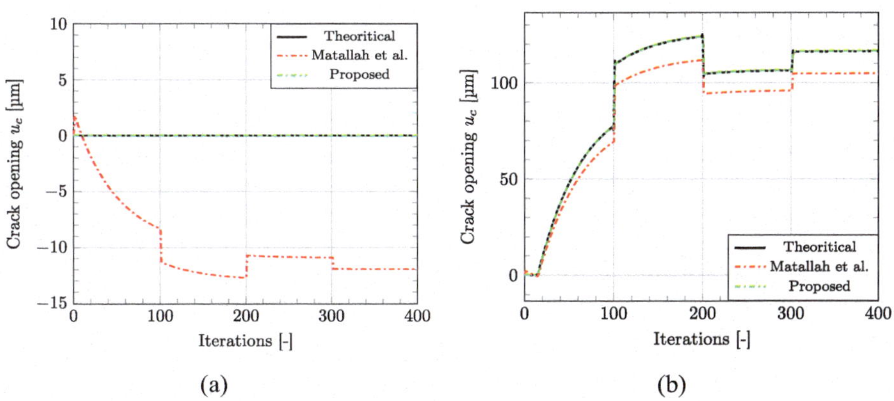

Fig. 3 Crack opening along the axial displacement: **a** free to deform and **b** restrained

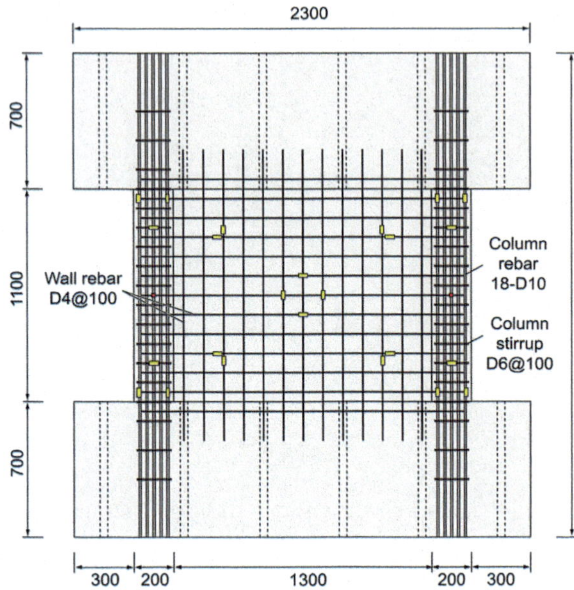

Fig. 4 Schematic representation of the shear wall extracted from [7]

3.2 *Shear Wall Under Drying*

In this application, a semi-structural modelling of a shear wall submitted to drying is undertaken. The modelling relies on the experimental investigation conducted by Sasano et al. [7]. In this study two shear walls are exposed to a dried and sealed environment for approximately 1 year. The ambient conditions regarding dry environment is an average temperature and relative humidity of 18.4°C and 58.7% RH respectively. The dimension of the wall are depicted on Fig. 4.

To model the ageing of the shear wall a previous model developed by the authors is used [8]. It is composed of 4 main stages: the modelling of the drying process with a diffusion equation including the capillary pressure as a driving potential "Eq. (9)".

$$\frac{dS_w}{dp_c}\frac{dp_c}{dt} = div(\frac{k_{rl}(S_w)K}{\mu_w \phi} grad(p_c)) \tag{9}$$

where S_w, p_c, ϕ, K, k_{rl} and μ_w are, respectively, the saturation degree, the capillary pressure, the porosity, the intrinsic permeability, the relative permeability and the viscosity of the liquid water. Saturation and relative permeability are computed according to van Genuchten's relations [9]. Then the consideration of the drying shrinkage "Eq. (10)":

$$\epsilon_{ds} = (1-2\nu)b_w S_w \int \left(\frac{1}{E}\frac{dp_c}{dt'} + J\left(t-t'\right)\frac{dp_c}{dt'}\right)dt' \tag{10}$$

with b_w (-) the Biot coefficient and J (Pa^{-1}) the creep compliance. Thus, the creep is considered according to "Eqs. (11)" and "(12)"

$$\frac{\dot{\sigma}}{k_{kv}} = \tau\varepsilon_{kv} + \left(1 + \frac{\dot{k}_{kv}}{k_{kv}}\right)\dot{\varepsilon}_{kv} \text{ and } \sigma = \eta_{am}\dot{\varepsilon}_{am} \text{ with } \dot{\varepsilon} = \dot{\varepsilon}_{kv} + \dot{\varepsilon}_{am} \quad (11)$$

$$\dot{\epsilon}_{dc} = \lambda_{dc}|\dot{\epsilon}_{ds}|\sigma \quad (12)$$

where k_{kv} (Pa), τ (s), η_{am} (Pa.s) and λ_{dc} (-) are material parameters and a constant respectively. Creep is extended to multiaxial state of stress thank to the creep Poisson ratio. The mechanical model (presented in Sect. 2.1) is slightly modified to consider the pore pressure which will strengthen the material according to desaturation "Eq. (13)".

$$\sigma = (1 - D)\,\tilde{\sigma} + b_m S_w p_c \mathbf{1} \text{ with } \dot{\tilde{\sigma}} = \mathbb{C} : (\dot{\epsilon} - \dot{\epsilon}_c - \dot{\epsilon}_{ds}) \quad (13)$$

with $\tilde{\sigma}$ the effective stress, b_m a material parameter and **1** the unit tensor. Finally, to model the ageing of shear wall, the identification of the model parameter has to be done. There were informations regarding drying, drying shrinkage and mechanical parameters in Sasano et al. [7] work. However, there was no consideration of creep. To overcome this limit an assumption on creep parameters was done (parameters taken from another identification on a concrete close to the one used in Sasano study). The identified parameters are gathered in the Tables 2, 3 and 4. Regarding mechanical parameters, it is assumed that damage in compression never occurs. As a result compressive damage parameters are unused. Moreover, an energetic regularization is introduced according to "Eq. (3)" in order to prevent mesh dependency.

Table 2 Set of drying parameters identified on Sasano et al. [7] experimental tests

β (-)	α (-)	n_k (-)	K_{int} (m^2)	φ (-)	ρ ($kg.m^{-3}$)
0.48	13	−0.23	6.4. 10–21	0.143	2365

Table 3 Set of creep parameters extracted from [8]

τ_{am} (d)	η_∞ (GPa.s)	k_{kv} (GPa)	τ_{kv} (d)	ν_{bc} (-)	λ_{dc} (MPa^{-1})
90	130	135	15	0.1	7.82. 10^{-2}

Table 4 Identified mechanical parameters for the shear wall under drying

f_t (MPa)	E (GPa)	ν (-)	G_f (Jm^{-2})	At (-)	b_m (-)
3.1	30.4	0.23	100	0	0.01

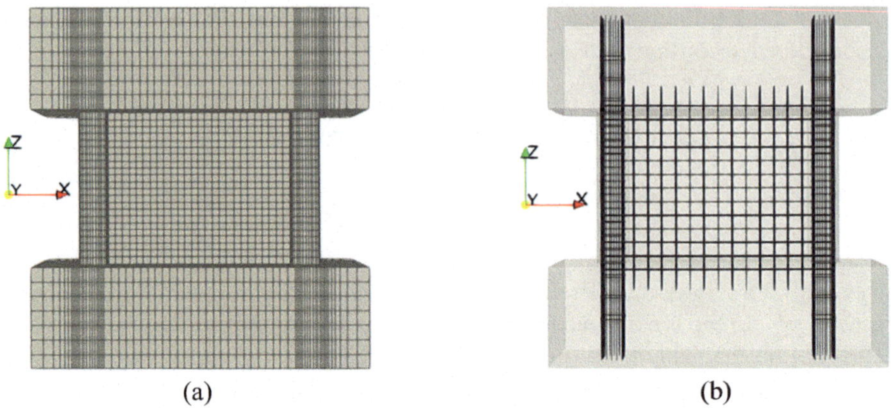

Fig. 5 Modelling of the RC wall specimen **a** 3 dimensional modelling and **b** modelled rebar reinforcements

The shear wall is modelled using a 3 dimensional modelling. The mesh is composed of 35576 cubic elements and 3644 linear elements with a linear formulation (Fig. 5a and b). Rebar reinforcements and volumetric element are linked with a kinematic relationship that imposes a perfect adhesion. Moreover the boundaries conditions are implemented to block the symmetry axes allowing a free deformation. As a first attempt, a random field on the tensile strength is generated using the Turning Band Method [10]. A coefficient of variation equal to 10% and a correlation length equal to three times the biggest aggregate (15 mm) are considered.

For a sake of clarity, the results of this study focus only on crack openings. However, in order to highlight the hydro-mechanical modelling some insights on the relative humidity fields resulting of one year of drying and on the mechanical variable are given such as damage and the crack openings.

Thus, Fig. 6 shows the distribution of the relative humidity fields after 372 days of drying. One can note that the wall reached the equilibrium state ($h_r = 58.7\%$). However, it is not the case concerning the columns and the stubs. These are close to 75% and 100% relative humidity in the core respectively. As a result in addition to the rebar reinforcement, these elements will prevent the shrinkage of the wall and favor cracks at the interface between the wall and them.

Concerning mechanical results, Fig. 7a draws the damage variable. This variable is mainly concentrated superficially on the edges of the structure. The stubs are the most damaged parts of the structure. If one focus on the wall, the damage variable does not exceed the value of 0.45. It is noticeable that damage concentration are occurring on the corner of the wall. One of this concentration goes through the wall. Then, the post-processing technic is applied. The results are depicted on Fig. 7b. The crack opening field induced by drying shrinkage is very similar to the damage field. A concentration of cracks is found on the stubs. The crack openings are in the range of 100 μm with the biggest value equal to 150 μm. One can note that the crack

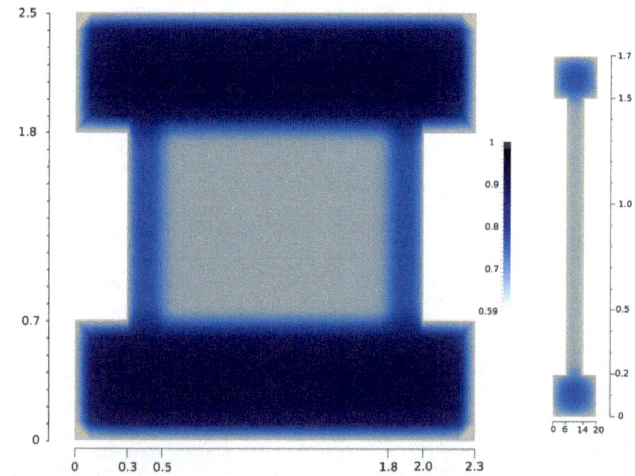

Fig. 6 Relative humidity field after 372 days of drying. (Left) Cross section at the middle of the depth, (right) Cross section at the middle of the height

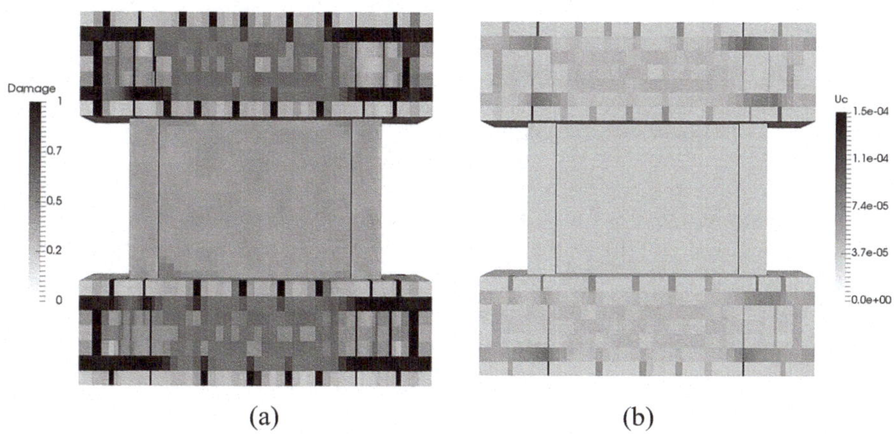

Fig. 7 Computed mechanical fields after 372 days: **a** Damage and **b** crack openings in meter

openings found in the wall are much smaller than that. They are invisible at with the range scale used in Fig. 7.

Figure 8 highlights the cracks on the wall. Monitored cracks are placed in front of computed cracks. These experimental cracks openings ranged from a few tenth of micrometer to 150 μm. Corresponding numerical openings are close to few micrometer. Wide deviations between them exist. This is probably due to both the assumption on the identified parameters (creep for instance) and the modelling hypothesis. Indeed, perfect adhesion between concrete and rebar leads to a slight overestimation of the structure behaviour. As the interface between both materials is perfect there is no slippage and the resulting behaviour is stiffer. Regarding the localization

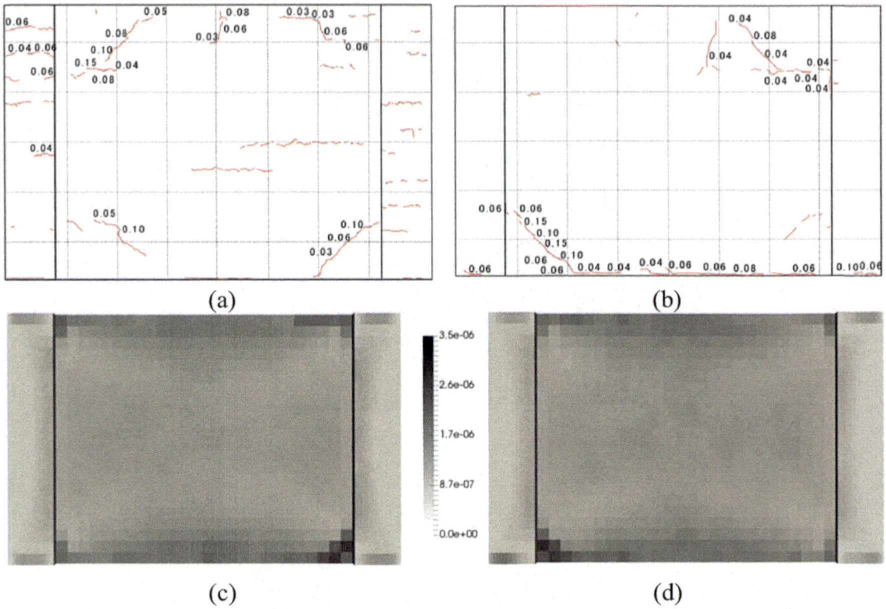

Fig. 8 Experimental and numerical crack opening induced by drying shrinkage at the end of drying (372 days) **a** Face 1 **b** Face 2 **c** Numerical Face 1 and **d** Face 2. Crack opening measured in mm for experimental results and in meter in modelling

of cracks, computed values found agreements with experimental findings. Diagonal cracks at four corners of the wall are partially reproduced and cracks at the interface between stubs and wall are captured. Conversely, it is harder to model the cracks on the columns. These are controlled by material variability, damage model parameters and steel-concrete binding assumptions.

4 Conclusion

Crack is an important feature of concrete material. Often model with damage approach, crack is not explicitly available. Post-processing methods were developed to overcome this limit and roughly estimate the crack openings. They were designed to consider only cracks induced by mechanical loadings.

To extend these methods to multi-physics loadings, a post processing routine was developed. It was tested on academical tests and compared to the analytical solutions. This step yielded satisfactory results. Further investigations were undertaken with a semi-structural application based on Sasano et al. work. With the use of a chained hydro-mechanical modelling, cracks induced by drying were characterized.

The routine managed to locate the area where cracks developed but partially due to model assumptions the cracks openings were slightly underestimated.

The proposed method is thus a good alternative to sophisticated approaches to capture cracks in large structures.

References

1. Bottoni, M., Dufour, F., Giry, C.: Topological search of the crack pattern from a continuum mechanical computation. Eng. Struct. **99**, 346–359 (2015)
2. Matallah, M., La Borderie, C., Maurel, O.: A practical method to estimate crack openings in concrete structures. Int. J. Numer. Anal. Meth. Geomech. **34**, 1615–1633 (2010)
3. Dufour, F., Pijaudier-Cabot, G., Choinska, M., Huerta, A.: Extraction of a crack opening from a continuous approach using regularized damage models. Computers and Concrete **5**, 375–388 (2008)
4. Mazars, J.: A description of micro- and macroscale damage of concrete structures. Eng. Fract. Mech. **25**, 729–737 (1986)
5. Bazant, Z.P., Oh, B.H.: Crack band theory for fracture of concrete. Mater. Struct. **16**, 155–177 (1983)
6. Hillerborg, A., Modéer, M., Petersson, P.-E.: Analysis of crack formation and crack growth in concrete by means of fracture mechanics and finite elements. Cem. Concr. Res. **6**, 773–782 (1976)
7. Sasano, H., Maruyama, I., Nakamura, A., Yamamoto, Y., Teshigawara, M.: Impact of drying on structural performance of reinforced concrete shear walls. J. Adv. Concr. Technol. **16**, 210–232 (2018)
8. Soleilhet, F., Benboudjema, F., Jourdain, X., Gatuingt, F.: Role of pore pressure on cracking and mechanical performance of concrete subjected to drying. Cem. Concr. Compos. **114**(103727) (2020)
9. Van Genuchten, M.T., Nielsen, D.: On describing and predicting the hydraulic properties. Ann. Geophys. **3**(5), 615–628 (1985)
10. Matheron G.: The intrinsic random functions and their applications. Adv. Appl. Probab. 439–468 (1973)

Sustainability of Internally-Cured Concrete for Mitigating Shrinkage Cracking Using Service Life Prediction Models

Sara Kalantari, Mohammad Ali Dastan Diznab, and Fariborz M. Tehrani

Abstract Internal curing using fine lightweight aggregate is an effective method to address early-age cracking. This study focuses on the sustainability of using fine lightweight aggregates for internally curing concrete. Application of fine lightweight aggregates has a positive influence on transport properties of concrete and enhances the durability and service life of concrete infrastructure. These properties contribute to the performance of those infrastructures in various environments and climate zones. This paper highlights the sustainability of internal curing application through cost, energy, and emission analyses based on existing service life prediction models and using transport properties of internally-cured concrete obtained through experimental investigations. Presented methodology covers data acquisition and refinement procedures to perform such analyses. Analysis includes a range of concrete applications to simulate current trends in concrete market. Results highlight quantitative sustainability performance measures, such as energy and emissions in addition to cost and time for selected applications in parking structures and bridge decks.

Keywords Concrete cracking · Internal curing · Lightweight aggregate · Service life · Sustainability

S. Kalantari
Amirkabir University, 15916-34311 Tehran, Iran

M. A. D. Diznab
Arak University, 38481-77584 Arak, Iran

F. M. Tehrani (✉)
California State University, Fresno, CA 93740-8030, USA
e-mail: ftehrani@csufresno.edu

F. Kanavaris et al. (eds.), *International RILEM Conference on Early-Age and Long-Term Cracking in RC Structures*, RILEM Bookseries 31,
https://doi.org/10.1007/978-3-030-72921-9_23

1 Introduction

1.1 Concrete Cracking and Durability

Concrete materials have a broad range of applications in built infrastructure, including pavements, buildings, and environmental structures [1]. Production of concrete is the main market for cement consumption, which is associated with substantial environmental footprints [1, 2]. Hence, the sustainability and resilience of concrete infrastructure are essential elements to reduce the environmental footprints of concrete materials [3]. Durability of concrete contributes to lower maintenance, repair, and upgrade costs of various applications, like roads, bridges and ports that are generally exposed to corrosive environments and chemical agents [4]. Vulnerability of concrete to cracking determines the service life of concrete which has substantial impact on the lifecycle analyses of infrastructure [5].

Mitigating early-age cracking is a preventive measure to reduce the effect of corrosive environments on the degradation of concrete throughout its service life [6]. Proper curing of concrete contributes to hydration of cementitious contents and recues or eliminates volume changes associated with chemical shrinkage at early ages [6, 7]. However, conventional curing, also known as external curing is not necessarily an efficient method for curing high cementitious content mixtures, large areas or volumes, vertical surfaces, thick applications or agile constructions with limited access time to the concrete surface [8–11]. Hence, internal curing is promoted as an efficient method to cure cementitious contents within the body of the concrete using reservoirs of water embedded in voids of high absorptive fine lightweight aggregates [12, 13].

1.2 Internal Curing and Sustainability

Internal curing using fine lightweight aggregate addresses early-age cracking and volume changes and enhances the strength of the concrete due to higher hydration rate of cementitious materials and quality of interfacial transition zone (ITZ) between cement paste and lightweight aggregate [14–17]. Existing literature indicates the value of internal curing for lowering lifecycle cost of concrete applications [18]. Further, application of lightweight aggregate enhances transport properties of concrete [19–24] as major contributors to enhancing the durability and extending the service life of concrete [25]. Different models are available for predicting the service life of concrete using transport properties [26–29] like Life-365™ [30–33].

Extended service life of concrete infrastructure has a direct influence on lifecycle cost, energy consumption and greenhouse gas emissions of infrastructure [34, 35]. These parameters facilitate an objective determination of the sustainability of applied materials, means, and methods of construction [36, 37]. The focus of this work is tracking the influences of internal curing on design, construction, and maintenance of concrete elements containing fine lightweight aggregates and incorporating production and manufacturing data into lifecycle analyses of cost, energy, and emissions. The presented work attempts to utilize experimental transport properties and, analytical service life modeling to cover selected applications within a practical range of environmental conditions and climate zones to portray a realistic perception of the value of internal curing for concrete industry in respect to sustainability performance measures.

2 Methodology

2.1 Experimental Transport Properties

Experimental data rely on a series of testing on internally-cured concrete mixtures (IC) in comparison with conventionally cured concrete (CC) as witness samples using ASTM [38–44] and NT [45] standards. Data includes mixture proportioning as well as physical and mechanical properties. Mixtures contained 390 kg/m^3 Portland cement, with water-to-cementitious materials ratio of 0.37, air content of 6%, and volumetric aggregate contents of 38% and 29% for coarse and fine aggregates, respectively (Table 1).

Figure 1 exhibits changes in concrete properties due to internal curing using average values of 3 specimens for each test. The fine aggregate portion in internally-cured mixture contained 10% lightweight and 19% normalweight aggregates by volume. This substitution reduced the equilibrium density of concrete by 3%. Internal curing raised the 28-day compressive strength of concrete by 24% from 37.7 to 46.6 MPa. Results of the restrained shrinkage ring test revealed that the internally-curing also extended the cracking age from 12 to 22-day and reduced stress rate by nearly 50%. Other major influences of internal curing include increasing diffusion by 25% from 3.6 to 4.5 μm^2, increasing chloride threshold by mass by 20% from

Table 1 Concrete mixture proportions [25]

Description	Internally-cured concrete	Conventional concrete
Cement (kg/m^3)	390	390
Normal weight fine aggregate (kg/m^3)	502	768
Normal weight coarse aggregate (kg/m^3)	1068	1068
Light weight fine aggregate (kg/m^3)	180	0
Water (kg/m^3)	144	144
Air (%)	6	6

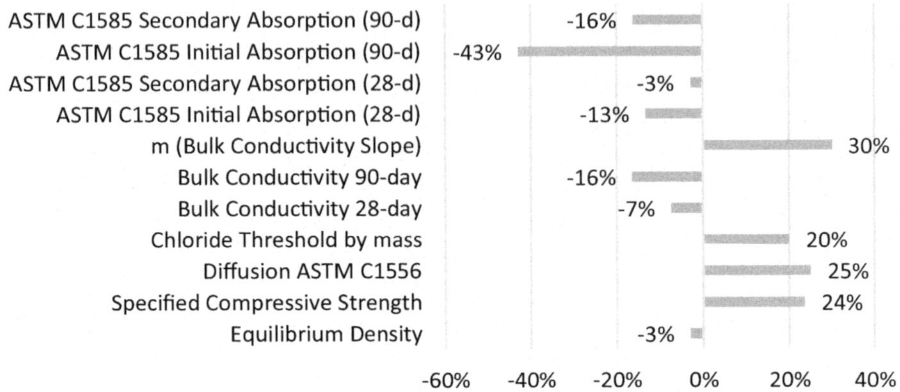

Fig. 1 The influence of internal curing on selected properties of concrete [25]

0.05 to 0.06%, and reducing bulk conductivity at 28- and 90-day by 7% and 16%, respectively. Average and standard deviation bulk conductivity in *ms/m* were 8.8 and 0.02 at 28-day and 5.6 and 0.015 at 90-day, respectively for internally-cured concrete. These changes indicate the long-term effect of internal curing which is responsible for 30% increase in the bulk density slope, m value. Moreover, both initial and secondary absorption values at 28- and 90-day have dropped with application of fine lightweight aggregates (Fig. 1). These changes hint toward slower transport of ions in the internally-cured concrete, resulting in enhanced durability [25].

2.2 *Service Life Prediction*

Service life prediction follows Life-365™ parameters and procedures. These procedures determine the total time for corrosion of steel to exceed the acceptable threshold. An assumed 6-year period is included in this total time to account for the corrosion process after initiation. Hence, the main purpose of service life prediction is determination of the initiation time, that is the time for diffusion of chloride as presented in the Fick's second law [32]:

$$\frac{dC}{dt} = D\frac{d^2C}{dx^2} \tag{1}$$

where, C is the chloride content, D is the diffusion coefficient, x is the depth of the reinforcing steel, and t is time. Diffusion coefficient is a function of time in Eq. (2) and temperature in Eq. (3), benchmarked at the 28-day and 293 K referenced time and temperature, respectively:

Table 2 Selected locations in United States and relevant climate zones [46]

Zone	Dry	Wet
Freeze	Denver (CO)	Detroit (MI)
No-freeze	San Francisco (CA)	New Orleans (LA)

$$D(t) = D_{ref}\left(\frac{t_{ref}}{t}\right)^m \tag{2}$$

$$D(T) = D_{ref}exp\left[\frac{U}{R}\left(\frac{1}{T_{ref}} - \frac{1}{T}\right)\right] \tag{3}$$

where U is the activation energy of the diffusion process (35000 J/mol) and R is the gas constant (8.3145 J/mol/K). Life-365^{TM} does not consider the direct effect of cracking on the performance of concrete, and hence, results indicate the indirect effect of internal curing on the service life of concrete through diffusion testing of uncracked samples in the laboratory.

Modeled prototypes represent a range of concrete applications, including buildings, bridges, pavements and curtain structures in three major types of structures, including frame members (building columns and bridge piers), horizontal surfaces (pavements and bridge decks), and vertical surfaces (walls and curtains). Analysis report service life prediction for four general climate zones of dry-freeze (Denver, CO), wet-freeze (Detroit, MI), dry-no-freeze (San Francisco, CA), and wet-no-freeze (New Orleans, LA) as listed in Table 2. For simplicity, all applications included a 75 mm concrete cover and a single reinforcement ratio of 1.5% with 300 mm thickness or width [46].

2.3 Cost, Energy, and Emissions Analyses

The cost analysis considers a zero-interest financial environment to simplify comparison of results. This analysis includes initial cost of construction, relative maintenance and repair costs, and the demolishing cost at the end of the lifecycle based on available industry trends and specifications [47, 48]. Summarized values in Table 3 provide a comparative base to understand the weight of each item on the overall impact of activities on cost, energy, and emissions, as elaborated further down.

Energy and emissions relate to cost analysis using productivity calculations and determination of required crew, machine and equipment for each task. The input energy and emissions associated with materials and operations are based on available environmental product declarations and engineering analysis of machines and equipment [49–52].

Repair and demolition of concrete products require additional energy and emissions throughout the life cycle of the project. Typical activities related to cracking

Table 3 Resource breakdown analysis for concrete placement in a small site (m^3) [48]

Items	Column and pier	Slab and deck	Walls and curtains
Concrete with 3% waste (m^3)	1.03	1.03	1.03
Burlap for external curing (m^2)	0.6	0.6	0.6
Curing water (l)	300	300	300
Foreman 1 (d)	0.036	0.036	0.036
Foreman 2 (d)	0.15	0.048	0.11
Mason assistant (d)	0.15	0.048	0.11
Labor (d)	0.35	0.42	0.27
Vibrator and operator (h)	1.01	0.38	0.86
Shovel (#)	0.005	0.005	0.005

in concrete materials include construction and maintenance of joints and demolition of damaged materials at the end of their service life. Estimation of labor and equipment for these activities follow standard construction practices for productivity calculations [47, 48]. Joint volume is estimated to be 0.015% of concrete volume for a typical depth of 1/4–1/3 of the concrete thickness, equal portion of 3 mm unsealed and 6 mm sealed joints, over an average 3 m spacing for a 0.3 m-thick concrete (Table 4). Frequency of repair is assumed between 4 and 8 years [53, 54].

Determination of energy and emissions for construction activities utilizes productivity analyses to provide a bottom-up estimate of energy and emissions for labor, equipment and consumed fuel [34, 35]. Input energy includes 143.2 MJ/kg of equipment mass, 47.8 MJ/l of fuel consumption, and 2.28 MJ/h of labor. Estimated emissions include 0.081 kg/MJ of energy input for equipment and 2.68 kg/l of fuel [34, 35]. Demolition rate is assumed to be 1% of the area during major maintenance activities every 5–15 years in addition to salvaging at the end of the service life

Table 4 Resource breakdown analysis for joint repairs (dm^3) [53, 54]

Items	Cut	Seal
Sealant (dm^3)		1.061
Foreman 1 (d)	0.21	0.023
Sealcoat assistant (d)		0.023
Operator (d)		0.00255
Compressor (h)		0.0204
Blade (#)	0.007	

Table 5 Resource breakdown analysis for reinforced concrete demolishing during corrective maintenance (m^3) [54]

Items	Bridge
Operator (d)	0.636
Labor (d)	1.273
Assistant (d)	1.273
Compressor (h)	5.855
70-lb Breaker hammer (h)	5.855
Shovel (#)	0.0172
Pickaxe (#)	0.0172

Table 6 Estimation of energy and emissions per unit of work or material [49–52, 55]

Items	Energy (MJ)	Emissions (kg-CO_2e)
Concrete 34.5–41.4 MPa (m^3)	3601	551.5
Concrete 41.3–55.1 MPa (m^3)	4126	642.0
Fine normal weight aggregate (tonne)	79.05	4.89
Fine light weight aggregate (tonne)	2512	175
Joint maintenance (dm^3)	0.623	0.0357
Sealant (dm^3)	111.9	5.138
Demolishing—parking (m^3)	513.8	29.17
Demolishing—bridge (m^3)	395.2	22.44

(Table 5). Productivity of demolishing in open sites of bridge works is 30% higher than confined areas of buildings, such as parking garages [18, 53, 54].

Summarized values in Table 6 represent the production stage of each components and hence cannot be used for the evaluation of activities at life cycle scales, which is presented in results later [49–52, 55].

3 Results

Service life prediction modeling indicates that internal curing increases the service life of concrete applications in all selected locations (Table 7). The increase varies between 6% in a wet-freeze region (Detroit, MI) to 24% in a wet-no-freeze region (New Orleans, LA). The rate of increase is almost identical for parking garage

Table 7 Predicted service life (year) of conventional (C) and internally-cured (IC) concrete

Conditions	Garage slab		Bridge deck	
Location	CC	IC	CC	IC
Denver (CO), dry freeze	36	39	38	41
Detroit (MI), wet freeze	31	33	33	35
San Francisco (CA), dry no freeze	84	91	90	98
New Orleans (LA), wet no freeze	102	123	111	138

slab and bridge deck applications, with maximum 3% difference observed for New Orleans, LA (Fig. 2).

Annual estimation of cost, energy and emissions values reveal the overall benefits of internal curing in selected cities (Fig. 3). These values indicate that internal curing contributes the most toward sustainability performance measures, including natural resources and environmental footprints, in addition to saving construction costs.

Internal curing contributes to the lifecycle cost of concrete by 18% in wet and no freeze zone and 6–8% for other zones.

The gained strength due to internal curing provides a significant saving in energy as it would allow using lower class concrete with less cement. Further, internal curing reduces required energy for curing operations. The total energy also includes the difference between energy input values of normal weight and lightweight fine aggregates. Energy saving due to internal curing is 7–20%. Estimation of emissions follows the same procedure and has resulted in 14–26% saving due to internal curing. In freeze zones, savings in energy and emission are up to 1% higher for parking garages. These savings are up to 3% higher for bridge decks in non-freeze zones.

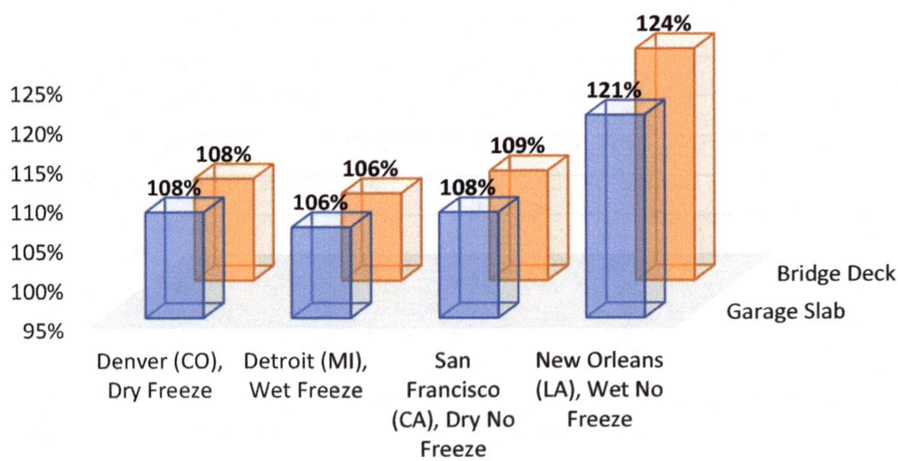

Fig. 2 Influence of internal curing on predicted service life of concrete prototypes

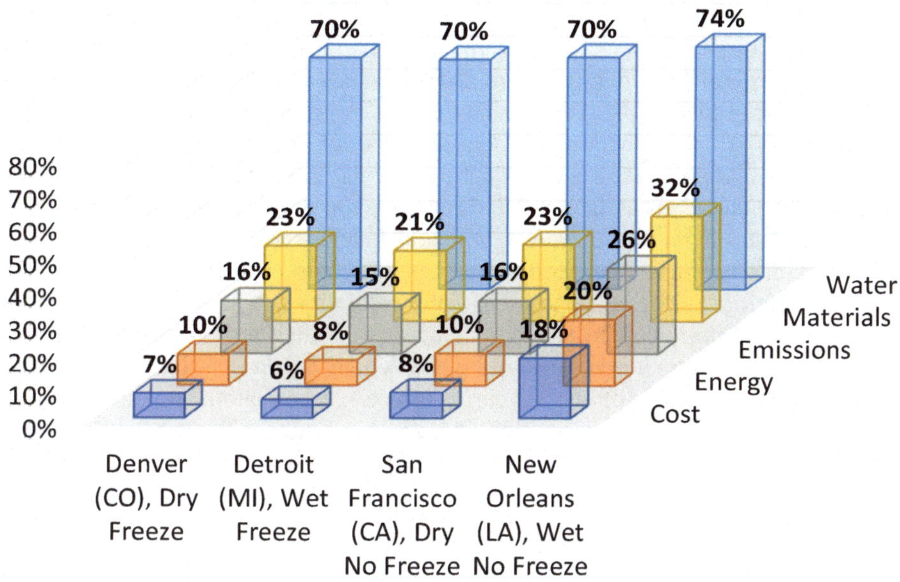

Fig. 3 Savings in cost, energy and emissions of concrete by internal curing

Moreover, replacement of portion of fine lightweight aggregates with manufactured expanded aggregates reduces the consumption of natural resources, which are mined materials. Internal curing reduces mined materials by 21–32%.

Furthermore, the application of internal curing optimizes consumption of water, which is another valuable resource. The total water reported includes water needed for saturating aggregates (considering different saturation rates for normal weight and lightweight aggregates), proportioning concrete mixture (based on saturated surface dry conditions) and curing placed concrete (internal or external, as applicable). The substantial saving in water is due to two major factors; first, the internal curing has no effect on water-to-cementitious material ratio; and second, the water consumption for external curing involves substantial waste due to evaporation and limited mobility of water inside concrete and therefore it is several fold more than water needed for saturating lightweight aggregate for the purpose of internal curing, which is done in controlled environment of aggregate or concrete production plants. Water saving is nearly 70–74%.

4 Conclusion

Internal curing using fine lightweight aggregate enhances physical, mechanical and transport properties of concrete. The main purpose of internal curing is mitigating shrinkage cracking in concrete. However, internal curing has broader impact on the sustainability of concrete. An analysis of cost, energy and emissions indicate substantial contribution to sustainability performance measures. Numerical results for four selected zones indicate reductions in mined materials (21–32%), water consumption (70–74%), input energy (8–20%), greenhouse gas emissions (15–26%) and life-cycle cost (6–18%). These reductions are consistent for all zones with maximum reductions in a wet and no-freeze zone. Reductions are also consistent for bridge deck and garage slab applications with up to 3% variation favoring bridge deck in freeze zones and garage slabs in no freeze zones. These conclusions rely on standard practices in concrete construction and hence represent a realistic approach to evaluating sustainability of internal curing, as opposed to using cradle-to-gate performance measures of energy and emissions. The accuracy of the service life determination is an important factor in the reliability of results. Enhancement in service life predictions to incorporate the full range of internal curing effects, such as elimination of long-term cracks will provide a better perception of presented results in this study.

References

1. Kosmatka, S.H., Panarese, W.C.: Design and Control of Concrete Mixtures. Portland Cement Association (PCA), Skokie, Illinois (2016). www.cement.org
2. PCA.: Environmental Product Declaration. Portland Cement Association (PCA), Skokie, IL (2016). www.cement.org
3. Tehrani, F.M.: Deploying and rating sustainable practices for resilient bridge infrastructure. Keynote Lecture. In: Proceedings of The Fifth International Conference on Bridges, December 17–18, 2019, MS05. Amirkabir University of Technology, Tehran, Iran (2019). www.ibc.aut.ac.ir
4. PCA.: The Real Value of Resilient Constriction. 05.2019. Portland Cement Association (PCA), Skokie, IL (2019). www.cement.org
5. Nazari, M., Tehrani, F.M., Ansari, M., Jeevanlal, B., Rahman, F., Farshidpour, R.: Green strategies for design and construction of non-auto transportation infrastructure. Report 19–17, Mineta Transportation Institute, San Jose, CA (2019). www.transweb.sjsu.edu
6. ACI 231R-10.: Report on early-age cracking: causes. In: Measurement and Mitigation. American Concrete Institute (ACI), Farmington Hills, MI (2010). www.concrete.org
7. ACI 308R-16.: Specification for Curing Concrete. American Concrete Institute (ACI), Farmington Hills, MI (2016). www.concrete.org
8. Bentz, D.P.: Internal curing of high-performance blended cement mortars. ACI Mater. J. **104**(4), 408–414 (2007). www.concrete.org/publications/internationalconcreteabstractsportal/m/details/id/18831
9. Lane, D.S.: Effect of wet curing duration on curability parameters of hydraulic cement concretes. FHWA/VTRC 10-R11. Virginia Transportation Research Council, Richmond, VA, March (2010). vtrc.virginiadot.org/PubDetails.aspx?PubNo=10-R11
10. Khayat, K.J., Mehdipour, I.: Economical and crack-free high-performance concrete for pavement and transportation infrastructure construction. cmr 17–007. Missouri Department of Transportation, Jefferson City, MO, May (2017). spexternal.modot.mo.gov/sites/cm/CORDT/cmr17-007.pdf
11. Tankasala, A. and Schindler A. K. Early-age cracking of lightweight mass concrete. ACI Mater. J. **117**(1), 223–232 (2020). www.concrete.org/publications/internationalconcreteabstractsportal.aspx?m=details&ID=51719082
12. ACI 308–213R-13.: Report on internally cured concrete using pre-wetted absorptive lightweight aggregate. American Concrete Institute (ACI), Farmington Hills, MI (2013)

13. Bonyadian, S., Mohammadi, M., Foroutanmehr, B., Tehrani, F.M.: An experimental investigation of internally-cured concrete application for bridge decks. In: Proceedings of The Fifth International Conference on Bridges, (December 17–18, 2019), MS02, Amirkabir University of Technology, Tehran. Iran (2019). ibc.aut.ac.ir
14. Henkensiefken, R.: Internal curing in cementitious systems made using saturated lightweight aggregate. Thesis, Purdue University, West Lafayette, IN (2008). docs.lib.purdue.edu/dissertations/AAI1469671
15. Castro, J.: Moisture transport in cement-based materials: application to transport tests and internal curing. Ph.D. dissertation. Purdue University, West Lafayette, IN (2011). docs.lib.purdue.edu/dissertations/AAI3475407
16. Roberts, J.W.: The 2004 practice and potential of internal curing of concrete using lightweight sand. In: Advanced in Concrete through Science and Engineering. RILEM Publications S.A.R.L., Bagneux, France, March (2004). https://doi.org/10.1617/2912143926.035
17. Hoff, G.C. Internal curing of concrete using lightweight aggregate. In: Theodore Bremner Symposium on High Performance Lightweight Concrete, Sixth CANMET/ACI International Conference on Durability, June, pp. 185–203, Thessaloniki, Greece (2003). www.concrete.org
18. Vosoughi, P., Tritsch, S., Ceylan, H., Taylor, P.C.: Lifecycle cost analysis of internally cured jointed plain concrete pavement. Part of IHRB Project TR-676, Iowa Highway Research Board, Ames, IA (2017). lib.dr.iastate.edu/intrans_reports/222
19. Thomas, M. D. A. Chloride diffusion in high-performance lightweight aggregate concrete. In: ACI Symposium Paper 234, vol. 50, pp. 797–812 (2006). www.concrete.org/publications/internationalconcreteabstractsportal/m/details/id/15974
20. Maltais, Y., Ouellet, E., Marchand, J., Samson, E., Burke, D.: Prediction of the long-term durability of lightweight aggregate concrete mixtures under severe marine environment. Mater. Struct. **39**, 911–918 (2006). https://doi.org/10.1617/s11527-006-9127-7
21. Liu, X., Chia, K.S., Zhang, M.: Development of lightweight concrete with high resistance to water and chloride-ion penetration. Cement Concr. Compos. **32**, 757–766 (2010). https://doi.org/10.1016/j.cemconcomp.2010.08.005
22. Ferrer, B., Bogas, J.A., Real, S.: Service life of structural lightweight aggregate concrete under carbonation-induced corrosion. Constr. Build. Mater. **120**(1), 161–171 (2016). https://doi.org/10.1016/j.conbuildmat.2016.05.108
23. Real, S., Bogas, J.A., Ferrer, B.: Service life of reinforced structural lightweight aggregate concrete under chloride-induced corrosion. Mater. Struct. **50**, 101 (2017). https://doi.org/10.1617/s11527-016-0971-9
24. Zhen, N., Qian, X., Justnes, H., Martius-Hammer, T.A., Tan, K.H., Ong, K.C.G.: Durability of lightweight aggregate concrete in marine structures. ISOPE-I-18-649. In: The 28th International Ocean and Polar Engineering Conference, 10–15 June. International Society of Offshore and Polar Engineers Sapporo, Japan 2018. onepetro.org/ISOPEIOPEC/ISOPE18/conference/All-ISOPE18
25. Tehrani, F.M., Service life prediction of structural lightweight concrete using transport properties. ESCSI Report 4363, October 2020. Expanded Shale, Clay and Slate Institute, Chicago, IL (2020). www.escsi.org
26. Bentz, E.C.: Probabilistic modeling of service life for structures subjected to chlorides. ACI Mater. J. **100**(5), 391–397 (2003). www.concrete.org/publications/internationalconcreteabstractsportal/m/details/id/12814
27. Alexander, M., Thomas, M.: Service life prediction and performance testing—current developments and practical applications. Cement Concr. Res. **78**(Part A), 155–164 (2015). https://doi.org/10.1016/j.cemconres.2015.05.013
28. fib Bulleting 34.: Model code for service life design. fédération internationale du béton Lausanne, Switzerland (2006). www.fib-international.org
29. Papworth, F., Matthews. S.: fib Model Code 2020—Durability design and through life management of new and existing structures. In: Proceedings of Sixth International Conference on Durability of Concrete Structures, Paper Number KN04, 18–20 July 2018, Leeds, pp. 53–62. University of Leeds, West Yorkshire, United Kingdom (2018). eps.leeds.ac.uk

30. Violetta, B.: Life-365 Service Life Prediction Model. Concrete International. December, pp. 53–57 (2002). www.concrete.org/publications/internationalconcreteabstractsportal.aspx?m=details&ID=12376
31. Ehlen, M.A., Thomas, M.D.A., Bentz, E.C.: Life-365 service life prediction model^{TM} Version 2.0. Concrete International (May), pp. 41–46 (2009). www.concrete.org/publications/internationalconcreteabstractsportal.aspx?m=details&ID=56508
32. Life-365.: Life-365TM service life prediction model^{TM}. Life-365™ Consortium III (2013). www.life-365.org
33. Ehlen, M.A., Kojundic, A.: Life-365^{TM} v2.2. Concrete International (May), pp. 41–44 (2014). www.concrete.org/publications/internationalconcreteabstractsportal.aspx?m=details&ID=51686933
34. Tehrani, F.M., Alexandrou, A., Machoney, M., Adhikari, D., Raymond, M.: Energy inputs and carbon dioxide emissions from construction equipment during construction of a golf course. Int. J. Eng. Res. Innov. **6**(2), 78–86 (2014). ijeri.org/IJERI-Archives/issues/fall2014/fall2014.htm
35. Tehrani, F.M., Dadkhah, M.: A case study on the analysis of energy and emissions for sustainability rating. Int. J. Clim. Change: Impacts Responses **10**(3), 13–23 (2018). https://doi.org/10.18848/1835-7156/cgp/v10i03/13-23
36. Tehrani, F.M., Farshidpour, F., Pouramini, M., Mousavi, M., Esfahani, A.N.: Sustainability rating of lightweight expanded clay aggregates using energy inputs and carbon dioxide emissions in life-cycle analysis. In: Proceedings of Sixth International Symposium on Life Cycle Civil Engineering, (October 2018), pp. 2989–2993. IALCCE, Ghent, Belgium (2018). ialcce2018.ugent.be
37. Tehrani, F.M., Nazari, M., Truong, D., Farshidpour, R.: Sustainability of tire-derived aggregate concrete: a case study on energy, emissions, economy, and Envision. In: Proceedings of International Conference on Sustainable Infrastructure, November 6–9, 2019, pp. 399–408. American Society of Civil Engineers (ASCE), Los Angeles, CA (2019). https://doi.org/10.1061/9780784482650.043
38. ASTM C39/C39M-16.: Standard Test Method for Compressive Strength of Cylindrical Concrete Specimens. ASTM International, West Conshohocken, PA (2016). www.astm.org. https://doi.org/10.1520/c0039_c0039m-16
39. ASTM C1760-12.: Standard Test Method for Bulk Electrical Conductivity of Hardened Concrete. ASTM International, West Conshohocken, PA (2012). www.astm.org. https://doi.org/10.1520/c1760-12
40. ASTM C1202-17.: Standard Test Method for Electrical Indication of Concrete's Ability to Resist Chloride Ion Penetration. ASTM International, West Conshohocken, PA (2017). www.astm.org. https://doi.org/10.1520/c1202-17
41. ASTM C1556-11a.: Standard Test Method for Determining the Apparent Chloride Diffusion Coefficient of Cementitious Mixtures by Bulk Diffusion. ASTM International, West Conshohocken, PA (2016). www.astm.org. https://doi.org/10.1520/c1556-11ar16
42. ASTM C1585-13.: Standard Test Method for Measurement of Rate of Absorption of Water by Hydraulic-Cement Concretes. ASTM International, West Conshohocken, PA (2013). www.astm.org. https://doi.org/10.1520/c1585-13
43. ASTM C1581/C1581M-18a.: Standard Test Method for Determining Age at Cracking and Induced Tensile Stress Characteristics of Mortar and Concrete under Restrained Shrinkage. ASTM International, West Conshohocken, PA (2018). www.astm.org. https://doi.org/10.1520/c1581_c1581m-18a
44. ASTM C567/C567M-14.: Standard Test Method for Determining Density of Structural Lightweight Concrete. ASTM International, West Conshohocken, PA (2014). www.astm.org. https://doi.org/10.1520/c0567_c0567m-14
45. NT BUILD 492.: Concrete, Mortar and Cement-Based Repair Materials: Chloride Migration Coefficient from Non-Steady State Migration Experiments. Nordtest, Espoo, Finland (1999). www.nordtest.info

46. Schwartz, C.W., Elkins, G.E., Li, R., Visintine, B.A., Forman, B., Rada, G.R., Groeger, J.L.: Evaluation of long-term pavement performance (LTPP) climatic data for use in mechanistic-empirical pavement design guide (MEPDG) calibration and other pavement analysis. FHWA-HRT-15-019. Office of Infrastructure Research and Development Federal Highway Administration, McLean, VA (2015). www.fhwa.dot.gov
47. Peurifoy, R.B., Schexnayder, C.J., Shapira, A., Schmitt, R.L.: Construction Planning, Equipment, and Methods. McGraw-Hill, New York, NY (2018)
48. ACCO.: The Applied Cost Analysis of Construction Works in Persian. Association of Construction Companies, Tehran, Iran (2004). www.acco.ir
49. ESCSI.: Embodied Energy to Manufacture Expanded Shale, Clay and Slate (ESCS) Lightweight Aggregate. Information Sheet 9153. Expanded Shale, Clay, and Slate Institute, Chicago, IL. (2015). www.escsi.org
50. NRMCA.: Environment Product Declaration. National Ready Mixed Concrete Associate, Silver Spring, MD (2020). www.nrmca.org
51. PCA.: Environmental Product Declaration. Portland Cement Association, Skokie, IL (2016)
52. Vulcan.: Environmental Product Declaration. Vulcan Materials Company, Western Division, Glendale, CA (2016). www.vulcanmaterials.com
53. SUDAS.: Roadway Design. Chapter 5, In SUDAS Design Manual. Statewide Urban Design and Specifications Program, IA (2021).www.iowasudas.org/manuals/design-manual
54. PBO.: Cost Analysis of Referenced Units in Persian. Plan and Budget Organization. Tehran, Iran (2018). www.sama.mporg.ir
55. Sika.: Sealing and Bonding Environmental Product Declaration—Cradle-to-Gate Sikaflex®-1a. Sika Corporation, Lyndhurst, NJ (2019). www.usa.sika.com

Validation of Thermo-hygro-Mechanical FEM Analysis of Thick Restrained RC Members by Comparison with Experiments

Cláudio Ferreira, Dirk Schlicke, Carlos Sousa, José Gomes, Rui Faria, and Miguel Azenha

Abstract Crack width control is an important design task to achieve the required durability and serviceability of reinforced concrete (RC) structures. The thickness of the member is an important contributing factor for the mode of cracking—and thus for the required reinforcement for crack control. While in thin RC elements, e.g. industrial floors or tank walls, visible cracks usually penetrate through the entire element thickness, in thick elements with common reinforcement configuration (reinforcement is located near the surface) the cracking process is much more complex: initially, a primary crack occurs and separates the element over its entire thickness; and in addition, secondary cracks may occur due to the transfer of the steel force in the primary crack back into the concrete by bond action. These secondary cracks do not penetrate through the entire thickness, but end somewhat behind the surface zone. This paper presents the validation of a thermo-hygro-mechanical framework for simulation of the cracking process in thick restrained RC members. The study focusses on the short-term experimental campaign on restrained reinforced concrete panels made at TU Graz. The experimental measurements are compared with the results of the finite element simulation. The comparison is made for the crack patterns, the crack width development for primary and secondary cracks and the development of the restraint force in function of the imposed elongation. In addition, a parametric study is performed to investigate the drying shrinkage effect on the long-term crack width evolution of primary and secondary cracks.

Keywords Cracking · Thick RC member · Shrinkage · Imposed deformation · Non-linear analysis

---

C. Ferreira (✉) · C. Sousa · R. Faria
CONSTRUCT, University of Porto, Porto, Portugal
e-mail: cfbferreira@fe.up.pt

D. Schlicke
Institute for Structural Concrete, Graz University of Technology, Graz, Austria

J. Gomes · M. Azenha
ISISE, University of Minho, Guimarães, Portugal

F. Kanavaris et al. (eds.), *International RILEM Conference on Early-Age and Long-Term Cracking in RC Structures*, RILEM Bookseries 31,
https://doi.org/10.1007/978-3-030-72921-9_24

1 Introduction

Crack control is one of the criteria for serviceability design of RC structures. Efforts have been made to understand the physical principles beyond cracking, and to develop engineering solutions to be used in the design, namely experimental campaigns [1], analytical approaches [2] and numerical analyses [3]. Thermo-hygro-mechanical (THM) finite element modelling (FEM) is a powerful tool to assess the long-term behaviour of complex crack sensitive structures, and to develop engineering solutions for typified design problems.

In thick, or massive, RC structures, THM FEM is particularly useful, owing to the explicit simulation of temperature and moisture gradients, triggered by cement hydration, concrete shrinkage and varying ambient conditions. With increasing element thickness, these effects give rise to self-equilibrated concrete stresses within the member. In addition, usually those types of structures are also externally restrained since casting (at least partially), which enhances especially the risk of separating cracks. For most RC structures avoidance of cracking is not a viable design strategy. Instead, design solutions focus on limitation of the crack widths at the surface, or at the level of the reinforcement, to acceptable values, whilst larger crack widths in the interior of the element are tolerated. The typical reinforcement configuration for RC members, namely arrangement of reinforcement near the surface, leads to the formation of two types of cracks (see Fig. 1): primary cracks, which fully penetrate through the element thickness; and secondary cracks, which are formed in the vicinity of the primary ones, but do not cross the entire member thickness. Secondary cracks are the consequence of the transfer of the steel force in the primary crack into the surface region of the member. In the sense of crack control, the occurrence of secondary cracks is beneficial once the resulting widths of the superficial cracks are smaller than the widths in an identical member with only primary cracks.

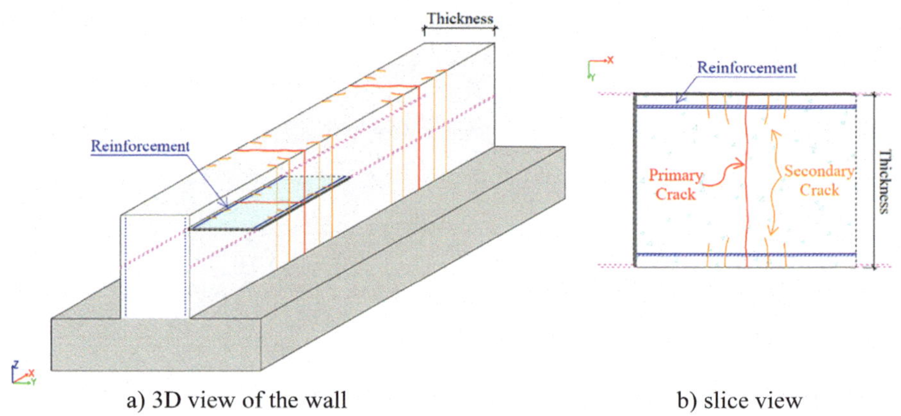

Fig. 1 Primary and secondary cracks in a thick, axially retrained, RC wall

This crack formation process can be assessed through THM analyses. Azenha [4] and co-workers have developed, calibrated and validated, at the material and structural levels, a THM FEM framework. However, validation of this framework was missing for prediction of the outlined crack process in thick restrained RC members, by comparison with experiments. The purpose of this work is to give a contribution in this regard.

At Graz University of Technology a laboratory system was developed to investigate the progression of cracking and restraining forces in thick RC elements [1, 5]. Several experimental campaigns were performed to investigate two features of the behaviour of thick RC elements: (i) the timely development of restraint stresses and crack widths during cement hydration, plus interaction with additionally imposed deformations at later age, which was achieved with tie-shaped specimens; and (ii) the crack mechanism of thick elements consisting of primary and secondary cracks, which was achieved with panel-shaped specimens. This work considers the experimental campaign focused on issue (ii). The campaign consisted on the imposition of an axial deformation in the specimen, as represented in Fig. 1b. The experiment simulated in a simplified manner the crack formation process in thick RC members, where temperature and shrinkage deformations are externally restrained. It should be noted that the experiment was designed to exclusively investigate the crack formation process in thick elements under uniform tension. The effect of self-equilibrated stresses, which is always present under real conditions due to temperature and moisture gradients across the cross-section, was deliberately omitted in these experiments.

This paper presents the FEM simulation of this short-term test. The influence of the bond-slip model, which describes the transfer of stresses at the steel-concrete interface, is investigated, once this has a major impact in the development of restraint forces, crack patterns and crack widths. For this set of analyses, no thermal or hygral model is used, because the imposed deformations are applied to the specimen ends, according to the experiment. The objective of this task consists in the validation of the adopted FEM analysis approach.

Then, a parametric study is performed to investigate the influence of self-equilibrated stresses due to non-uniform drying shrinkage strains upon the crack pattern formation.

2 Experimental Campaign

2.1 Description

The principle of the experimental campaign is to impose an axial deformation in a panel with only one reinforcement bar at each side, leading to the generation of one primary crack at half-length of the specimen, plus secondary cracks at both sides of the primary crack. Restraint force, elongation of the specimen and crack widths

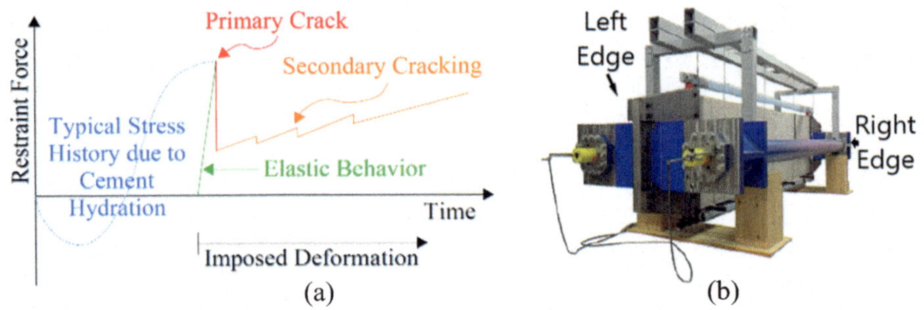

Fig. 2 Adjustable restraining frame developed at TU Graz for experimental assessment of the crack formation in thick restrained RC members: **a** schematic representation of the experimental procedure; **b** image of the test frame in the mode for panel-shaped specimens, adapted from [1]

were continuously monitored. The restraint procedure of the experiment and the photograph of the setup are illustrated in Fig. 2.

The solid steel frame includes load cells in both longitudinal rods, and the left edge can be moved by hydraulic jack cylinders (Fig. 2b). Further details on the configuration of the frame are given in [1]. The specimen is cast outside of the frame, and then it is cautiously mounted into it. A higher density of reinforcement is set at the edges of the specimen panel, allowing the full connection between the specimen and the frame (see Fig. 3). A notch is localized at half-length of the panel specimen, to induce location of the primary crack at the middle section.

2.2 Experimental Data

The concrete panel has a length of 3.5 m, a depth of 1.03 m and a width of 0.097 m, and it simulates the crack formation in a 1.03 m thick wall. The reinforcement consists of a single rebar with a diameter of 14 mm located by symmetry at each edge of the panel. The cover is 37 mm, measured from the concrete surface to the rebar centre.

The used concrete corresponds to the strength class grade C35/45. The concrete composition and mechanical properties, determined in tests in a parallel program, are presented in Table 1.

The total imposed deformation is 2.4 mm (corresponding to an average strain of $\approx 0.7‰$). LVDT as well as DD1-strain gauges are located at the panel surface and in the middle span, allowing to measure the crack widths of the primary and secondary cracks during the total duration of the experiment. Extra measurements are performed at the uncracked concrete zone for evaluation of the concrete strains. The LVDT and DD1 locations, and the base-length values for deformation measurements, are indicated in Fig. 3. The measurements were performed for the top and bottom cracks.

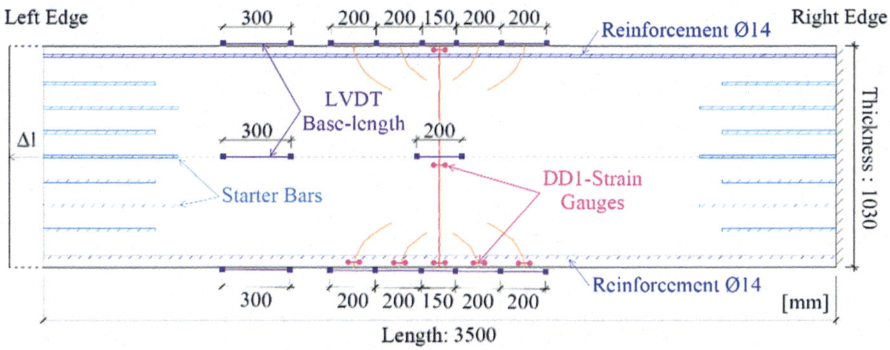

Fig. 3 Position of the DD1 and LVDT transducers for measurement of crack widths, base length for LVDT measurements (mm) and position of the reinforcement

Table 1 Concrete composition and mechanical properties

Concrete composition		Mechanical properties	
Component	Value (kg/m^3)	Property	Value
CEM III/A 32.5 N Holcim	300	Compressive strength	43 MPa
Water	145	Tensile strength	3.21 MPa
Sand and Aggregates: 0–2/2–8/8–16	620/720/630	Elastic modulus	34.1 GPa
BV ViscoCrete-1051 PCE	3		

3 Finite Element Model

3.1 General Modelling Aspects

The mechanical analyses are performed in the DIANA 10.3 software, using a 2D model. Concrete is modelled with 8-nodded quadrilateral isoparametric plane stress FE, with a 2×2 Gauss integration scheme. Reinforcement is modelled with 3-nodded truss FE with a two-point Gauss integration scheme. Bond between concrete and reinforcement is modelled using interface FE with 6-nodded and a 3-point Newton-Cotes integration scheme. The FE sizes vary between 20×18.5 mm^2 for the four superficial concrete FE layers, and 20×19.12 mm^2 for the remaining mesh. Only half of the panel is modelled due to symmetry about the horizontal axis. The total dimensions of the FE mesh are thus 3500×515 mm^2 (see Fig. 4). At mid-length of the mesh a vertical column of FE is considered with a lower tensile strength (1.5 MPa in the simulation of the experiment; $2/3 f_{ctm}$ in the parametric analyses), inducing

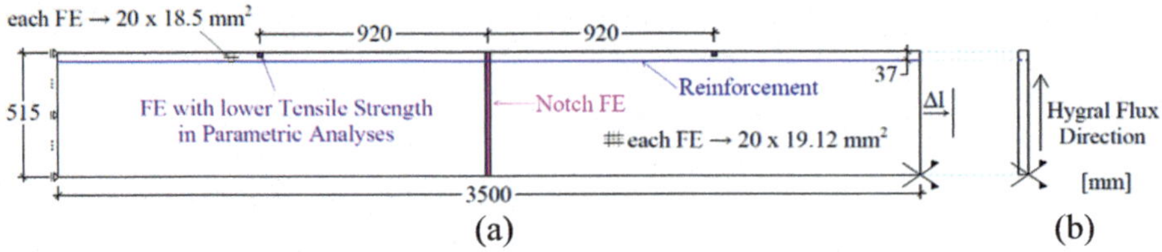

Fig. 4 FE mesh scheme: **a** mechanical analysis; **b** hygral analysis

the localization of the primary crack as in the experiment. The non-linear analysis is conducted by imposing deformation at the rightmost cross-section, in a total of 231 steps. The first step imposes a deformation of 0.1 mm (the structural behaviour is linear in this step), whereas the other 230 steps have a deformation of 0.01 mm each, reaching the total imposed deformation of 2.4 mm.

The hygral analysis is performed for a rectangular prism with a height equal to the panel thickness (1030 mm). Drying occurs from the top and bottom surfaces in a 1D flow, trough the height of the specimen, which resembles the behaviour of a continuous slab or wall. The mesh is generated with nodal compatibility with the one used in the mechanical analysis, allowing the allocation of a moisture history to all the nodes during the mechanical analysis. Two additional FE of lower tensile strength are considered in the long-term parametric analyses (see Fig. 4a), for an easier localization of the surface cracks due to drying.

3.2 Material Models for the Mechanical Analysis Simulating the Experiment

A smeared cracking model is used, based on the total strain approach, switching from rotating to fixed cracks at a threshold crack strain of 5×10^{-3} (crack width of approximately 0.01 mm). The mechanical properties of concrete, such as the compressive and tensile strength and the elastic modulus are constant. They are the same as the experimentally reported values in Table 1. The Poisson's ratio of uncracked concrete is 0.20. A damage function is applied to decrease this ratio with increasing damage due to cracking. The fracture energy is calculated from *fib* Model Code [6] based on the concrete compressive strength (Table 1), which leads to 0.144 MPa.mm. The adopted tension softening model is the one from *fib* Model Code [6], with bilinear ascending and descending paths. The constant shear retention approach is adopted, with a shear retention factor β of 0.01.

The actual bond-stress slip relationship is not known. Given that this property significantly influences the mechanical behaviour of the specimen, a parametric study is presented simulating the bond-slip law as

$$\tau = k_{bs}\sqrt{f_{cm}}\left(\frac{s}{1mm}\right)^{0.4} \tag{1}$$

where τ and f_{cm} are the bond stress and the concrete compressive strength in MPa, s is the slip in mm and k_{bs} is a calibration parameter. For $k_{bs} = 2.5$ and $k_{bs} = 1.25/1.8^{0.4} = 0.988$, Eq. (1) corresponds to the *fib* [6] proposals for "good bond conditions" and for "all other bond conditions", respectively. An intermediate value, $k_{bs} = 2.0$, was also considered because it provides the best agreement with the experimental results. The bond-slip loading and unloading stiffness corresponds to the secant stiffness for a slip value of 0.01 mm.

3.3 *Material Models for the Parametric Study of Non-uniform Drying Shrinkage Effects*

The thermo-hygral analysis framework, developed by Azenha [4], based on FEM and using MatLab software, was used to calculate the relative humidity in the concrete pores and the potential drying shrinkage strain for each material point at each age. The drying model and the parameters used are based on the work of Azenha et al. [7]. The moisture transport is ruled locally by the simplified moisture diffusion equation:

$$\frac{\partial H}{\partial t} = div(D_{H*}\nabla H) \tag{2}$$

where H is the local moisture content for a specific time t. The diffusion coefficient, D_{H*}, is given by:

$$D_{H*} = D_1\left[\frac{D_0}{D_1} + \frac{1 - D_0/D_1}{1 + ((1 - H)/(1 - H_c))^m}\right] \tag{3}$$

where H_c is the relative humidity corresponding to $D_{H*} = 0.5(D_1 + D_0)$, D_1 is the value of D_{H*} for $H = 1$, D_0 is a lower bound for the diffusion parameter and m is a material property.

The moisture flux through boundaries is simulated by a Neumann-type formulation, $q_m = h_m(H - H_{env})$, where h_m is the moisture emissivity coefficient and H_{env} is the relative humidity of the environment. The local drying shrinkage is computed by the Kwak equation [8]:

$$\varepsilon_{sh}(H) = \varepsilon_{sh,ult}\left[0.97 - 1.895(H - 0.2)^3\right] \tag{4}$$

where $\varepsilon_{sh,ult}$ is the ultimate (or asymptotic) drying shrinkage.

Table 2 Drying and creep models for the parametric analyses

Drying model		Creep model	
Parameter	Value	Parameter	Value
D_1	3.08×10^{-10} m^{2}s^{-1}	q_1	19.33
D_0	0.0967 D_1	q_2	107.78
H_c	0.8	q_3	1.72
m	2	q_4	5.37
h_m	4.81×10^{-8} ms^{-1}	η_{dc}	5 GPa/day
h_{env}	0.60	k_{dc}	25 GPa
$\varepsilon_{sh,ult}$	539×10^{-6}		

While the simulation of the experimental test is done through deformation increments, the analyses of the parametric study are performed with time increments, as detailed in Sect. 4.2.

The viscoelastic behaviour of concrete is simulated using an ageing Kelvin chain, fitted to the B3 creep compliance model [9]:

$$J\left(t,t'\right) = q_1 + q_2\,Q\left(t,t'\right) + q_3\ln\left[1+\frac{\left(t-t'\right)^{0.1}}{1day}\right] + q_4\ln\left(\frac{t}{t'}\right) \tag{5}$$

The additional creep due to simultaneous drying is given by a Kelvin-Voigt element, related to the local moisture ratio as used by [3] and Benboudjema et al. [10]:

$$\eta_{dc}\dot{\varepsilon}_{dc} + \theta\left|\dot{H}\right|k_{dc}\varepsilon_{dc} = \theta\left|\dot{H}\right|\sigma \tag{6}$$

where $\eta_{dc}/\theta\left|\dot{H}\right|$ is the viscosity of the dashpot and k_{dc} is the elasticity of the spring.

The creep parameters q_1, q_2, q_3 and q_4 are estimated based on the concrete composition and strength presented in Table 1, following model B3 guidelines [9]. Table 2 summarizes the drying and creep model parameters adopted in the parametric analysis.

4 Results

4.1 Simulation of the Experiment

In Fig. 5 the experimental restraint force is compared with the calculated one for different bond conditions ("good" and "other bond conditions", according to the *fib* Model Code [6]), and an intermediate bond model characterized by $k_{bs} = 2.0$, as explained in Sect. 3.2. The first peak value (labelled as 1 in Fig. 5) corresponds to the

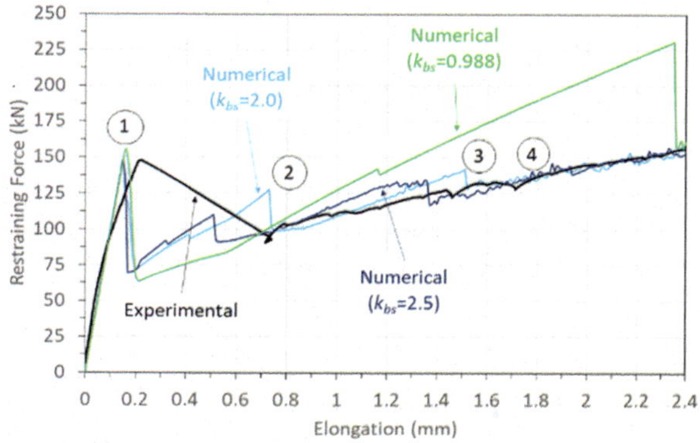

Fig. 5 Comparison of the numerical and experimental evolutions of the restraining force

formation of the primary crack. Then, the restraining force evolves in the experiment until the next position of equilibrium, which occurred for an elongation of ~0.7 mm, owing to the release of energy in the hydraulic actuators used to apply the imposed deformation. The numerical analysis with the softer bond model ($k_{bs} = 0.988$) provides a higher slope of the evolution of the restraining force, as a consequence of a weaker bond-slip activation of the reinforcement, which limits the generation of secondary cracks. In contrast, the restraint force evolutions for the analyses using the stiffer bond models ($k_{bs} = 2.0$ and 2.5) show similar slopes, which agree with the experimental results. However, whilst in the analysis with $k_{bs} = 2.5$ the formation of the first pair of secondary cracks occurs for an elongation of ~0.5 mm, in the analysis with the intermediate bond model ($k_{bs} = 2.0$) the formation of the first pair of secondary crack occurs for an elongation of ~0.75 mm.

Figure 6 presents the crack pattern for the analysis considering the bond model which provided the best global agreement with the experiment results ($k_{bs} = 2.0$). The order of crack formation is also shown, using the same labels as in Fig. 5. The first crack (the primary one), labelled as 1, penetrates through the total depth of the specimen. Subsequently, secondary cracks are formed, symmetrically, in the vicinity

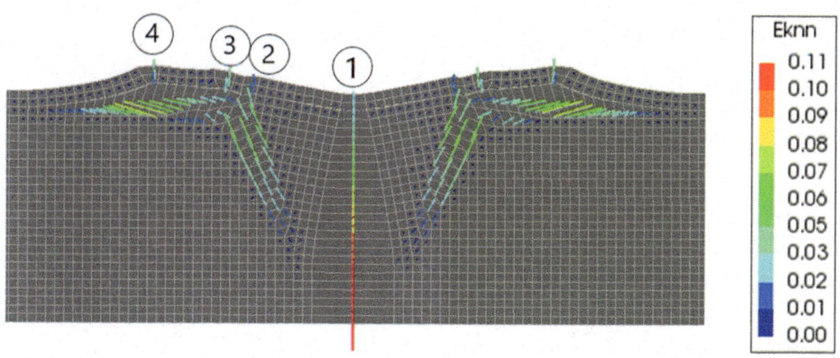

Fig. 6 Cracking pattern for the analysis with $k_{bs} = 2.0$, with identification of the sequence of crack formation, as well as the deformed mesh at the end of analysis

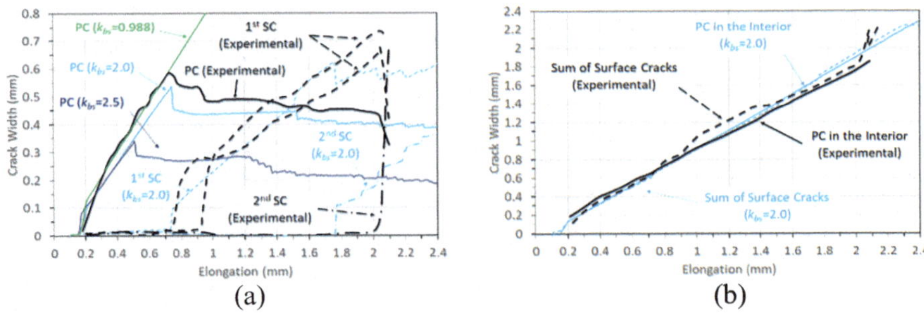

Fig. 7 Experimental and calculated crack widths, for primary cracks (PC) and secondary cracks (SC): **a** experimental versus calculated widths of primary cracks for three scenarios ($\boldsymbol{k_{bs}} = 2.5$, 2.0 and 0.988) and secondary cracks for $\boldsymbol{k_{bs}} = 2.0$ (best-fit scenario); **b** sum of the surface crack widths, and the crack width of the primary crack at the interior of the panel

of the primary crack. In the numerical analyses, each pair of symmetrical secondary cracks occurs simultaneously, and with an equal crack width on each. The first pair of secondary cracks, labelled as 2, starts to develop vertically, but once they advance in-depth, the cracks start to rotate, approaching the primary crack. With the increase of the imposed deformation, additional crack strains develop near the first pair of secondary cracks, at the position labelled as 3. These crack strains are considered to belong to the first pair of secondary cracks, given their geometrical proximity. The bond stresses in the activation area led to the generation of crack strains along the line of the reinforcement. At the same time, the second pair of secondary cracks (labelled as 4) was formed.

Figure 7 presents a comparison between the numerical and experimental crack widths. The numerically obtained crack widths are computed by the horizontal length variation between two points at a distance equal to the base-length of the LVDT used in the experiment (see Fig. 3).

In Fig. 7a, it is perceptible that the bond-slip model influences significantly the width of the primary crack. A stiffer bond model ($k_{bs} = 2.5$) leads to smaller crack widths. The stiffer bond model also leads to the generation of secondary cracks for lower imposed deformations. As soon as the first pair of secondary cracks is formed, the width of the primary crack drops, and stabilizes, or even slightly decreases, during the rest of the test. As the imposed deformation increases, the width of the first secondary cracks rises until the formation of the new pair of secondary cracks. Once a new pair of secondary cracks occurs, the width of the previous pair of secondary cracks stabilizes, as observed for the primary crack, and the new pair of secondary cracks retains the additional imposed deformation till the end of the test, or until a new pair of secondary cracks is formed. Figure 7b shows that the sum of the crack widths at the surface of the element is equal to the width of the primary crack in the core of the element. Also, the width of the primary crack in the core is slightly smaller than the imposed deformation due to the elongation of the uncracked concrete along the specimen length.

4.2 Parametric Study of Non-uniform Drying Shrinkage Effects

In order to assess the consequences of self-equilibrated stresses due to non-uniform shrinkage effects, a parametric analysis was conduced by applying an external imposed deformation at certain ages (7 days, 14 days, 28 days or 90 days after casting), after a period of drying which starts at 7 days after casting in all of the analyses. The imposed deformation ratio is 0.01 mm at every 50 s, until a total of 0.4 mm. The analyses stop 10 years after casting.

A typical crack pattern, at long-term, is presented in Fig. 8. The primary crack penetrates the total depth of the panel, while secondary cracks develop in the vicinity of the primary one. Drying-induced cracks develop vertically, starting from the concrete surface, as drying progresses. They are visible in Fig. 8, all over the length of the element. After formation of the primary crack, the drying-induced cracks near the primary one perform like secondary cracks, i.e., they rotate in-depth approaching the primary one. The level of rotation of the secondary cracks depends on the current state of cracking damage when the primary crack is formed. As drying progresses, the drying-induced cracks continue to develop vertically through the depth of the element, even after the primary and secondary crack system is created. This occurs due to the non-linearity of the drying shrinkage process and the resulting internal restraint provided by concrete.

Figure 9 presents the crack width evolution for the primary crack, at the interior and at the surface of the structure. Figure 9a shows the results during the application of the imposed deformation. As the duration of drying before imposing the external deformation is increased, the externaly imposed deformation that induces the formation of the primary crack decreases. Initially, the crack width is larger at the surface, because of drying effects. But the crack width at the interior increases at a higher rate than at the surface, the latter being overpassed at the end of application of the imposed deformation. This is explained by the contribution of the secondary cracks and drying cracks, responsible for the deformation compatibility at the surface.

Figure 9b shows the evolution of the width of the primary crack over 10 years of analysis. The crack width at the surface has a small increase, while at the interior it increases significantly with the progression of drying. All analyses show similar long-term crack widths, regardless of the time of formation of the primary crack.

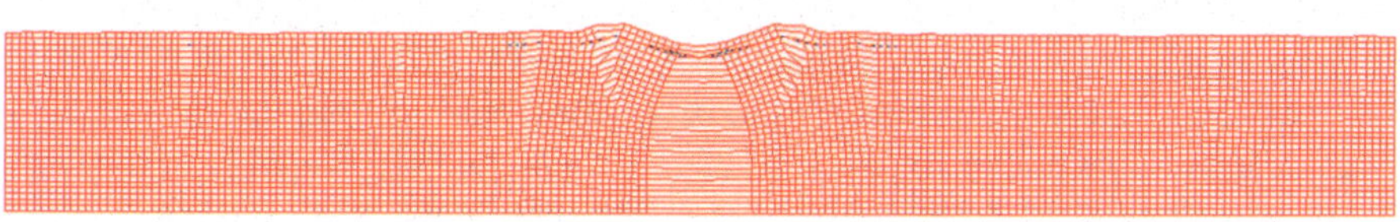

Fig. 8 Crack pattern at 10 years (analysis for imposed deformation at the age of 14 days)

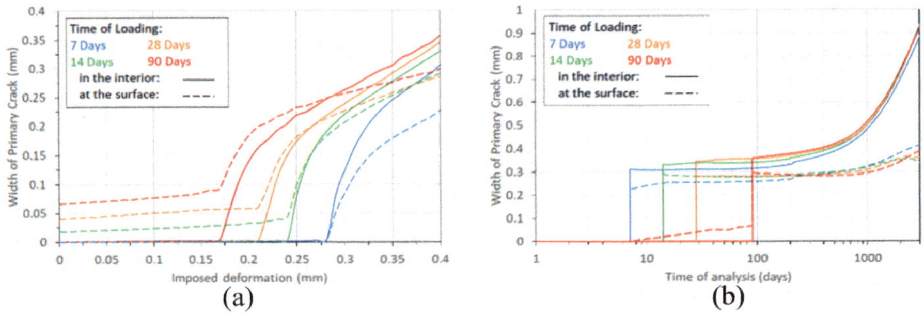

Fig. 9 Width of primary cracks at the surface and in the interior of the structure: **a** during the application of the externally imposed deformation; **b** long-term evolution

5 Conclusions

This research contributes to a better understanding of the cracking phenomena in thick RC members, with a common arrangement of reinforcement near the surface, and its FE simulation. The following conclusions are drawn:

- The crack pattern in the FE simulation is analogous to the one reached in the experiment considered as benchmark for validation and calibration of the FE model. It is composed by a primary crack that penetrates through the entire thickness of the element, and neighbouring secondary cracks. The secondary cracks rotate in-depth, in the direction of the primary one. The primary crack width stabilizes, or even decreases, once a pair of secondary cracks is formed. The width of the latter increases until a new pair of secondary cracks occurs. This process is repeated when a new pair of secondary cracks is generated.
- The bond stress-slip relationship significantly affects the calculated crack widths. By introducing a calibration coefficient into a reference bond-slip law, it was possible to obtain a good agreement with the experiment, both in terms of crack widths and stiffness of the RC member after cracking.
- In the absence of non-uniform drying shrinkage effects, the crack width at the core of the element is equal to the sum of the crack widths at the surface, both in the experiment and in the calculations.
- In the parametric analyses to study the influence of drying shrinkage, it was observed that the secondary cracks in the vicinity of the primary one have a shape configuration similar to the one in the benchmark experiment. Additional superficial cracks are formed, due to drying progressing from the surface. In the calculations, the long-term crack widths are the same independently of the degree of advancement of drying shrinkage at the instant when the primary crack is formed.

Acknowledgements This work is financed by national funds through FCT—Foundation for Science and Technology, under Grant agreement 2020.04450.BD attributed to the first author.

This work was also partly financed by FCT/MCTES through national funds (PIDDAC) under the R&D Unit Institute for Sustainability and Innovation in Structural Engineering (ISISE), reference UIDB/04029/2020, and under the R&D Unit CONSTRUCT—Instituto de I&D em Estruturas e Construções, reference UIDB/04708/2020.

References

1. Schlicke, D., Hofer, K., Tue, N.V.: Adjustable restraining frames for systematic investigation of cracking risk and crack formation in reinforced concrete under restrained conditions. In: Serdar, M., et al. (eds.) Advanced Techniques for Testing of Cement-Based Materials, pp. 211–239. Springer International Publishing, Cham (2020)
2. Schlicke, D., Tue, N.V.: Minimum reinforcement for crack width control in restrained concrete members considering the deformation compatibility. Struct. Concr. **16**(2), 221–232 (2015)
3. Ferreira, C., et al.: Thermo-hygro-mechanical simulation of cracking in thick restrained RC members: application to a 50 cm thick slab. J. Adv. Concr. Technol. **17**(8), 489–505 (2019)
4. Azenha, M.: Numerical Simulation of the Structural Behaviour of Concrete Since its Early Ages. Ph.D. Thesis, University of Porto, Faculty of Engineering (2009)
5. Turner, K.: Ganzheitliche Betrachtung zur Ermittlung der Mindestbewehrung für fugenlose Wasserbauwerke (EN: An Integrated Approach to Determination of Minimum Reinforcement for Jointless Hydraulic Structures). Ph.D. Thesis, Graz University of Technology (2020)
6. CEB-FIP: Model Code for Concrete Structures 2010. (2013)
7. Azenha, M., et al.: Experimental validation of a framework for hygro-mechanical simulation of self-induced stresses in concrete. Cem. Concr. Compos. **80**, 41–54 (2017)
8. Kwak, H.-G., Ha, S.-J., Kim, J.-K.: Non-structural cracking in RC walls: Part I. Finite element formulation. Cem. Concr. Res. **36**(4), 749–760 (2006)
9. Bažant, Z.P., Baweja, S.: Creep and Shrinkage Prediction Model for Analysis and Design of Concrete Structures: Model B3. ACI Special Publications **194**, 1–84 (2000)
10. Benboudjema, F., Meftah, F., Torrenti, J.-M.: Interaction between drying, shrinkage, creep and cracking phenomena in concrete. Eng. Struct. **27**(2), 239–250 (2005)

Binders, Additions, and Admixtures

Bleeding of Cement Pastes Modified with the Addition of Granite Powder

Adrian Chajec and Łukasz Sadowski

Abstract The article presents the results of bleeding tests for cement pastes modified with the addition of granite powder. Bleeding is the process of dispensing the water from the inside of the mixture towards the top. Bleeding in cementitious mixes is an important process related to early-age properties of composites (like cracking of concrete slabs). There are known examples of damage to horizontally formed concrete structures due to the uncontrolled bleeding process. The research was performed for four series of modified cement slurries of the addition of granite powder. The test was carried out with the use of a method allowing to obtain high accuracy of measurement (ASTM C243-95 Standard Test Method for Bleeding of Cement Pastes and Mortars (Withdrawn 2001), [1]) of the released water over time, with particular emphasis on the first 30 min of the process. Three phases of the bleeding process were observed during the investigation (initial bleeding, accelerated bleeding, stable bleeding) into cement pastes. An additional division of the first phase on three phases (start of the process, components ordering, appropriate bleeding) was also proposed. The results of the research also present that thanks to the addition of granite powder, it is possible to reduce the volume of dispensed water in the bleeding process. The main issues were recognized and directions of future research area are indicated.

Keywords Bleeding · Granite powder · Bleeding rate · Bleeding capacity · Grains morphology

A. Chajec (✉) · Ł. Sadowski
Wroclaw University of Science and Technology, Wroclaw, Poland
e-mail: adrian.chajec@pwr.edu.pl

Ł. Sadowski
e-mail: lukasz.sadowski@pwr.edu.pl

F. Kanavaris et al. (eds.), *International RILEM Conference on Early-Age and Long-Term Cracking in RC Structures*, RILEM Bookseries 31,
https://doi.org/10.1007/978-3-030-72921-9_25

1 Introduction

The properties of fresh cement paste have usually a significant impact on the properties of hardened composites. One of the important fresh properties of cement pastes is bleeding. Cement pastes are the mix of cement (or different types of binders or else powder additives) with water. Granite powder is a material produced during the process of cutting or crushing granite rocks. It is an additive that is increasingly used to improve the properties of cement composites [2, 3]. Currently, the amount of this material in heaps is so significant that it has become a hazardous waste causing pneumoconiosis, lung cancer and environmental pollution. One of the ways to reduce the amount of deposited material is to use it in cement-based materials. Bleeding is a key phenomenon related to fresh properties of the cementitious mixes used in civil engineering. Bleeding is the process of dispensing the water from the inside of the mixture towards the top.

Bleeding is a property that could determine the characteristics of hardened composites [4–6]. Too large a volume of dispensed water may reduce the mechanical properties of cement pastes, on the other hand, too small a volume of dispensed water may result in cracking of the surface of cement pastes due to shrinkage [7, 8]. Bleeding is a well-known process, although the ability to control the bleeding process is therefore one of the research goals of many researchers [9–12]. In the past, researchers believed that bleeding was a homogeneous process, but research conducted in recent years have shown that it is a heterogeneous process [13]. Bleeding shows a variable course, which means that to control this process, precise research should be performed. Until now, the precision of measuring the volume of dispensed water was impossible due to outdated measuring methods (based on manual harvesting water from the surface of the mixture and are subject to significant uncertainty of measurement (±1 mL)). However, now we can use more precise measurement methods [1] that allow us to accurately determine the course of the bleeding process and the volatility of the volume of dispensed water (based on the measurement of the separated water as a result of the gravity separation process of two liquids, allowing for very accurate measurement (±0.2 mL)).

Previously, bleeding was thought to depend only on the force of gravity and is a process related to the cement pastes compacting process [5, 6]. Recent studies indicate that the process is much more complicated and it may vary depending on the morphological properties of grains of ingredients of cement pastes (e.g. type of grains [14], particle size distribution [15] or shape of grains [13]).

The bleeding process is particularly problematic in horizontally concreted structures. Inadequately controlled bleeding is a frequent cause of cracks, voids and peeling of concrete on concrete floors. A frequent problem is also the appearance of scratches in concrete slabs, equated with the uncontrolled bleeding process. Early-age stage of curing concrete (first 30 min) is especially suitable for bleeding cracks. Research has shown [5, 7, 9] that it is possible to control the bleeding process by modifying the cement pastes with the addition of mineral powders. One of them is

Table 1 Compositions of the research series

Series	Cement (kg)	Water (kg)	Granite powder (kg)	w/c (-)	w/b (-)
REF	2.626	1.200	0	0.457	0.457
GP10	2.363	1.200	0.263	0.508	0.457
GP20	2.101	1.200	0.525	0.571	0.457
GP30	1.838	1.200	0.788	0.623	0.457

granite powder. However, there are no examples of studies on the effect of granite powder on bleeding of cementitious mixes in the literature.

Considering the above, the main aim of the article is to understand the bleeding process, especially in its initial stage (first 30 min of the process), and to check the influence of granite powder additive on the bleeding process in cement pastes.

2 Materials and Methods

2.1 Materials Used in the Research

As part of the research, 4 research series of cement pastes were prepared. Table 1 presents the composition of tested series of cement pastes. To mix the mixtures used cement CEM I 42.5R (Górażdże, Poland), water (water supply in Wroclaw, Poland) and granite powder (Strzegom, Poland). Figure 2 shows the particle size distribution of cement and granite powder which were used in the test. Compositions of cement pastes were established according to the guidelines in ASTM C 243 [1].

2.2 Preparation of the Cement Mixes

Preparation of cementitious pastes began with measuring the appropriate amount of dry ingredients (cement, granite powder, with an accuracy of 0.1 g). Then, a measured amount of water was placed in the prepared bucket, and then the measured amount of cement was added and mixed for 180 s at a mixer speed of 140 rpm. In the first 15 s after switching off the mixer, debris was scraped off the bottom of the bucket. After the initial mixing, the paste was left for 3 min. After this time the paste was mixed for 3 min at a speed of 60 rpm. After mixing instantly started bleeding measurement procedure described in Sect. 2.3.

2.3 Bleeding of the Cement Pastes

The bleeding test was performed according to the procedure described in the standard ASTM C 243 [1]. The test is used in the USA, but it is not widely used in Europe. Immediately after mixing procedure began to pour of pastes in the sample container. The sample container was filled to a height of 127 mm (up to the ring). Refilling of sample containers was completed in 15 s. The stacking end time of pastes in the sample container is taken as the start time for the test. Then the collecting ring is placed in the centre of the paste surface and embed it to a depth of approximately 19.1 mm using the three support brackets. The surface of pastes has been covered with 1,1,1-trichloroethane to a depth of 31.8 mm. Funnel-buret assembly to within 6.4 mm and directly over the collecting ring set. After that, pulled up 1,1,1-trichloroethane drown up into the buret and closed the stopcock. The device uses the gravitational sinking effect of the heavier liquid (1,1,1-trichloroethane) as a result of which the water travels up the pipette and it is possible to accurately read the amount of dispensed water in the bleeding process. The first measurement was performed 4 min after the starting time. Then measurements were taken at 2 min intervals until 30 min after starting time, and then at 10 min intervals until bleeding is not over. Figure 1 shows bleeding measurement procedure.

Bleeding in cement pastes is characterized by the following indicators: bleeding rate (R_B) and bleeding capacity (C_B). The following formulas are described in the

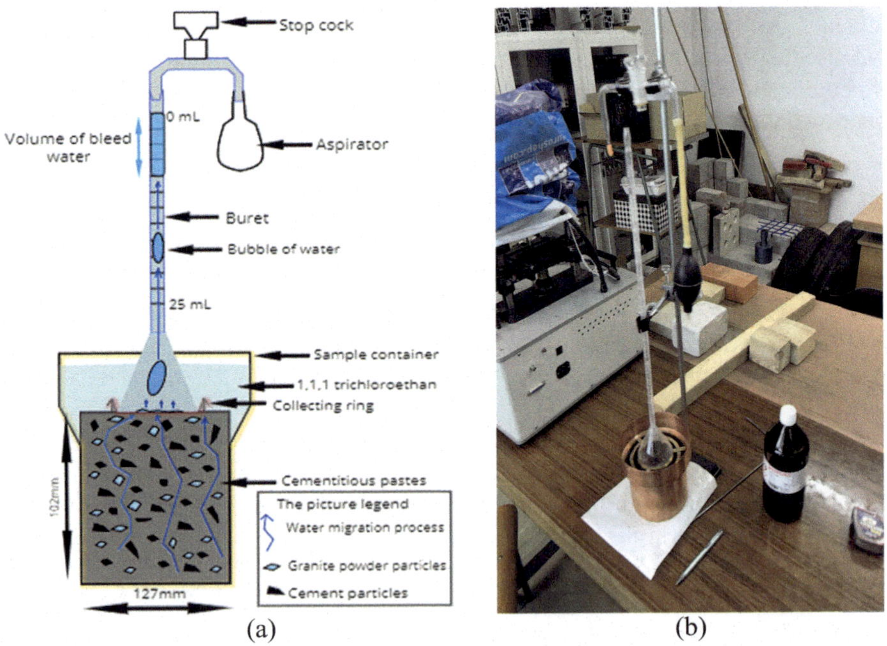

Fig. 1 Cement pastes bleeding process research **a** test diagram **b** photo from research conducting

standard [1].

$$R_B = \frac{V_1}{A \cdot t}\,(\mathrm{mL/cm^2\,s}) \tag{1}$$

R_B bleeding rate (mL/cm^2 s)
V_1 the volume of bleeding water (mL), measured during the time interval (t)
A area covered by the 76.2 mm diameter collecting ring, (cm^2)
t time during which bleeding occurs at a uniform rate (s).

$$C_B = \frac{V_2}{V_3}\,(\mathrm{mL/mL}) \tag{2}$$

C_B bleeding capacity (mL/mL)
V_2 the total volume of bleeding water (mL)
V_3 volume of the cylinder of paste directly under the 76.2 mm diameter collecting ring, (mL).

3 Results and Analysis

3.1 Morphological Properties of Cement and Granite Powder

The materials used in the production of cement pastes have determined morphological properties. The most important of these can be the graining curve and the Blaine surface specific area.

Figure 2 shows the particle size distribution of the cement and granite powder used in the tests. The curves presented in the diagram are dissimilar. Granite powder consists of more particles smaller than 0.063 mm as cement. It consequently has a

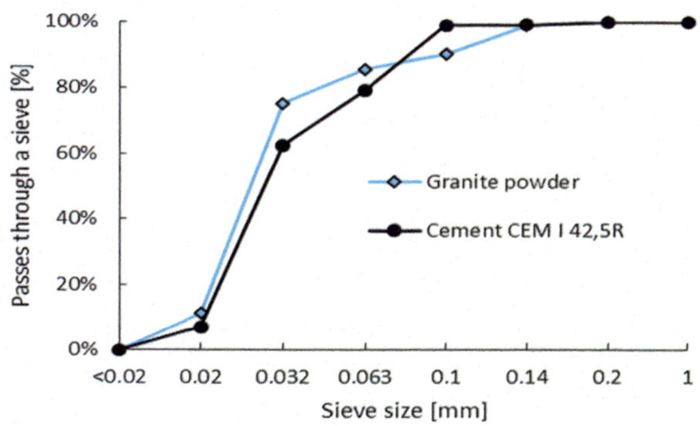

Fig. 2 The particle size distribution of the cement and granite powder used in research

Table 2 Blaine specific surface area of ingredients

Ingredient	Blaine specific surface area (cm^2/g)
Cement CEM I 42, 5R	3590
Granite powder	3920

finer graining compared to cement. The materials are characterized by the similar grain size of parts >0.1 mm.

The particle size distribution of cement systems has an impact on the properties of cement pastes: compactness, consistency, air-void content [16]. Additionally, it was observed [17] that optimized particle size distribution may be able to influence on the permeability or physical properties of cement composites.

Table 2 shows the Blaine specific surface area of cement and granite powder used in the research. Blaine specific surface area of that two materials are dissimilar. Granite powder has a higher value of Blaine specific surface area compared to cement. It may be related to the results of a sieve size development test (granite powder characterised of a higher amount of smaller parts of grains <0.063 mm).

Blaine specific surface area is important properties of raw materials. In research [18] authors observed the relationship between the specific surface area of cement and pozzolanic reactivity of this material. Also in research [19] authors are suggesting, that specific surface area of raw materials is properties that could have an impact on the hydration process in cement. An interesting observation was presented by researchers [20]. They concluded that the shape factor (morphological properties of cement grains) and specific surface area are deeply related to each other.

3.2 *Bleeding of the Cement Pastes*

Bleeding is a process also related to the rheological properties of pastes [12, 21]. The preparation of the ingredients and the mixing procedure influences the bleeding process [22, 23].

Figure 3 shows the volume of the bleeding water V_b that is dispensed from the cement pastes overtime of the bleeding process. The graph shows the averaged results of 4 measurements made for each test series. The uncertainty in the results between measurements for each series was less than 5%. For the sake of clarity of the chart, the uncertainty has not been shown in the diagram. Water volume dispensed from cement pastes modified with granite powder addition is significantly smaller (approximately 10%) than the water volume measured in the reference mix. Moreover, it can be seen that the amount of dispersed water is dependent on the quantity of granite powder addition—with increasing of granite powder addition, the amount of dispersed water in the bleeding process decreases. The characteristic of the bleeding process tends to change over time. The chart shows a clear division into 3 zones in which bleeding can change dynamic of water volume dispensed in that process: initial bleeding (0–30 min from mixing), accelerated bleeding (30–60 min from mixing) and stable

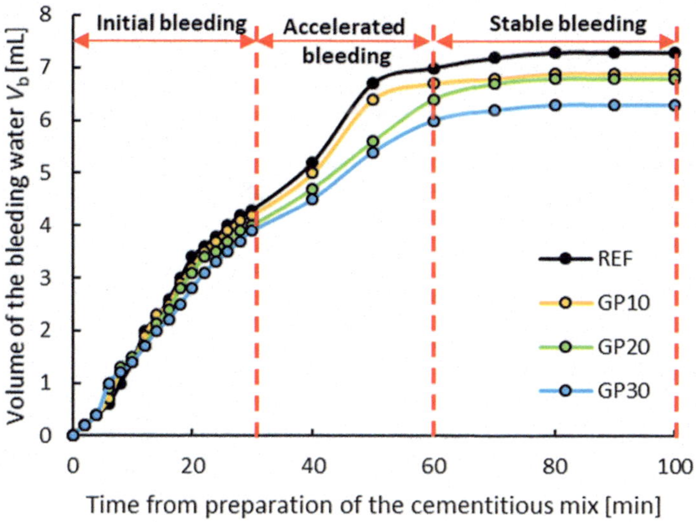

Fig. 3 The bleeding process of the cement pastes

bleeding (60 min from mixing—end of bleeding). Initial bleeding is characterized by an almost linear increase in the volume of dispersed water from cement pastes. It was also noticed that in this zone the volume of dispensed water in time is similar for all research series. In this zone, the impact of modification of cement pastes with the addition of granite powder is not noted. However dynamic of changes in that phase needed more observation (Fig. 3). Accelerated bleeding occurs approximately 30 min after the mixture has been mixed. After this time, the volume of dispensed water in cement pastes series can be easily distinguished. Moreover, in this zone, bleeding may have a differential cycle for each investigated series. Significant variances were noticed in the process of the bleeding and it can depend on the material composition of the research series. Stable bleeding begins approximately 60 min after the mixture has been mixed. It is characterized by small changes in the volume of dispensed water and a slight dynamic variation in the process. There are variances in the volume of dispensed water. The plotline is close to a straight line. The differences in the cycle of the bleeding process (related to the material composition) are also clearly visible on the chart.

The described properties allow us to conclude that the bleeding process is heterogeneous. Similar observations were made during the research [13]. After the investigation of bleeding nature, the authors proposed to divide the bleeding process into 5 phases. There was also a thesis that bleeding depends on the formation of water extraction channels, which depend on the material composition of cement pastes and the mixing conditions of pastes. Moreover, authors of research [4] noticed that bleeding was dependent on the w/c rate used. They also proposed a division into materials in which bleeding occurs and not due to the w/c rate. It was also shown that the bleeding process for pastes is dependent on the yield stress of those materials.

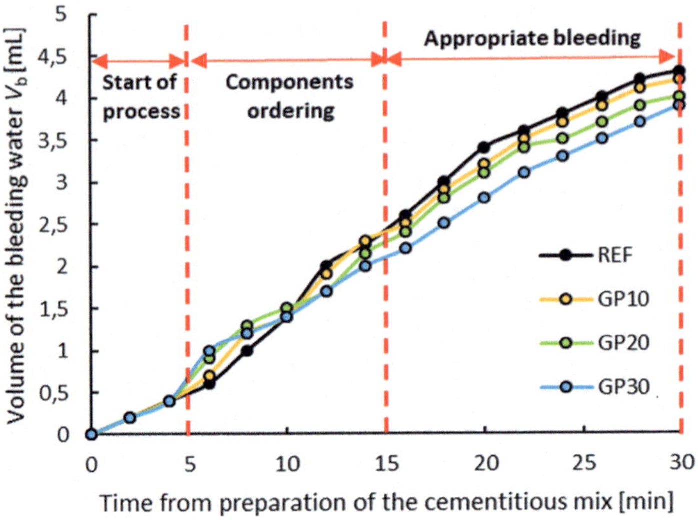

Fig. 4 The process of initial bleeding phase in the cement pastes

The range of bleeding occurrence was also determined depending on the yield stress value. The analysis of the test results of other researchers allows for the conclusion that they indicate the right research path related to the bleeding process in cement pastes, however, they may not be accurate due to the use of an inaccurate measurement method (thickness of the water layer on top of pastes in the plastic cylinder—measurement with high inaccuracy).

Figure 4 shows the volume of the bleeding water V_b that is dispensed from the cement pastes over the first phase of bleeding—initial bleeding (0–30 min after mixing the mixture). The graph shows the averaged results of 4 measurements made for each test series. The uncertainty in the results between measurements for each series was less than 5%. For the sake of clarity of the chart, the uncertainty has not been shown in the diagram. Water volume dispensed from cement pastes modified with granite powder addition is significantly smaller (approximately 10%) than the water volume measured in reference mix, however, in that phase it is not so evident compared with Fig. 4. The characteristic of the initial bleeding process (first 30 min after mixing) has the tendency to change over time. The chart shows a clear division into 3 phases in which bleeding is able to change dynamic of water volume dispensed in that process: phase 1—the start of the process (0–5 min from mixing), phase 2—components ordering (5–15 min from mixing) and phase 3—appropriate bleeding (+15 min from mixing). Phase 1 is characterized by an almost linear increase in the volume of dispersed water from cement pastes. It was also noticed that in this zone the volume of dispensed water in time is similar for all research series. In this zone, the impact of modification of cement pastes with the addition of granite powder is not noted. Phase 2 occurs approximately 5 min after the mixture has been mixed. After this time, the volume of dispensed water in cement pastes series can be easily distinguished. Moreover, in this zone, bleeding may have a differential cycle for each investigated series. This phase is characterized by the reordering of the internal structure of cement pastes. The grains of the materials begin to arrange themselves following the action of gravity and intermolecular interactions. Phase 2 is directly related to the type of used materials and size of grains that materials. Appropriate bleeding begins approximately 15 min after the mixture has been mixed. It is characterized by almost linear changes in the volume of dispensed water and variances in the volume of dispensed water.

There are many examples of bleeding being defined in cement pastes in the literature [4, 5, 24, 25]. Nevertheless, most of them were determined as a result of the method of measuring of thickness of water layer in a plastic tube. Due to the characteristics of this method, the measurements may be inaccurate. The description of the bleeding process, especially in the initial phase, is extremely important due to the functional properties of cement pastes. Thanks to the method used in these tests, it is possible to define the course of bleeding in the first 30 min after mixing the mixture.

Figure 5 shows the bleeding rate R_B and the bleeding capacity C_B of cement pastes modified with the addition of granite powder. Bleeding rate R_B is an indicator that describes the process of bleeding in the first 30 min. It can allow investigating the main relationship between the volume of dispensed water in the bleeding process and composition of the research series. Bleeding capacity C_B is an indicator that describes the relationship between the total volume of bleeding water (V_2) and the volume of the cylinder of paste directly under the collecting ring (V_3). It may be used to describe the changes of volume of bleed water in the whole bleeding process,

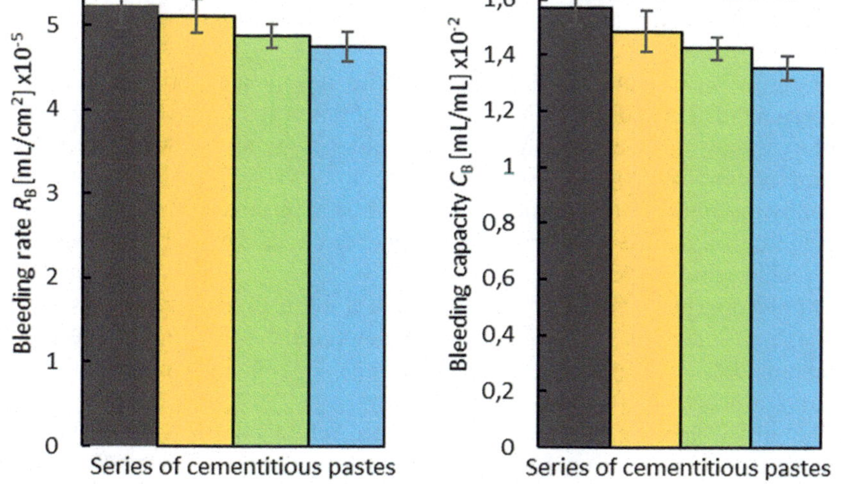

Fig. 5 The bleeding rate R_B (**a**) and the bleeding capacity C_B (**b**) in cement pastes

mainly used to compare the impact of the composition of research series on the bleeding process.

Figure 5a shows the bleeding rate R_B of cement pastes modified with the addition of granite powder. The addition of granite powder reduces the bleeding rate R_B of cement pastes. Bleeding rate of series with the addition of 30% of granite powder (GP30) is approximately 10% lower compared to the reference series (REF).

Figure 5b shows the bleeding capacity C_B of cement pastes modified with the addition of granite powder. The addition of granite powder allows on decrease the bleeding capacity C_B of cement pastes. The bleeding capacity of series with the addition of 30% of granite powder (GP30) is approximately 15% lower compared to the reference series (REF).

Noticeable changes of bleeding rate R_B and bleeding capacity C_B may be related to the difference in specific surface area of both materials (Blaine specific surface area of granite powder is higher compared to cement). Additionally, granite powder has a dissimilar grain structure as cement, which may also affect the bleeding rate [5]. Granite powder and cement have a different grain size curve. More fine particles can help reduce the number and the size of water extraction channels in cement pastes structure [13].

4 Conclusions

Based on performed research, authors can conclude that the addition of granite powder has a significant impact on the bleeding process in cementitious pastes. It was observed that with the increase in the amount of granite powder the bleeding rate of cement pastes was decreased. Additionally, the following relations were observed:

- The morphological properties of raw materials have a significant impact on the bleeding process,
- The influence of shape factor on the properties of cement materials is an important issue in the field. Additionally, the authors of this article believe that the bleeding process may be dependent on the shape factors of grains of raw materials used in the production of cement pastes. Research related to this issue is currently carried out by the authors of this article and should be published in forward,
- Addition of granite powder to cement pastes could reduce bleed water in the bleeding process about 10%,
- The bleeding process is heterogeneous and it is able to divide that process into three phases: 1—initial bleeding (0–30 min after mixing), 2—accelerated bleeding (30–60 min after mixing) and 3—stable bleeding (+60 min after mixing), the phases are significantly different from each other,
- The initial bleeding phase could be divided into 3 phases: 1—the start of the process (0–5 min from mixing), 2—components ordering (5–15 min from mixing) and 3—appropriate bleeding (+15 min from mixing),

- The bleeding rate R_B and the bleeding capacity C_B are dependent on the composition of the research series.

The conducted research indicates significant shortcomings in the research on the bleeding process. The morphological properties of particles of raw materials on the bleeding process should be performed. Besides, tests related to changes in the microstructure of the cement paste related to the bleeding process should also be carried out. This may help to establish the proper nature of the bleeding process.

Acknowledgements The authors received funding from the project supported by the National Centre for Research and Development under the program LIDER XI [grant no. LIDER/35/0130/L-11/19/NCBR/2020 "The use of granite powder waste for the production of selected construction products"].

References

1. ASTM C243-95: Standard test method for bleeding of cement pastes and mortars (Withdrawn 2001). https://compass.astm.org/Standards/WITHDRAWN/C243.htm. Accessed 06 Dec 2020
2. Asadi Shamsabadi, E., Ghalehnovi, M., de Brito, J., Khodabakhshian, A.: Performance of concrete with waste granite powder: the effect of superplasticizers. Appl. Sci. **8**(10), 1808 (2018). https://doi.org/10.3390/app8101808
3. Gupta, L.K., Vyas, A.K.: Impact on mechanical properties of cement sand mortar containing waste granite powder. Constr. Build. Mater. **191**, 155–164 (2018). https://doi.org/10.1016/j.conbuildmat.2018.09.203
4. Perrot, A., Lecompte, T., Khelifi, H., Brumaud, C., Hot, J., Roussel, N.: Yield stress and bleeding of fresh cement pastes. Cem. Concr. Res. **42**(7), 937–944 (2012). https://doi.org/10.1016/j.cemconres.2012.03.015
5. Peng, Y., Jacobsen, S.: Influence of water/cement ratio, admixtures and filler on sedimentation and bleeding of cement paste. Cem. Concr. Res. **54**, 133–142 (2013). https://doi.org/10.1016/j.cemconres.2013.09.003
6. Wainwright, P.J., Ait-Aider, H.: The influence of cement source and slag additions on the bleeding of concrete. Cem. Concr. Res. **25**(7), 1445–1456 (1995). https://doi.org/10.1016/0008-8846(95)00139-4
7. Topçu, I.B., Elgün, V.B.: Influence of concrete properties on bleeding and evaporation. Cem. Concr. Res. **34**(2), 275–281 (2004). https://doi.org/10.1016/j.cemconres.2003.07.004
8. Peng, Y., Lauten, R.A., Reknes, K., Jacobsen, S.: Bleeding and sedimentation of cement paste measured by hydrostatic pressure and Turbiscan. Cem. Concr. Compos. **76**, 25–38 (2017). https://doi.org/10.1016/j.cemconcomp.2016.11.013
9. Mirza, J., Mirza, M.S., Roy, V., Saleh, K.: Basic rheological and mechanical properties of high-volume fly ash grouts. Constr. Build. Mater. **16**(6), 353–363 (2002). https://doi.org/10.1016/S0950-0618(02)00026-0
10. Ji, Y., Pel, L., Sun, Z.: The microstructure development during bleeding of cement paste: an NMR study. Cem. Concr. Res. **125**, 105866 (2019). https://doi.org/10.1016/j.cemconres.2019.105866
11. Poon, C.S., Kou, S.C., Lam, L.: Influence of recycled aggregate on slump and bleeding of fresh concrete. Mater. Struct. Constr. **40**(9), 981–988 (2007). https://doi.org/10.1617/s11527-006-9192-y
12. Ben aicha, M., Burtschell, Y., Alaoui, A.H., El Harrouni, K., Jalbaud, O.: Correlation between bleeding and rheological characteristics of self-compacting concrete. J. Mater. Civ. Eng. **29**(6), 05017001. https://doi.org/10.1061/(asce)mt.1943-5533.0001871

13. Massoussi, N., Keita, E., Roussel, N.: The heterogeneous nature of bleeding in cement pastes. Cem. Concr. Res. **95**, 108–116 (2017). https://doi.org/10.1016/j.cemconres.2017.02.012
14. Han, J., Wang, K.: Influence of bleeding on properties and microstructure of fresh and hydrated Portland cement paste. Constr. Build. Mater. **115**, 240–246 (2016). https://doi.org/10.1016/j.conbuildmat.2016.04.059
15. Olorunsogo, F.T.: Particle size distribution of GGBS and bleeding characteristics of slag cement mortars. Cem. Concr. Res. **28**(6), 907–919 (1998). https://doi.org/10.1016/S0008-8846(98)00042-8
16. Sevim, Ö., Demir, İ: Optimization of fly ash particle size distribution for cementitious systems with high compactness. Constr. Build. Mater. **195**, 104–114 (2019). https://doi.org/10.1016/j.conbuildmat.2018.11.080
17. Sevim, Ö., Demir, İ: Physical and permeability properties of cementitious mortars having fly ash with optimized particle size distribution. Cem. Concr. Compos. **96**, 266–273 (2019). https://doi.org/10.1016/j.cemconcomp.2018.11.017
18. Cordeiro, G.C., Toledo Filho, R.D., Tavares, L.M., Fairbairn, E.D.M.R., Hempel, S.: Influence of particle size and specific surface area on the pozzolanic activity of residual rice husk ash. Cem. Concr. Compos. **33**(5), 529–534 (2011). https://doi.org/10.1016/j.cemconcomp.2011.02.005
19. Mantellato, S., Palacios, M., Flatt, R.J.: Reliable specific surface area measurements on anhydrous cements. Cem. Concr. Res. **67**, 286–291 (2015). https://doi.org/10.1016/j.cemconres.2014.10.009
20. Hunger, M., Brouwers, H.J.H.: Flow analysis of water-powder mixtures: application to specific surface area and shape factor. Cem. Concr. Compos. **31**(1), 39–59 (2009). https://doi.org/10.1016/j.cemconcomp.2008.09.010
21. Sonebi, M., Abdalqader, A., Fayyad, T., Perrot, A., Bai, Y.: Optimisation of rheological parameters, induced bleeding, permeability and mechanical properties of supersulfated cement grouts. Constr. Build. Mater. **262**, 120078 (2020). https://doi.org/10.1016/j.conbuildmat.2020.120078
22. Pan, J., Gao, X., Ye, H.: Influence of rheological behavior of mortar matrix on fresh concrete segregation and bleeding. Iran. J. Sci. Technol. Trans. Civ. Eng. 1–15. Springer Science and Business Media Deutschland GmbH (2020). https://doi.org/10.1007/s40996-020-00450-5
23. von Bronk, T., Haist, M., Lohaus, L.: The influence of bleeding of cement suspensions on their rheological properties. Materials (Basel) **13**(7), 1609 (2020). https://doi.org/10.3390/ma13071609
24. Josserand, L., Coussy, O., de Larrard, F.: Bleeding of concrete as an ageing consolidation process. Cem. Concr. Res. **36**(9), 1603–1608 (2006). https://doi.org/10.1016/j.cemconres.2004.10.006
25. Sun, Z., Young, C.: Bleeding of SCC pastes with fly ash and GGBFS replacement. J. Sustain. Cem. Mater. **3**(3–4), 220–229 (2014). https://doi.org/10.1080/21650373.2013.876373

Effect of Expansive Additives on the Early Age Elastic Modulus Development of Cement Paste by Ambient Response Method (ARM)

Mayank Gupta, Igarashi Go, Jose Granja, Miguel Azenha, and Tetsuya Ishida

Abstract Expansive additives are being widely used as shrinkage reducing/compensating admixture to reduce the early-age cracking and chemical prestressing. Such expansive additives influence the composition of cement hydration products and mechanical properties (stiffness and strength) of the cement and concrete, which ultimately affects the early age deformation. In the current study, the EMM-ARM (Elastic Modulus Measurement by Ambient Response Method), is used because of its ability to continuously monitor cement's real-time stiffness development paste with expansive additives. Stiffness development of the cement paste with expansive additives is captured in the unrestrained condition by allowing the free expansion during the stiffness measurement. An expansive additive, based on free lime, with a 10% (by weight) cement replacement ratio is studied in the current work. Results from the EMM-ARM are compared to the elastic modulus measurement from the compression testing of cylindrical specimens, and the mechanisms of expansive action on the microstructure in early-age are discussed.

Keywords Expansive additives · Elastic modulus · EMM-ARM · Compressive strength

1 Introduction

Concrete has the inherent property of shrinking due to drying, and cracking may occur due to the combined effect of restrained shrinkage deformations and the low tensile strength of concrete. These cracks affect the mechanical properties of concrete, such as strength and stiffness, and also negatively affect the durability. Shrinkage reducing admixture (SRAs) or expansive additives (EAs) have been extensively used

M. Gupta (✉) · I. Go · T. Ishida
Department of Civil Engineering, The University of Tokyo, Tokyo 113-8656, Japan
e-mail: mayank@concrete.t.u-tokyo.ac.jp

J. Granja · M. Azenha
ISISE, Department of Civil Engineering, University of Minho, Guimarães, Portugal

F. Kanavaris et al. (eds.), *International RILEM Conference on Early-Age and Long-Term Cracking in RC Structures*, RILEM Bookseries 31,
https://doi.org/10.1007/978-3-030-72921-9_26

to mitigate the effects of concrete shrinkage and ultimately to reduce the risk of cracking [1, 2], or at least to reduce the extent of cracking. EAs operate by causing early age expansion, and depending on the amount and type of expansive additives, such additives also find their application in chemical prestressing [3].

Different EAs include calcium sulfoaluminate, free lime, and MgO based. Such EA's differ in terms of different parent species, expansive product, rate of hydration, and generated expansive forces [4–6]. The current study is focused on expansive additives based on free lime. Due to the very high reactivity of free lime, such expansive additives reacts very rapidly and causes high early-age expansion [7]. Literature shows that the addition of such EA affects the mechanical properties of the cement concrete [8, 9]. Since rate of hydration of such expansive additives is very fast, it is essential to study the effect of such EA on the early age mechanical properties of cement and paste with expansive additives namely strength and stiffness.

The Elastic Modulus Measurement through Ambient Response Method (EMM-ARM) has been developed a decade ago, and it is used to measure the stiffness development of cement-based materials since early ages [10]. EMM-ARM enables the measurement of the elastic modulus of the tested material continuously right after cement paste mixing/casting [11]. In the version of EMM-ARM applied in this research work, cement paste is poured into an acrylic tube, and the fundamental frequency of the resulting composite tube is measured continuously. The measured resonant frequency is further related to the elastic stiffness of the cement paste. The normally applied procedure of filling the tube completely might find issues in the current application to EA's, because of the potential cross-sectional restraints that might come around and induce artifacts, such as eigenstresses due to restrained expansion, and possibly result in a denser microstructure. Thus to capture the effect of expansive additives on the early age stiffness of the cement paste under free expansion, the acrylic tube is filled only half, which enables expansion to happen in a less restrained manner inside the tube.

2 Experimental Materials and Methods

For the current study, Ordinary Portland Cement (OPC) and free lime (CaO based) based expansive additive (EA) was used. The chemical and mineral composition of OPC and expansive additive is given in Tables 1 and 2. Mineral composition of the OPC and EA is calculated by internal standard method, by mixing 10% corundum (Al_2O_3) of the weight of the cement powder. Sample preparation for XRD was done by back-loading method and Rietveld analysis was done using TOPAS (Bruker) software.

To determine the mechanical properties of the cement paste with EA two different mixes were prepared, with 0 and 10% cement replacement (percentage with the total weight of the solids) by expansive additive, and using the water to cement ratio (w/c) of 0.4. After dry mixing the cement and EA for 90 s at 140 rpm, further mixing is done according to ASTM C 305-06 [12].

Table 1 Chemical composition of expansive additive and ordinary Portland cement used

Material	Density	Chemical composition (Weight %)						
	g/cm^3	LOI	SiO2	Al_2O_3	Fe_2O_3	CaO	MgO	SO_3
Expansive additive	3.10	3	1.2	5.0	1.2	69.5	0.9	19.5
OPC	3.16	0.6	21.8	4.5	2.9	64	1.8	2.26

2.1 Compression Test

Compression test was performed on cylindrical specimens of 5 cm diameter and 10 cm height. In order to allow free expansion cylindrical specimens are demolded after 8–9 h of casting. After demolding, specimens are wrapped with polythene for sealed curing and kept in controlled environment chamber with temperature of 20 °C. Measurement for compression tests were done after 1, 3 and 7 days of age. For the measurement of elastic modulus, two Linear variable differential transducers (LVDT) are used, which are supported by two steel rings with a gauge length of 5cm. Testing was performed with a displacement rate of 0.5 mm/min. Elastic modulus is calculated by calculating the slope of the stress-strain plot up to 30% of the maximum load.

2.2 EMM-ARM Measurement

EMM-ARM is used for the continuous measurement of the elastic modulus right after the cement paste being poured to the tube. In this methodology, the fundamental flexural frequency of the composite cantilever beam of cement paste and acrylic tube is determined continuously. This frequency is then related to the elastic modulus of the composite beam by using the dynamic equation of motion. The experimental setup for the measurement is shown in Figs. 1 and 2, details of which can be found elsewhere [8, 9].

A high-resolution ceramic shear piezoelectric accelerometer (PCB Piezotronics 352B, sensitivity 1000 mV/g; frequency range: 2–10,000 Hz and mass of 24.25 gm) is attached at the free end of the cantilever tube to measure the vibration in the vertical direction. The beam is subjected to naturally occurring environmental vibration which is magnified by a fan placed in the vicinity. Acceleration data was recorded at the data acquisition frequency of 1000 Hz for a time interval of 132 s after every 10 min. Measurements were done for seven days. The fundamental frequency is calculated using the Welch procedure [13] by applying fast Fourier transform (FFT) to subsets of 8192 points and Hanning window with 50% overlaps. The fundamental frequency is calculated by identifying the highest peak in the power spectral density (PSD) plot.

For the measurement of elastic modulus of cement paste with expansive additives under free expansion, the tube is filled half by keeping the tube vertically and then

Table 2 Mineral composition in expansive additive and ordinary Portland cement used

	Mineral phases (Weight %)								
	C_3A	C_2S	C_3S	C_4AF	f-CaO	$C_4A_3\overline{S}$	$C\overline{S}H_2$	$C\overline{S}H_{0.5}$	$C\overline{S}H$
Expansive additive	–	–	–	–	44.0	12.2	–		31.0
OPC	5.6	18.3	62.2	9.0	–	–	1.6	2.0	

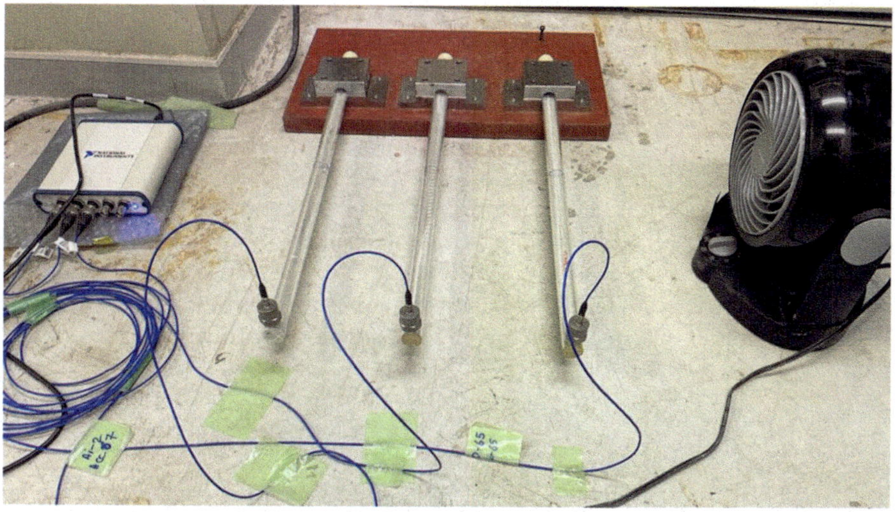

Fig. 1 Experimental setup for elastic modulus measurement by EMM-ARM

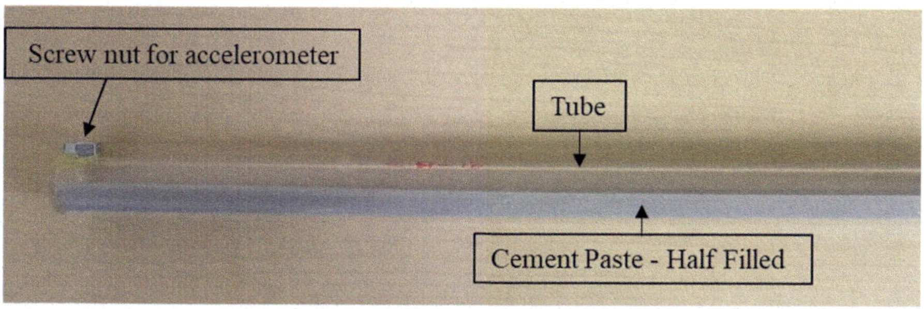

Fig. 2 Experimental setup for elastic modulus measurement by EMM-ARM

vibrated by holding the tube horizontally such that the tube is filled up to the diameter and paste has space for expansion. Although authors do acknowledge that, due to use of the expansive additives the height of the cement paste might slightly increase with the hydration. However, with 10% addition of free lime based additives the expansion was observed to be around 0.2% and thus such change in the height of the cement paste with time in currently ignored [14]. All the measurements are done at 20 °C under sealed conditions. Elastic modulus of the cement paste is calculated from the resonant frequency by the Eq. (1):

$$a^3[\cosh(aL)\cos(aL)+1]+\frac{\omega^2 m_p}{EI}[\cos(aL)\sinh(aL)-\cosh(aL)\sin(aL)]=0 \tag{1}$$

where,

$$a = \frac{\sqrt[4]{\omega^2 \overline{m}}}{EI} \tag{2}$$

Here, ω is the first angular frequency of the beam calculated as $\omega = 2\pi f$, f is the fundamental frequency (Hz), $\overline{m}$ is the mass per unit length of the beam (kg/m), m_p is the mass at the free end of the beam (accelerometer and screw nut), L is the length of the beam (450 mm for the current study). EI is the flexural rigidity (Pa m^4) of the composite cross-section (includes the acrylic tube and cement paste filled up to the diameter of the tube) calculated along the neutral axis (Fig. 3), which can be calculated as per the Eq. 3.

$$Y_{NA} = \frac{Y_{tube} \times E_{tube} \times A_{tube} + Y_{CP} \times E_{CP} \times A_{CP}}{E_{tube} \times A_{tube} + E_{CP} \times A_{CP}} \tag{3}$$

Here, E_{tube}, E_{CP}, $A_{tube,}$ and A_{CP} are the elastic modulus of the tube, the elastic modulus of cement paste, area of cross-section of tube and cement paste, respectively. Y_{NA}, Y_{tube}, Y_{CP} represents the depth of the neutral axis, central of gravity (CG) of tube, and CG of cement paste, respectively from the central axis of the tube. The elastic modulus and density of the acrylic tube were determined by calculating the frequency of vibration for the empty tube and measuring the weight and dimensions of the tube before each test (Table 3).

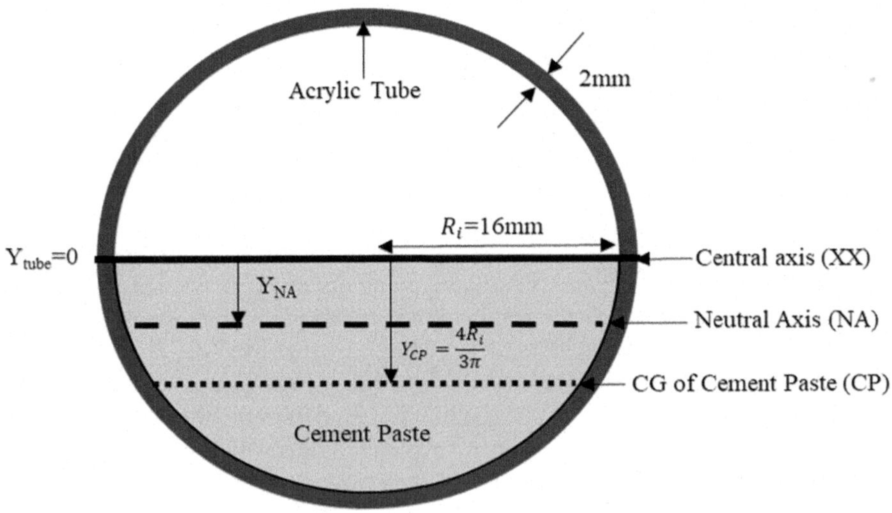

Fig. 3 Cross Section of the half-filled tube for the EMM-ARM

Table 3 Physical and mechanical characteristics of the cantilever beam for EMM-ARM

Cement Mix	OPC	OPC + 10% EA
Tube geometry		
Inner diameter (mm)	16	16
Outer diameter (mm)	20	20
Span length (mm)	450	450
Density of mould (kg/m^3)	1144.9	1147.1
Elastic modulus of mould (GPa)	4.22	4.31
Density of cement paste (kg/m^3)	1797.5	1798.6

3 Results and Discussion

Elastic modulus development curves were obtained for both the cement mixes and plotted together in Fig. 4. Both the curves start with some dormant stage up to the percolation limit. Further, the elastic modulus development is rapid and much similar up to 1 day of age, whereas after such age, the slope of the curve decreases, and both the curves separate. It is evident from the plot that with the addition of 10% of expansive additive elastic modulus decreases significantly. Due to ease of expansion inside the tube bulk volume of the cement paste increased, and it may potentially have resulted in a higher porosity and less stiff cement matrix with lower elastic modulus, similarly to what has been reported in [15, 16].

Elastic modulus values calculated from the compression test follow a similar trend as EMM-ARM, hence confirming the validity of the current methodology. The observed decrement in the elastic modulus is in agreement with previous work on EA with concrete specimens [8, 9]. For the case of OPC, values of elastic modulus from two methods of measurement (EMM-ARM and Compression test) are much similar, but values of elastic modulus from compression test for cement paste with 10% EA are slightly lower. Although due to free space, there is an ease in expansion still, due to bonding between the tube and the cement paste, there is still some amount of restraint in the longitudinal direction with may result in a denser microstructure. This might be the reason for a little higher modulus of elasticity from EMM-ARM as compared to the compression test.

The addition of expansive additive has also affected the compressive strength of the cement paste. Figure 5 illustrates the results of compressive strength for the two cement mixes: similar trends to those already observed in E-modulus are observable herein. It is evident that with the addition of 10% free lime-based EA, strength also decreased significantly. Such decrement was also observed in previous works on concrete specimens [8, 9, 16].

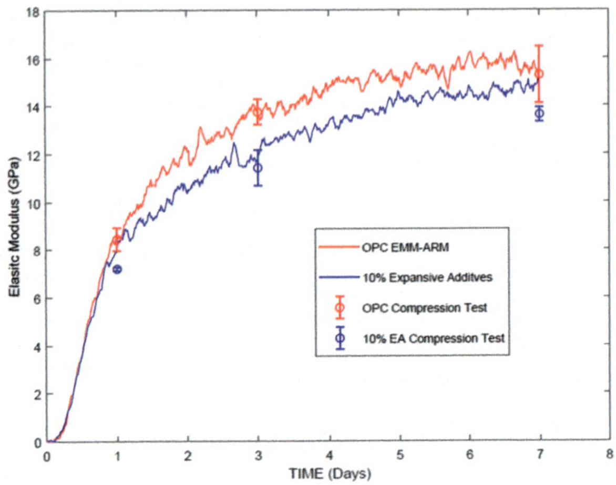

Fig. 4 Elastic modulus measurements by EMM-ARM and compression test

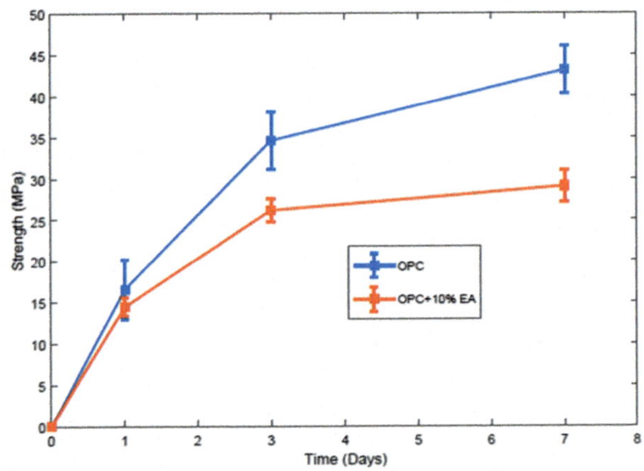

Fig. 5 Comparison of compressive strength for OPC and OPC + 10% EA

4 Conclusion

In the current work, the effect of expansive additive, based on free lime, on the early age mechanical properties, namely strength, and stiffness is studied. Both the properties are tested under conditions of unrestrained expansion. EMM-ARM results have effectively captured the stiffness development of OPC and OPC + 10% EA paste. The addition of expansive additive has reduced the stiffness of the cement paste. Elastic modulus calculated from the stress-strain plot of the compression test shows a similar trend. Both the test methods show a good correlation for the OPC cement paste; however, for the EA case, mean values are slightly smaller for the compression test. The bonding between the paste and the tube might have caused some restraint, which resulted in a slightly higher modulus from the EMM-ARM. The compressive strength also decreased with the addition of expansive additive.

In the current work, unrestrained mechanical properties of cement paste were evaluated. However due to the effect reinforcement and molds, expansion is restrained in the real structures. Depending upon the degree of restrained, microstructure of cement paste can be denser its mechanical properties can be significantly improved [16].

References

1. Nagatki, S., Tsuji, Y.: Recommended practice for expansive concrete (1994)
2. Nmai, C., Tomita, R., Hondo, F., Buffenbarger, J.: Shrinkage-reducing admixtures. Concr. Int. **20**(4), 31–37 (1998)
3. Chaunsali, P., Mondal, P.: Physico-chemical interaction between mineral admixtures and OPC–calcium sulfoaluminate (CSA) cements and its influence on early-age expansion. Cem. Concr. Res. **80**, 10–20 (2016)
4. Mo, L., Deng, M., Tang, M.: Effects of calcination condition on expansion property of MgO-type expansive agent used in cement-based materials. Cem. Concr. Res. **40**(3), 437–446 (2010)
5. Nagataki, S., Gomi, H.: Expansive admixtures (mainly ettringite). Cem. Concr. Compos. **20**(2–3), 163–170 (1998)
6. Higuchi, T., Eguchi, M., Morioka, M., Sakai, E.: Hydration and properties of expansive additive treated high temperature carbonation. Cem. Concr. Res. **64**, 11–16 (2014)
7. Van, N.D., Choi, H., Hama, Y.: Modeling early age hydration reaction and predicting compressive strength of cement paste mixed with expansive additives. Constr. Build. Mater. **223**, 994–1007 (2019)
8. Meddah, M.S., Suzuki, M., Sato, R.: Influence of a combination of expansive and shrinkage-reducing admixture on autogenous deformation and self-stress of silica fume high-performance concrete. Constr. Build. Mater. **25**(1), 239–250 (2011)
9. Wyrzykowski, M., Terrasi, G., Lura, P.: Expansive high-performance concrete for chemical-prestress applications. Cem. Concr. Res. **107**, 275–283 (2018)
10. Granja, J.L.D.: Continuous characterization of stiffness of cement-based materials: experimental analysis and micro-mechanics modelling (2016)
11. Dunant, C.F., Granja, J., Muller, A., Azenha, M., Scrivener, K.L.: Microstructural simulation and measurement of elastic modulus evolution of hydrating cement pastes. Cem. Concr. Res. **130**, 106007 (2020)
12. ASTM C 305-06: Standard practice for mechanical mixing of hydraulic cement pastes and mortars of plastic consistency. ASTM International (2006)
13. Welch, P.: The use of fast Fourier transform for the estimation of power spectra: a method based on time averaging over short, modified periodograms. IEEE Trans. Audio Electroacoust. **15**(2), 70–73 (1967)
14. Li, H., Liu, J., Wang, Y., Tian, Q.: Simplified chemo-mechanical model of hydration/expansion of calcium-oxide-type expansive agent in cement-based materials. Mag. Concr. Res. **72**(19), 1006–1015 (2020)
15. Shen, P., Lu, L., He, Y., Wang, F., Lu, J., Zheng, H., Hu, S.: Investigation on expansion effect of the expansive agents in ultra-high performance concrete. Cem. Concr. Compos. **105**, 103425 (2020)
16. Nguyen, T.B.T., Chatchawan, R., Saengsoy, W., Tangtermsirikul, S., Sugiyama, T.: Influences of different types of fly ash and confinement on performances of expansive mortars and concretes. Constr. Build. Mater. **209**, 176–186 (2019)

Effect of MgO-Based Expansive Agent on the Cement-Based Mortar Behavior

Lina Ammar, Kinda Hannawi, and Aveline Darquennes

Abstract As shrinkage can affect the concrete durability, several solutions were designed to limit this deformation. One of them consists to add expansive agents like MgO-based products. The aim of the present research work is to better understand the effect of MgO-based expansive agent on the self-healing potential of cementitious materials. At 28 days old, Portland cement mortars with different contents of expansive agent (0%, 5% and 10%) are pre-cracked by means of splitting test and stored under water. Self-healing is monitored using 2D and 3D tests: optical microscope observations and water permeability measurements respectively. Mechanical resistances and deformations of the studied mortars are also monitored and their microstructures are characterized by means of thermogravimetric analysis. The results show that mortars containing MgO-based expansive agent present a better self-healing capacity. It can be due to its swelling capacity decreasing/deleting shrinkage deformations and the formation of supplementary products like brucite.

Keywords Brucite · MgO-based expansive agent · Self-healing · Swelling

1 Introduction

The cracking of concrete structures has become a serious problem for the durability of civil engineering constructions. Shrinkage is one of the main reasons for mortar and concrete failures [1]. It appears when water containing into concrete decreases due to hydration process and exchanges with its external environment, increasing capillary

L. Ammar (✉) · K. Hannawi · A. Darquennes
Laboratoire de Génie Civil et Génie Mécanique (LGCGM), Institut National des Sciences Appliquées de Rennes (INSA Rennes), Rennes, France
e-mail: lina.ammar@insa-rennes.fr

K. Hannawi
e-mail: kinda.hannawi@insa-rennes.fr

A. Darquennes
e-mail: aveline.darquennes@insa-rennes.fr

F. Kanavaris et al. (eds.), *International RILEM Conference on Early-Age and Long-Term Cracking in RC Structures*, RILEM Bookseries 31,
https://doi.org/10.1007/978-3-030-72921-9_27

depression, the main shrinkage mechanism. When this deformation is restrained externally or internally, tensile stresses appear. When they are superior to the material tensile strength, concrete cracks reducing its durability. To minimize this problem, shrinkage reducers have been used over the last 30 years [2]. The shrinkage reducers are designed to reduce autogenous and drying shrinkage of concrete by reducing the surface tension that develops when concrete loses moisture [3]. Several solutions are available to reduce shrinkage [3] such as internal curing, use of Shrinkage Reducing Agents (SRA), expansive products, etc.... Due to crystallization pressures, this last solution leads to a swelling of cementitious materials at the macroscale reducing strongly their shrinkage. Several types of expansive agents are available such as iron powder, alumina powder, calcium oxide, calcium sulfoaluminate and magnesia [3]. Among them, MgO has recently drawn attention because of its good expansive behavior [2].

The first observations on the expansive behavior of MgO-based agents date back to 1884 [2, 4], when several bridges in France were damaged by delayed expansion of concrete containing from 16 to 30% of MgO. Therefore, they recommended to limit the MgO content to 5%/6% to prevent a concrete failure risk. Being an industrial by-products [5–7], MgO is also more environmental-friendly than Portland cement. Its chemical reaction with water given at Eq. (1) leads to macroscopic volume expansion due to the formation of brucite "$Mg(OH)_2$" at early age [3]. Crystallization and swelling pressures of this product act as the driving force of concrete expansion [8–10] reducing their autogenous and drying shrinkage [3, 8–10].

$$MgO + H_2O \rightarrow Mg(OH)_2 \tag{1}$$

Another phenomenon—affecting significantly crack reduction—is the "Autogenous Healing". Concrete presents this behaviour occurring usually in wet conditions thanks to the ongoing hydration and carbonation of particles along the crack [10–13]. This process contributes to close partially or totally the cracks [11–13]. Moreover, several studies [15–18] showed that shrinkage reducers can be benefit to strengthen the autogenous healing. With 5 and 7.5% of MgO-based expansive agent, cracks with an initial width equal to about 500 μm were closed up to 95% after 28 days of water curing [15, 16]. They found that magnesium oxide leads to the formation of brucite and other magnesium hydro-carbonate products along cracks. Other authors [17, 18] agree to the fact that supplementary products such as brucite could enhance the autogenous healing if they expand properly in the matrix. This healing mechanism is called "Autonomous Healing". Notice that methods used to determine self-healing performances have not been yet standardized [14]. The crack monitoring on specimen surfaces by visual observations is the more common method. However, it is limited to a 2D description of the crack healing.

The main objective of this research work is to better understand the healing mechanisms of cement-based mortars containing MgO-based expansive agent. So, a water permeability test giving 3D information on the healing process is performed ant the material behavior at micro-(hydration products) and macroscale (deformations, strength) are characterized.

Table 1 Mixtures proportions (kg/m^3)—fresh mortar properties

	REF	MGA5	MGA10
Portland cement	650	619	588
MgO	–	33	65
Sand	1300	1300	1300
Steel fibers	50	50	50
Water	291	292	292
Superplasticizer	1.95	1.95	1.95
Average slump (cm)	11.5	11.0	11.0
Air content (%)	2.1	2.8	2.8

2 Materials and Mixtures Proportions

The MgO-based expansive agent used in this study contains 80% of magnesium oxide. It is named "MgO" hereafter. Mortars with three contents of MgO (cement mass substitution) are tested: 0, 5 and 10%. They are named *REF*, *MGA5* and *MGA10* respectively. The used cement is CEM I 52.5 N. The water/binder and sand/binder ratios are equal to 0.45 and 2 respectively. The superplasticizer—Sika ViscoFlow-800 Power is added by 0.3% of the cement mass. Table 1 gives the mixtures proportions and the fresh mortar properties. To better control the initial crack width, fibers are added only into specimens used for monitoring the healing process. Hence, these mixtures contain 50 kg/m^3 of steel fibers (25 mm length and 0.25 mm diameter).

3 Specimens Preparation and Test Methods

3.1 Water Permeability Test

Cylindrical specimens (Φ 110 mm × H 200 mm) are casted for the Water Permeability test "WP". After demoulding, specimens are stored in a humid chamber (RH = 95 ± 5%, 20 °C) until the testing age—8 days old. Three days before the test (at 25 days old), each cylindrical specimen is cut into 3 disks (H 50 mm). At 28 days old, all the disks are pre-cracked using a splitting test to obtain a crack width equal to 0.2 ± 0.05 mm. The loading speed is fixed at 0.05 mm/min. During the splitting test, the crack opening is visually controlled using a specific ruler as shown in Fig. 1a.

Before WP test, cracks width is measured on the both disk surfaces using a reflective microscope. The measurements are performed beyond 1 cm from the disk edges and on several points (from 3 to 10) over a 50 mm long area (Fig. 1b).

After crack width measurements, all specimens are stored at 40 °C and 20% R.H. during 24 h. This curing is applied to slightly dry concrete disks to ensure a good adhesion of the PVC tubes ($ø_{inner}$ = 100 mm, H = 200 mm) glued on the disk top.

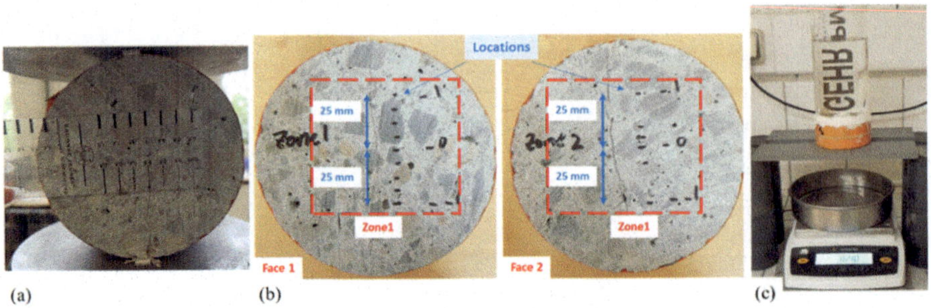

Fig. 1 Crack width controlled with a ruler during the splitting test (**a**)—Location of zones where the crack width is measured on specimen surfaces (**b**)—Water Permeability test (**c**)

After the glue drying, PVC tubes are filled with 1.55 L of tap water. To monitor the water flow through the crack, specimens are placed on a support above a balance (precision of 100 mg—Fig. 1c). The flowed water quantity is measured after 5, 10, 15, 20, 25 and 30 min to quantify a water flow (Liter/minute). Moreover, the water level is monitored during 3 h. After WP test, all disks are stored vertically in a tap water container. The water permeability test is made just after cracking (**0D**), after 7 days (**7D**) and 25 days (**25D**) of water curing. The water in the containers is renewed every month. From each test, a Healing Rate ($HR_{WP}(t)$) is calculated according to "Eq. 2".

$$HR_{WP}(t) = \frac{W_{unhealed}(0D) - W_{healing}(t)}{W_{unhealed}(0D)} x100 \quad (2)$$

where

$HR_{WP}(t)$ is the Healing Rate at time *t* determined with the Water Permeability test (%).

$W_{unhealed}(0D)$ is the initial water flow just after the cracking (at 0D) (l/min).

$W_{healing}(t)$ is the water flow after a healing period *t* under water curing (l/min).

3.2 Length Deformations

Length deformations are monitored on three prismatic specimens ($4 \times 4 \times 16$ cm^3) for each studied mortar. The length deformation measured is related to the hydration process and to the hydric exchanges between specimen and its environment. Until 28 days, specimens are stored in a humid chamber (RH = 95 ± 5%, 20 °C) and after, under water (Water Curing—*WC*). The length deformation measurements are made using a retractometer with 1 μm of precision and then calculated using Eq. (3):

$$\varepsilon(t) = \frac{L_i(t) - L_0}{L_0} \tag{3}$$

where L_0 is the initial length (*mm*) measured on 1 day old specimens (just after demoulding). Similarly, $L_i(t)$ is the length (*mm*) measured at time *t*.

3.3 Thermogravimetric Analysis

The hydration products (portlandite, brucite and calcite) are characterized on 28 days old specimens stored at 20 ± 2 °C and 95 ± 5 of % RH. Using ThermoGravimetric Analysis (TGA). Before the test, specimens are dried at 105 °C during 24 h, crushed and ground. During TGA, the powder (80 mg ± 10 mg) is placed in a nitrogen environment and temperature varies from 20 to 1000 °C with a rate of 10 °C/min. After the test, the hydration products (portlandite, brucite, and calcite) are calculated using Eqs. 4–6. The mass of water or carbon dioxide released at a given temperature range is used to calculate the mass of the corresponding hydration product.

- Brucite decomposes at 250 _ 400 °C [19];

$$Mg(OH)_2 \rightarrow MgO + H_2O \tag{4}$$

- Portlandite decomposes at 400 _ 600 °C [20];

$$Ca(OH)_2 \rightarrow CaO + H_2O \tag{5}$$

- Calcite decomposes at 600 _ 1000 °C [21];

$$CaCO_3 \rightarrow CaO + CO_2 \tag{6}$$

3.4 Mechanical Strengths

The mechanical strengths are measured on three prismatic specimens (4 × 4 × 16 cm^3) at each test age (at 3, 7, 28 and 90 days). They are stored in a humid chamber (RH = 95 ± 5%, 20 °C). The flexural and compressive strengths are measured according to the [EN 196-1] European standard. First, the bending test is carried out. Secondly, the compressive test is performed on the two half-specimens resulting from the previous test.

4 Results and Discussion

4.1 Self-healing Potential

Crack Width. Figure 2 shows the Crack Width (CW) evolution as function of their initial value after 0, 7 and 25 days of Water Curing "WC". These measurements are done on the both disk surfaces. The initial average crack width is 267 μm, 254 μm, 183 μm for *REF*, *MGA5* and *MGA10* respectively. With the water curing duration, crack width decreases for all compositions (Fig. 2). It is probably due to self-healing by the ongoing hydration leading to the formation of supplementary products on the crack surface. After 25 days of WC, a larger amount of *MGA5* and *MGA10*'s cracks are completely healed. To confirm these results and to obtain a 3D information on the healing process, a water permeability test is performed.

Water Flow. Figure 3 shows the Water Flow "WF" as function of the crack width at 0D, 7D and 25D for each composition. The initial WF increases with the crack width: the WF for *MGA5* characterized by an initial crack width located in the range [250–300] μm is equal to 0.054 l/min, while this value decreases to 0.027 l/min for *MGA10* characterized by an initial crack width located in the range [150–200] μm, e.g. To assess the healing capacity, the WF obtained for specimens having an initial crack width located in the range [200–250] μm are compared (Fig. 2). After 25 days of water curing, the WF decreases by 18% and 53% for *MGA5* and *MGA10* respectively. For REF, WF is characterized by a fast reduction during the first week, but no healing process is observed after 7 days of water curing.

Healing Rate. Figure 4 shows the average Healing Rate "HR" for each composition calculated with the Eq. 2. After 7 days of water curing (7D), the healing capacity of mortar is equal to 11%, 23% and 36% for *MGA5, REF* and *MGA10* respectively. As indicated previously, the healing stops after 7 days of water curing for *REF*, while it is increasing for mortars with MgO. After 25 days of water curing, HR reaches a value equal to 21% and 50% for *MGA5* and *MGA10* respectively. These results clearly show that *MGA10* has a more important capacity to healing crack. To understand these results, the hydration products are characterized as well as the mechanical performances (deformations, strengths).

4.2 Hydration Products

From TGA, brucite, portlandite and calcite contents are measured. The results are given in Table 2. Portlandite and calcite contents are measured for the three mortars. A supplementary product—Brucite is observed for *MGA5* and *MGA10*. It confirms that the growth of brucite is favored in solutions rich in magnesium Ions Mg^{2+} as suggested by [15–18]. Moreover, brucite content increases with the MgO content.

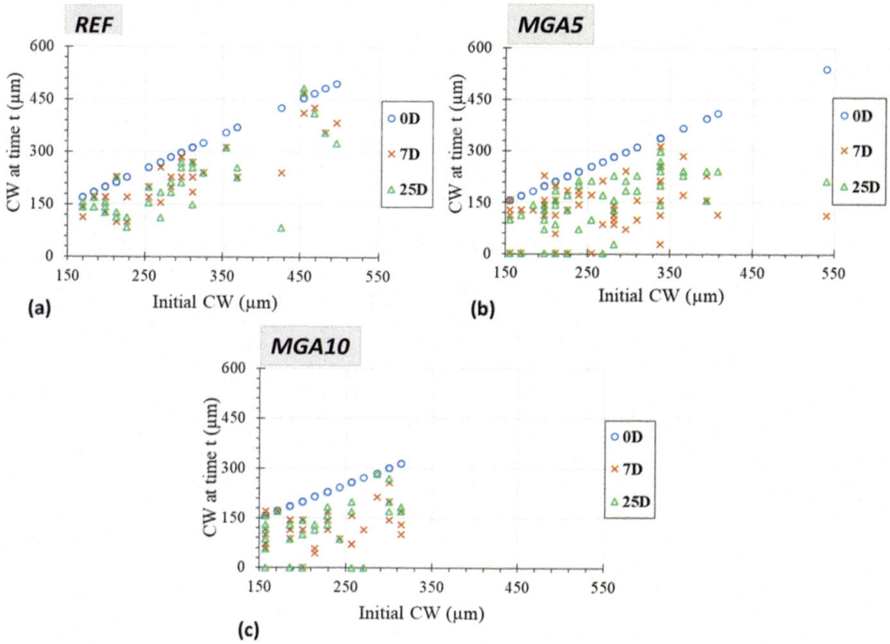

Fig. 2 Crack Width (CW) evolution after a water curing duration equal to 0, 7 and 25 days for *REF* **a**—*MGA5* (**b**)—*MGA10* (**c**)

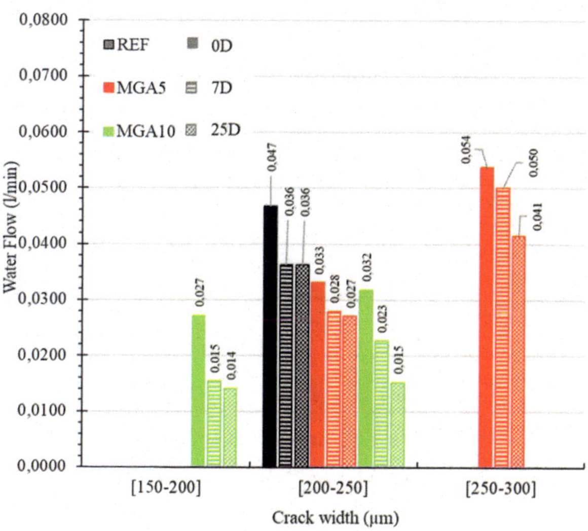

Fig. 3 Water flow as function of the crack width after a water curing duration of 0D, 7D and 25D for each composition

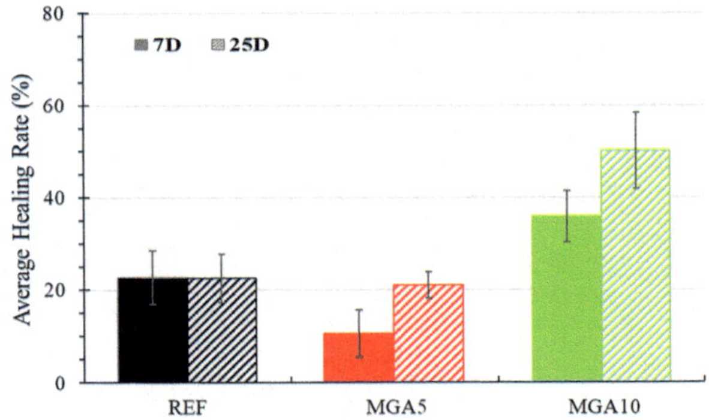

Fig. 4 Average healing rate after a water curing duration of 7D and 25D

Table 2 Brucite, portlandite, and calcite contents at 28 days (%)

Mass (%)	$Mg(OH)_2$	$Ca(OH)_2$	$CaCO_3$
REF	–	10.1	1.3
MGA5	5.3	11.6	1.7
MGA10	7.1	11.2	4.7

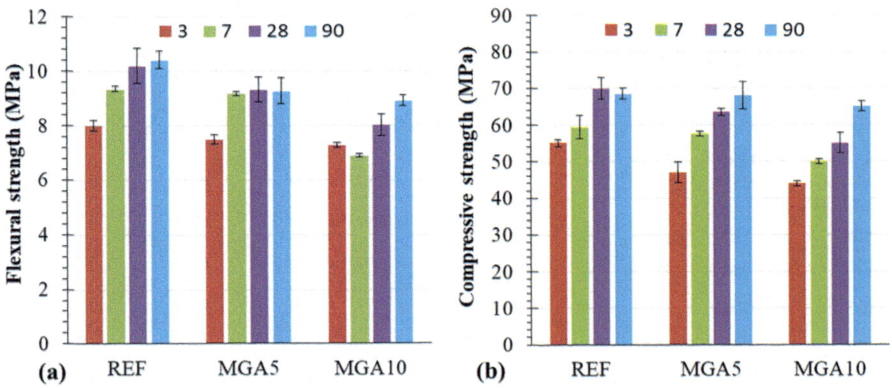

Fig. 5 Evolution of the flexural (**a**) and compressive strength (**b**) for all the studied mortars at 3, 7, 28 and 90 days

On the other hand, the calcite content is higher for *MGA10*. In the presence of MgO, magnesium carbonate can be formed besides of the calcite [22]. So, the calcite content in *MGA10* might be attributed to the magnesium carbonate as well.

4.3 Mechanical Performances

Figure 5 shows the evolution of the flexural and compressive strengths for all the studied mortars. They present the same trend. At 90 days, the flexural strengths vary between 8.9 MPa (for *MGA10)* and 11 MPA (for *REF*), while the compressive strengths vary between 65 MPa (for *MGA10*) and 68 MPa (for *REF*). *REF* shows slightly higher strengths than mortars with MgO. This strength decrease can be partially related to the weaker chemical bond between brucite crystals (Van der Waals force). Other research works [3, 11, 15] found that the brucite formation can lead to crystallization pressures and an expansion of the concrete matrix. These phenomena can induce cracks leading to a strength reduction.

Figure 6 shows the evolution of the length deformation for all the studied mortars. The main results are summarized hereafter:

- *REF* shrinks fast during the first 7 days. After, its deformation varies varies slightly around a constant value equal to −282 µm/m at 80 days.
- *MGA5* shrinks during the first 6 days to reach a value equal to about −128 µm/m. Then, its deformation is characterized by a swelling that cancels the shrinkage after 32 days. Beyond this period, *MGA5* presents a swelling and this deformation reaches a value of 85 µm/m at 80 days.
- *MGA10* shrinks slightly during the first days. Then, it swells very quickly and its shrinkage is cancelled before 3 days. It reaches a value superior to 1000 µm/m after 80 days.

During the first days from 1 to 7 days, shrinkage is mainly due to the hydration reactions (confirmed by the autogenous deformation measurements) for all the studied mixtures. Later, the formation of brucite in *MGA5* and *MGA10* leads to a swelling limiting/ cancelling shrinkage. It agrees with other research works [8–10]. This phenomenon is strengthened by the possibility of water exchanges from the environment to specimen. In this study, the swelling is more important for *MGA10* and no cracks are observed on the tested specimens after 80 days. Based on these observations and HR evolution (Fig. 4), it appears that the matrix swelling contributes to reduce the crack width as well as the formation of supplementary products like brucite.

5 Conclusion

In this research study, the effect MgO-based expansive agent on the self-healing potential of Portland-cement mortar is investigated. Based on the experimental results, mortars with MgO show a better self-healing potential. Two main mechanisms contribute to close partially or totally the crack. The first one is the formation of additional hydrates on the crack surface, which is confirmed by the 2D optical microscope observations after 7 days and 25 days of water curing. The second one

Fig. 6 Length deformation of the studied mortars stored under WC from 28 days old

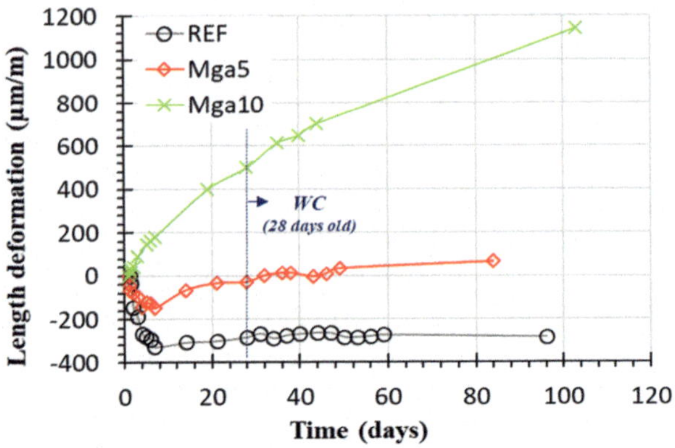

is the formation of brucite crystals. This last is showed by the thermogravimetric analysis. These products lead to the matrix expansion, which also may contribute to a reduction in the crack width. However, these mixtures present a lower strength related to the chemical bonds of brucite crystals and sometimes to internal cracks. This last will be confirmed by supplementary analyses at microscale.

References

1. Papayianni, I., Kesikidou, F., Alt., P-H.: The role of shrinkage reducing admixtures and supplementary cementitious materials in volume stability and strength development. In. MATEC Web of Conferences, vol. 289, p. 02005. Concrete Solutions (2019)
2. CaO, F.: Hydration characteristics and expansive mechanism of MgO expansive agents. Constr. Build. Mater. **183**, 234–242 (2018)
3. Yang, L.: Mitigation techniques for autogenous shrinkage of ultra-high-performance concrete—a review. Compos. Part B **178**, 107456 (2019)
4. Mehta, P.K.: History and status of performance tests for evaluation of soundness of cements. Cement Stand.-Evol. Trends **663**, 35–60 (1978)
5. GaO, P.: Production of MgO-type expansive agent in dam concrete by use of industrial by-products. Build. Environ. **43**, 453–457 (2008)
6. Xu, L.: Dolomite used as raw material to produce MgO-based expansive agent. Cem. Concr. Res. **35**, 1480–1485 (2005)
7. Temiz, H.: Influence of blast-furnace slag on behavior of dolomite used as a raw material of MgO-type expansive agent. Constr. Build. Mater. **94**, 528–535 (2015)
8. Mo, L.: Deformation and mechanical properties of the expansive cements produced by inter-grinding cement clinker and MgOs with various reactivities. Constr. Build. Mater. **80**, 1–8 (2015)
9. Mo, L.: Deformation and mechanical properties of quaternary blended cements containing ground granulated blast furnace slag, fly ash and magnesia. Cement Concr. Res **71**, 7–13 (2015)
10. Ding, W.: Utilization of light calcined magnesite tailings to compensate the autogenous shrinkage of steam-cured cement paste. Adv. Cement Res. **10**, 675–686 (2016)
11. Hearn, N.: Self-sealing, autogenous healing and continued hydration: what is the difference? Mater. Struct. **31**, 653–567 (1998)

12. Neville, A.: Autogenous healing, a concrete miracle? Concr. Int. (24) (2002)
13. Edvarsen, C.: Water permeability and autogenous healing of cracks in concrete. ACI Mater. J. 448–454 (1999)
14. Park, B.: Prediction of self-healing potential of cementitious materials incorporating crystalline admixture by isothermal calorimetry. Int. J. Concr. Struct. Mater. 13–36 (2019)
15. Qureshi, T.: Autogenous self-healing of cement with expansive minerals-I: impact in early age crack healing. Constr. Build. Mater. **192**, 768–784 (2018)
16. Qureshi, T.: Autogenous self-healing of cement with expansive minerals-II: impact of age and the role of optimised expansive minerals in healing performance. Constr. Build. Mater. **194**, 266–275 (2019)
17. Sherir, M.A.-A.: Self-healing and expansion characteristics of cementitious composites with high volume of fly ash and MgO-type expansive agent. Constr. Build. Mater. **127**, 80–92 (2016)
18. Sherir, M.A.-A.: The influence of MgO-type expansive agent incorporated in self-healing system of Engineered cementitious Composites. Constr. Build. Mater. **149**, 164–185 (2017)
19. Wang, J.-A.: Characterizations of the thermal decomposition of brucite prepared by sol-gel technique for synthesis of nanocrystalline MgO. Mater. Lett. **35**, 317–323 (1998)
20. Alonso, C.: Dehydration and rehydration processes of cement paste exposed to high temperature environments. J. Mater. Sci. 3015–3024 (2004)
21. Pei., Y.: Effets du chauffage sur les matériaux cimentaires—impact du “self-healing” sur les propriétés de transfert. Thèse de doctorat: Génie Civil, Ecole Centrale de Lille (2016)
22. Jin, F.: Strength and hydration properties of reactive MgO-activated ground granulated blast furnace slag paste. Cement Concr. Compos. **57**, 8–16 (2015)

Plasticizers as Dispersants for Carbon Nanotubes in Self-sensing Cement Composites

Pedro de Almeida Carísio, Oscar Aurelio Mendoza Reales, Eduardo de Morares Rego Fairbairn, and Romildo Dias Toledo Filho

Abstract Carbon nanotubes (CNT) modify not only the mechanical but also the electrical properties of the cement matrix, allowing the development of self-sensing cement composites (SSCC). These composites can sense their own deformation through changes in their electrical resistivity, acting as strain sensors, increasing the operational safety of concrete structures by indicating preventive or corrective maintenance. The efficiency of SSCC depends on the electrical properties of the conductive filler and its dispersion degree throughout the matrix. The most commonly employed method of ensuring adequate dispersion of CNT is using chemical dispersing agents. This work aims to compare the effectiveness of two types of plasticizers as dispersing agents for CNT, identifying how they affect the self-sensing capacity of SSCC. CNT contents of 0.10 and 0.20%, by mass of cement, were blended in a cement matrix using two types of commercial superplasticizers as dispersant agents, one naphthalene based and one polycarboxylate ether based. SSCC cement pastes were cast in 50 mm cubic molds with four copper embedded electrodes and cured for 30 days in a humid environment. After curing, an electrical potential was applied through the electrodes and the electrical resistivity of the SSCC was acquired during successive compressive load applications. A 10 V DC power supply was used for this purpose. The changes in electrical resistivity of each cube were correlated with deformations measured by LVDT. Results showed that both the type of superplasticizer used and the CNT content affect the piezoresistive response of SSCC. Naphthalene-based superplasticizer presented better performance as dispersing agent and led to SSCC with better self-sensing response. Results indicated that the chemical basis of the dispersing agent for CNT has great influence on the fabrication of efficient SSCC.

Keywords Self-sensing · Cement composites · Carbon nanotubes · Superplasticizer · Dispersions

P. de Almeida Carísio (✉) · O. A. Mendoza Reales · E. de Morares Rego Fairbairn · R. D. T. Filho
Universidade Federal Do Rio de Janeiro - UFRJ, Rio de Janeiro, Brazil
e-mail: pedro.carisio@numats.coc.ufrj.br

F. Kanavaris et al. (eds.), *International RILEM Conference on Early-Age and Long-Term Cracking in RC Structures*, RILEM Bookseries 31,
https://doi.org/10.1007/978-3-030-72921-9_28

1 Introduction

Concrete is a highly durable and versatile material for structural applications. Pozzolans [1], fibers [2] and nanoparticles [3], have been used to improve the performance concrete throughout its lifespan, modifying not only its mechanical performance but also its durability. Among these materials, carbon nanoparticles and metallic fibers have the capacity of generating intelligent materials for structural monitoring due to their electrical properties [4]. For this purpose, carbon nanotubes (CNT) are one of the most commonly used materials [5] due to their excellent electrical conductivity and mechanical properties [6], acting not only as reinforcement but also as conductive filler for self-sensing concrete fabrication [7].

A Self-Sensing cement composite (SSCC) is capable of monitoring its structural integrity by translating mechanical strain into electrical properties variations through a piezoresistive mechanism [8]. When a strain is imposed, the electrical resistivity of SSCC changes proportionally to it, allowing for a direct correlation between these two variables [9]. SSCC have been used to monitor traffic, as speed meter and vehicle counter [10]; applied in structural elements to monitor mechanical strain [11], and employed to detect crack opening and propagation [12].

Piezoresistivity is reached when conductive fillers such as carbon fibers [13], carbon black [14], CNT [15], or metallic fibers [16], are blended into conventional concrete. The addition of these conductive materials decreases the matrix electrical resistance and enhances electrical percolation. The self-sensing efficiency is directly linked to electrical percolation, which is consequence of the amount and spatial distribution of the conductive fillers throughout the matrix [17].

Although CNT is one of the most commonly used conductive fillers for SSCC, it is still a challenge to adequately disperse them in the concrete matrix. Due to their strong Van der Waals interactions, electromagnetic properties and hydrophobicity, CNT tend to agglomerate when mixed in water [18]. Physical and chemical processes have been developed in order to improve CNT wettability and dispersability in water. However, chemical agents such as surfactants, affect adversely cements hydration reaction [19]. On the other hand, superplasticizers, which already have compatibility with cement, are considered good candidates to work as dispersing agents for CNT in water [20].

Recent works have pointed out that the dispersing efficiency of superplasticizers is linked to their chemical nature; nevertheless, is not clear if second generation plastizicers (naphthalene based) [21] are more efficient dispersing agents for CNT than third generation superplasticizers (polycarboxylate ether based) [22]. A higher efficiency could be translated into more efficient SSCC. Therefore, the aim of this work is to study the self-sensing behavior of SSCC under compressive strength using two different types of superplasticizers as dispersing agents. The SSCC chosen was cement paste with CNT addition dispersed in water by ultrasonic pulses.

2 Materials and Methods

2.1 Materials

The materials used in this work were slag cement from Tupi (Brazil) ($d_{50} = 14{,}96\ \mu m$ and specific mass = 3,12 g/cm^3), powder multi walled carbon nanotubes from Nanocyl (Belgium) (average outer diameter = 9.5 nm, and an average length = 1.5 μm according to the manufacturer), a naphthalene based second generation plasticizer (SG) from Quartzolit (Brazil) with a solids concentration of 40%, and a polycarbolyxate ether based third generation (TG) superplasticizer from Aditibras (Brazil) with a solids concentration of 40%.

2.2 Methods

Fourier Transform Infrared spectroscopy (FTIR) was used to investigate the structural differences between the two plasticizers. A FTIR spectrometer was used in transmission mode with KBr pellets. The dispersing efficiency of each plasticizer was evaluated using UV-Vis spectroscopy and ζ-potential experiments. CNT aqueous dispersions with 0.40% of CNT solids content were fabricated using the two types of plasticizers in a 1:4 ratio of CNT to plasticizer. Each dispersion was placed in an ice bath and sonicated with a 550 W ultrasonic tip processor using 40% amplitude and in 20 s on/off cycles until reaching a total applied energy of 1000 $J/g_{dispersion}$. The UV-Vis absorbance spectrum of the dispersions after sonication was measured in a spectrophotometer. The same samples were used for ζ-potential measurements.

Cement pastes were manufactured with two CNT contents, 0.10 and 0.20% by mass of cement, and a fixed 0.45 water-to-cement ratio. The amount of CNT was chosen according to literature reports that indicated a better self-sensing performance of cement pastes containing 0.10% CNT by mass of cement [23]. Cement pastes were produced using a planetary mixer and cast in metallic cubic molds (50x50x50 mm^3). Three cubes were produced for each SSCC sample. Nomenclature and sample proportioning are presented in Table 1.

Table 1 Nomenclature and sample proportioning of SSCC for 3 cubes

Nomenclature	Plasticizer	CNT (%)	Cement (g)	Water (g)	CNT (g)	Plasticizer (g)
SG 0.20% CNT	SG	0.20	486.3	219.0	0.99	3.96
SG 0.10% CNT	SG	0.10			0.48	1.92
TG 0.20% CNT	TG	0.20			0.99	3.96
TG 0.10% CNT	TG	0.10			0.48	1.92

Fig. 1 Cured SSCC specimen with four copper wire electrodes

Three tangled 0.34 mm thickness copper wires were used as electrodes. Four electrodes were fixed at the bottom of the molds with instant glue in its central axis. The external electrodes were positioned at 1 cm from the cube edge, while the internal electrodes were positioned 0.5 cm from the cube center. After casting, samples were placed in a climate chamber for 7 days at a temperature of 23 ± 2 °C and 95% relative humidity. After this, specimens were demolded and placed back in the climate chamber for more 23 days, totalizing 30 days of curing. An example of the obtained SSCC is presented in Fig. 1.

A series circuit was adopted for self-sensing experiments. The external electrodes were connected to a DC voltage source with a 1 kOhm resistor welded to one of them. This was connected to a data acquisition system to register electric tension and to obtain electric current indirectly. The internal electrodes were also connected to the data acquisition system. An electric tension of 10 V was applied to the circuit. A scheme of the electrical circuit is shown in Fig. 2a. Constant voltage was applied during 30 min before load application. Mechanical stresses were applied in successive cycles of compression loads. The maximum compressive load applied was 9.0 kN with a constant deformation application rate of 0.3 mm/min. Strain was measured with two 50 mm LVDT, also connected to the data acquisition system. Fig. 2b shows the experimental setup. The same testing machine and experimental conditions were used to measure the compressive strength of the SSCC after 28 days of curing.

The electrical resistivity of each sample was obtained using Ohm's second law ($\rho = R.A/l$), where electrical resistivity (ρ—ohm.cm) is equal to the sample cross section area (A—cm^2) multiplied by the electrical resistance (R—ohm) and divided by the distance between the electrodes (l—cm). The changes in electrical resistivity due to load application (ΔR) were converted to fractional change in resistivity (FCR) as $FCR = \Delta R/R$ and compared with the average axial strain measured by the LVDT since it is known that these values are proportional [24].

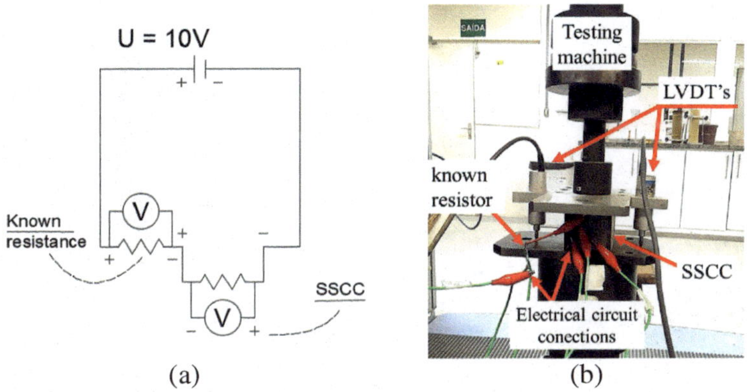

Fig. 2 Details of self-sensing testing. **a** Electrical circuit; **b** experimental set-up adopted

3 Experimental Results

3.1 CNT Dispersions Characterization

A detailed characterization of the carbon nanotubes used in this work can be found elsewhere [25]. FTIR spectra obtained for the two SP used are presented in Fig. 3. Janowska-Renkas [26] cite that specific functional groups from superplasticizers can be verified through their absorption bands in FTIR spectra as: OH- groups (3600–3200 cm^{-1}); aliphatic groups (3200–2500 cm^{-1}); carbonyl groups (1900–1500 cm^{-1}) and ether groups (1250–950 cm^{-1}). It can be observed that the chosen plasticizers presented absorption bands in the regions cited, with some differences among them in the ether groups. SG did not exhibit any band at 950 cm^{-1} (ether

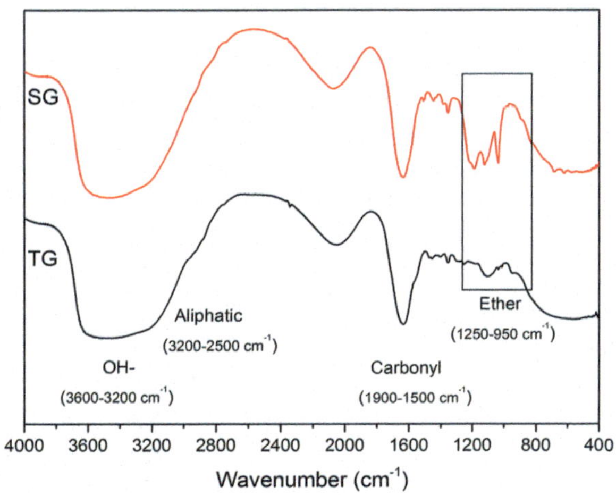

Fig. 3 FTIR spectra of plasticizers SG and TG used to disperse CNT

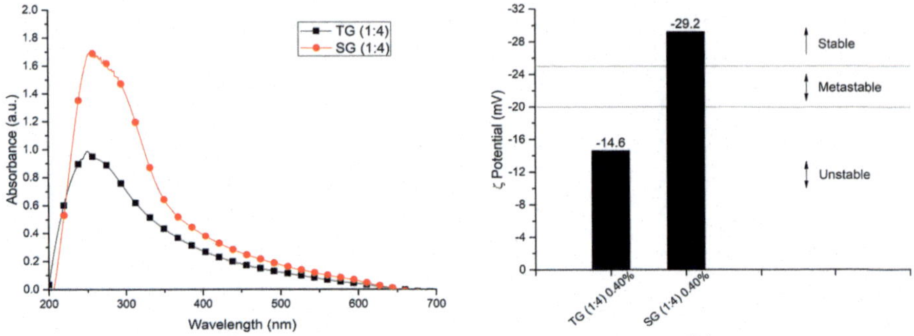

(a) (b)

Fig. 4 **a** UV-Vis absorbance and **b** ζ-potential for dispersions with TG and SG plasticizers

groups), and two additional bands at 1035 and 1190 cm^{-1}. This difference is associated with its naphthalene basis.

Figure 4 presents the absorbance spectra and ζ-Potential results of the CNT dispersions using a 1:4 CNT to plasticizer ratio and 0.40% CNT. It was found that the SG plasticizer showed higher absorbance value than the TG, indicating a higher dispersion degree and a higher dispersing efficiency for the same ultrasonic energy applied. ζ-potential results present negative values due to the presence of OH- groups in their structure, making the CNT negatively charged through adsorption [20]. Colloids with ζ-potential values in the region between +25 and −25 mV are considered of unstable electric potential, presenting more tendency to agglomerate [27]. Thus, according to the results found, only the SG plasticizer generated a stable dispersion, confirming its higher efficiency when compared with the TG plasticizer.

3.2 Mechanical Characterization of SSCC

Compressive strength of the SSCC was measured after 28 days of curing. The obtained results are presented in Table 2. It was found that both the presence of CNT and the plasticizer type affected the average compressive strength of the composite. A substantial strength decrease was identified for the TG 0.20% CNT sample. This can be associated with CNT agglomerations, as predicted by the ζ-potential results,

Table 2 Average compressive strength of SSCC after 28 days of curing

SSCC	Compressive strength (MPa)
SG 0.20% CNT	34,11 ± 2,16
SG 0.10% CNT	36,53 ± 4,72
TG 0.20% CNT	20,25 ± 3,65
TG 0.10% CNT	35,63 ± 3,59

that can act as stress concentration points [28]. Compressive strength results were used to keep the load applied in the self-sensing test under 45% of the peak load.

3.3 *Self-sensing Results*

The self-sensing results for the SG 0.20% CNT samples are presented in Fig. 5. It can be observed that the FCR results of cubes 1 and 2 presented peaks in phase with the strain peaks measured by the LVDT, indicating that a self-sensing response was achieved. It should be noticed that the FCR values of these two samples were similar, ranging between 0.00 and 0.25. The cube 3 did not show a conclusive self-sensing response, as can be observed in Fig. 5c.This can be associated with not uniform carbon nanotube dispersion or with a CNT amount under the electrical percolation threshold.

Figure 6 shows the self-sensing results for the SG 0.10% CNT samples. Similarly, to the SG 0.10% CNT samples, two of the three cubes presented a self-sensing response while one of them did not. It should be noticed that even tough cube 2 and cube 3 presented a self-sensing response, their FCR values differ by more than one order of magnitude, which can be indicative of an inadequate mixing procedure.

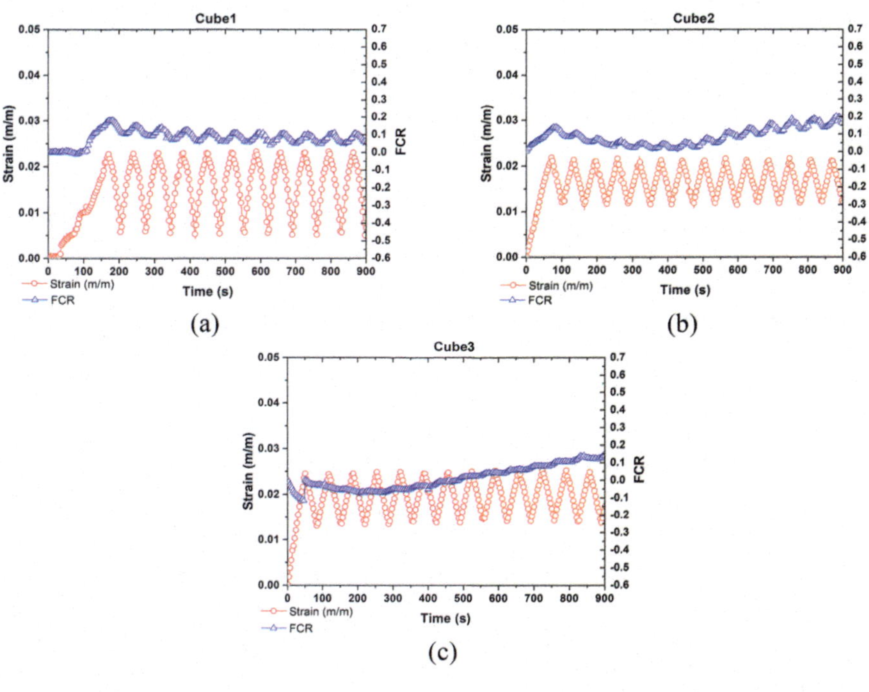

Fig. 5 FCR of SG 0.20% CNT samples compared to strain values registered by LVDT during cyclic loading. **a** Cube 1; **b** Cube 2; and **c** Cube 3

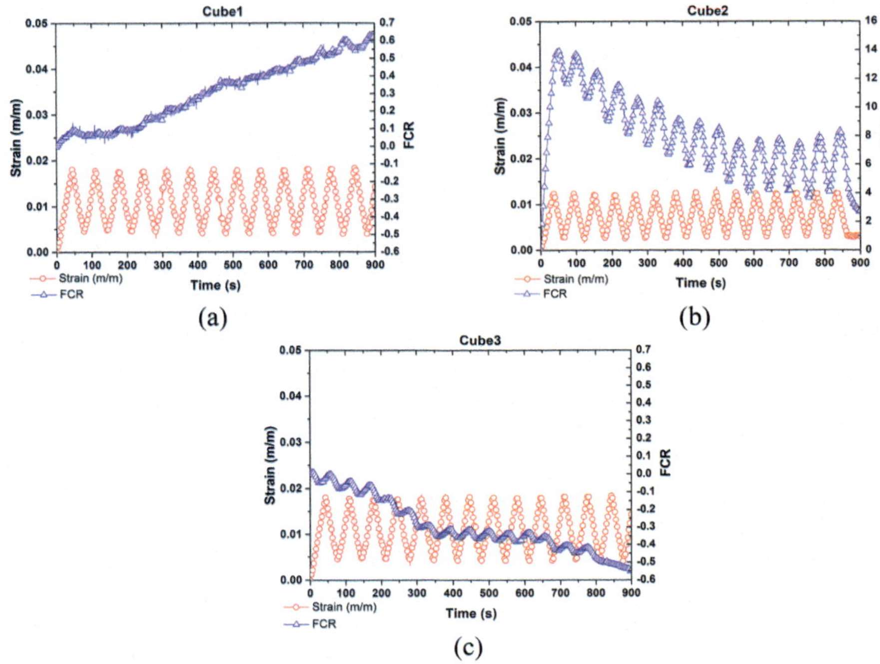

Fig. 6 FCR of SG 0.10% CNT samples compared to strain values registered by LVDT during cyclic loading. **a** Cube 1; **b** Cube 2; and **c** Cube 3

Additionally, the FCR results presented a non-horizontal tendency, which can be indicative of a charge up or polarization effect [29].

Self-sensing results for the TG 0.20% CNT and TG 0.10% CNT samples are presented in Fig. 7 and Fig. 8 respectively. For all cubes, the results showed a minimal self-sensing capacity. The FCR values did not show a response to the load cycles and the curves do not allow any correlation between the FCR values with LVDT strain values. This behavior can be seen as consequence of the low dispersing efficiency found for the TG plasticizer.

4 General Discussion and Conclusions

By comparing two plasticizers of different chemical nature it was found that they presented different efficiencies as dispersing agents for CNT when used in the same amounts. Structural differences between the naphthalene and polycarboxylate ether chemical bases, identified by FTIR, were found to be crucial when comparing the effectiveness of second generation and third generation plasticizers as dispersing agents. The naphthalene based plasticizer was found to generate a more disperse and stable CNT aqueous dispersion. This was found to translate into a SSCC with higher

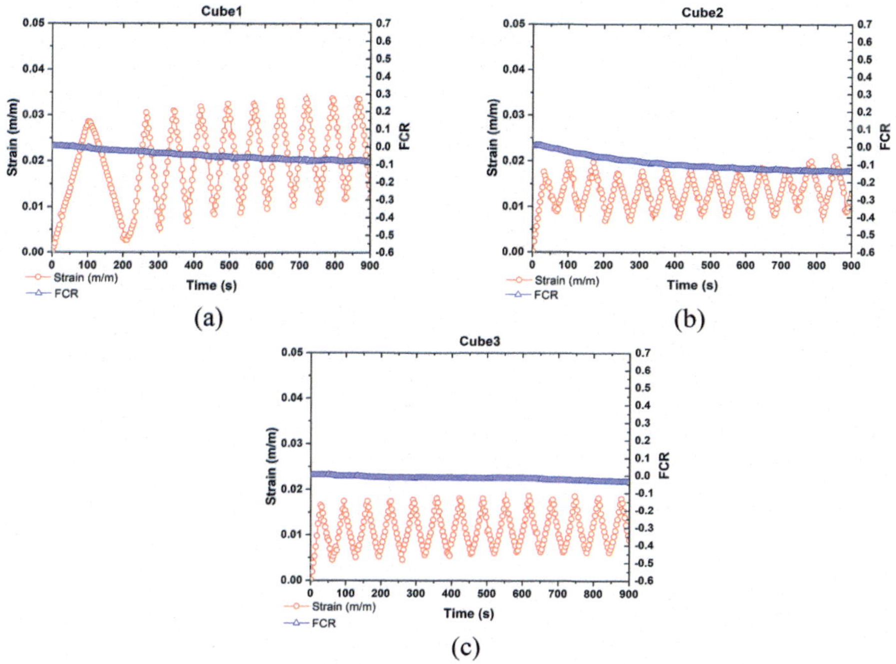

Fig. 7 FCR of TG 0.20% CNT samples compared to strain values registered by LVDT during cyclic loading. **a** Cube 1; **b** Cube 2; and **c** Cube 3

self-sensing capacity. These results agree with the findings of Al-rekabi et al. [30] and from Mendoza et al. [20], which concluded that polycarboxylates ether are not the most adequate superplasticizer to be used as a dispersant agent for CNT exposed to elevated pH environments.

A high variability was found between results. Differences in FCR of more than one order of magnitude, and absence of self-sensing effect within lots of SSCC, indicate that while CNT amount and dispersion degree are key parameters, an adequate mixing procedure that guarantees homogeneous distribution of CNT throughout the matrix is of vital importance to guarantee repeatability of the results.

The samples which presented self-sensing behavior also presented an increasing or decreasing deviation of their FCR. It is believed that this deviation is due to polarization acting in the cement paste combined with a capacitive behavior. Circuits that use direct current power supply have more chances to be affected by these effects [31]. The 30 min constant voltage application was not found to be sufficient to avoid these effects.

The use of a DC power supply was a limitation for this study. Polarization effects were found to induce slopes in the self-sensing response. Subsequent research can be done with a AC power supply and higher amounts of CNT in order to minimize the polarization effects and obtain more stable self-sensing curves.

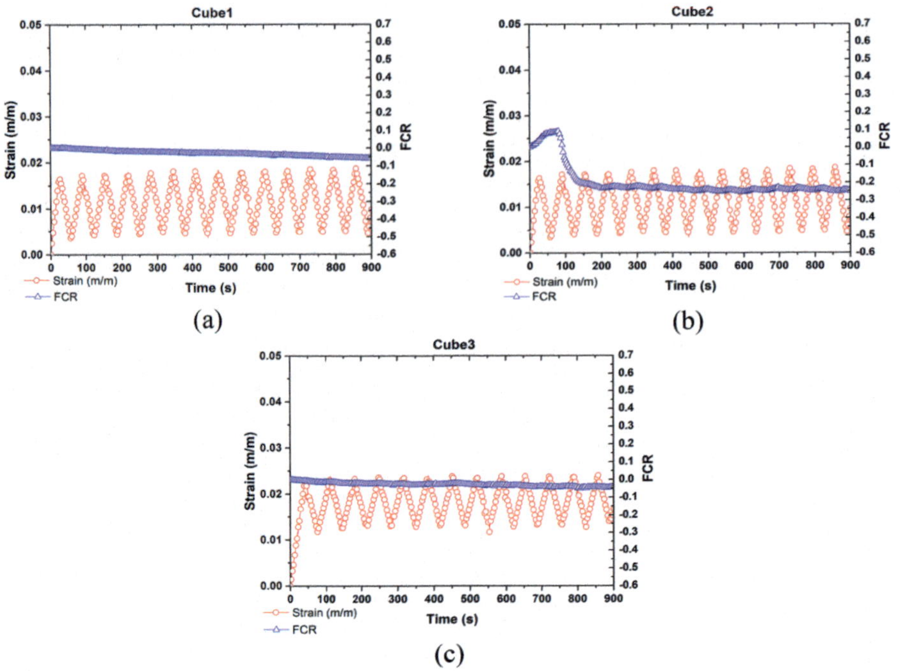

Fig. 8 FCR of TG 0.10% CNT samples compared to strain values registered by LVDT during cyclic loading. **a** Cube 1; **b** Cube 2; and **c** Cube 3

References

1. Lothenbach, B., Scrivener, K., Hooton, R.D.: Supplementary cementitious materials. Cem. Concr. Res. **41**(12), 1244–1256 (2011)
2. Islam, M.R., Picu, R.C.: Random fiber networks with inclusions: the mechanism of reinforcement. Phys. Rev. E **99**(6), 1–10 (2019)
3. de Almeida Carísio, P., Mendoza Reales, O.A., Toledo Filho, R.D.: Evaluation of mechanical properties of cement-based composites with nanomaterials. In Nanotechnology in Cement-Based Construction, Jenny Stanford Publishing, pp. 145–172 (2020)
4. Tian, Z., Li, Y., Zheng, J., Wang, S.: A state-of-the-art on self-sensing concrete: materials, fabrication and properties. Compos. Part B Eng. **177**, 107437 (2019)
5. Han, B., Yu, X., Ou, J.: Multifunctional and smart carbon nanotube reinforced cement-based materials. Nanotechnol. Civ. Infrastruct. 1–47 (2011)
6. Liew, K.M., Kai, M.F., Zhang, L.W.: Carbon nanotube reinforced cementitious composites: an overview. Compos. Part A Appl. Sci. Manuf. **91**(October), 301–323 (2016)
7. Mendoza Reales, O.A., Dias Toledo Filho, R.: A review on the chemical, mechanical and microstructural characterization of carbon nanotubes-cement based composites. Constr. Build. Mater. **154**, 697–710 (2017)
8. Han, B., et al.: Smart concretes and structures: a review. J. Intell. Mater. Syst. Struct. **26**(11), 1045389X15586452- (2015)
9. D'Alessandro, A., Ubertini, F., Materazzi, A.L., Laflamme, S., Porfiri, M.: Electromechanical modelling of a new class of nanocomposite cement-based sensors for structural health monitoring. Struct. Heal. Monit. An Int. J. **14**(2), 137–147 (Mar. 2015)

10. Han, B., Yu, X., Kwon, E.: A self-sensing carbon nanotube/cement composite for traffic monitoring. Nanotechnology **20**(44), 445501 (2009)
11. D'Alessandro, A., et al.: Static and dynamic strain monitoring of reinforced concrete components through embedded carbon nanotube cement-based sensors. Shock Vib. **2017** (2017)
12. Lim, M.-J., Lee, H.K., Nam, I.-W., Kim, H.-K.: Carbon nanotube/cement composites for crack monitoring of concrete structures. Compos. Struct. **180**, 741–750 (Nov. 2017)
13. Azhari, F., Banthia, N.: Carbon fiber-reinforced cementitious composites for tensile strain sensing. ACI Mater. J. (2017)
14. Deng, H., Li, H.: Assessment of self-sensing capability of carbon black engineered cementitious composites. Constr. Build. Mater. **173**, 1–9 (2018)
15. Parvaneh, V., Khiabani, S.H.: Mechanical and piezoresistive properties of self-sensing smart concretes reinforced by carbon nanotubes. Mech. Adv. Mater. Struct. **26**(11), 993–1000 (2019)
16. Demircilioğlu, E., Teomete, E., Schlangen, E., Baeza, F.J.: Temperature and moisture effects on electrical resistance and strain sensitivity of smart concrete. Constr. Build. Mater. **224**, 420–427 (2019)
17. Wang, H., Gao, X., Wang, R.: The influence of rheological parameters of cement paste on the dispersion of carbon nanofibers and self-sensing performance. Constr. Build. Mater. **134**, 673–683 (2017)
18. Konsta-Gdoutos, M.S., Metaxa, Z.S., Shah, S.P.: Highly dispersed carbon nanotube reinforced cement based materials. Cem. Concr. Res. **40**(7), 1052–1059 (2010)
19. Mendoza Reales, O.A., Arias Jaramillo, Y.P., Ochoa Botero, J.C., Delgado, C.A., Quintero, J.H., Toledo Filho, R.D.: Influence of MWCNT/surfactant dispersions on the rheology of Portland cement pastes. Cem. Concr. Res. **107**(August 2017), 101–109 (2018)
20. Mendoza, O., Sierra, G., Tobón, J.I.: Influence of super plasticizer and Ca(OH)2on the stability of functionalized multi-walled carbon nanotubes dispersions for cement composites applications. Constr. Build. Mater. **47**, 771–778 (2013)
21. Al-Rekabi, S.R.J.: Use of Carbon Nanofilaments in Producing Cementitious Composites with Improved Mechanical and Durability Performance. University of Brighton (2017)
22. Kim, G.M., Nam, I.W., Yoon, H.N., Lee, H.K.: Effect of superplasticizer type and siliceous materials on the dispersion of carbon nanotube in cementitious composites. Compos. Struct. **185**, 264–272 (2018)
23. Han, B., Yu, X., Kwon, E., Ou, J.: Effects of CNT concentration level and water/cement ratio on the piezoresistivity of CNT/cement composites. J. Compos. Mater. **46**(1), 19–25 (2012)
24. Zhang, L., et al.: Effect of characteristics of assembly unit of CNT/NCB composite fillers on properties of smart cement-based materials. Compos. Part A Appl. Sci. Manuf. **109**(February), 303–320 (2018)
25. Mendoza Reales, O.A., Duda, P., Dias Toledo Filho, R.: Effect of a carbon nanotube/surfactant aqueous dispersion on the rheological and mechanical properties of Portland cement pastes. J. Mater. Civ. Eng. **30**(10), 04018259 (2018)
26. Janowska-Renkas, E.: The effect of superplasticizers' chemical structure on their efficiency in cement pastes. Constr. Build. Mater. **38**, 1204–1210 (2013)
27. Srinivasan, S., Barbhuiya, S.A., Charan, D., Pandey, S.P.: Characterising cement-superplasticiser interaction using zeta potential measurements. Constr. Build. Mater. **24**(12), 2517–2521 (2010)
28. Collins, F., Lambert, J., Duan, W.H.: The influences of admixtures on the dispersion, workability, and strength of carbon nanotube-OPC paste mixtures. Cem. Concr. Compos. **34**(2), 201–207 (2012)
29. Zhang, L., Zheng, Q., Dong, X., Yu, X., Wang, Y., Han, B.: Tailoring sensing properties of smart cementitious composites based on excluded volume theory and electrostatic self-assembly. Constr. Build. Mater. **256**, 119452 (2020)
30. Al-rekabi, S.: "Influence of superplasticizer/surfactant aided aqueous dispersion of multi-walled Carbon nanotubes and its impact on workability and mechanical properties of cementitious composites. In: 25th Conferences and Laboratory Workshops (2016)

31. Downey, A., D'Alessandro, A., Ubertini, F., Laflamme, S., Geiger, R.: Biphasic DC measurement approach for enhanced measurement stability and multi-channel sampling of self-sensing multi-functional structural materials doped with carbon-based additives. Smart Mater. Struct. **26**(6) (2017)

Crack Healing

Autogenous Healing of High Performance Concrete Using Metakaolin

Reeja P. Soman and Nivin Philip

Abstract Concrete incorporates a natural self-healing capability that is primarily created by hydration and carbonation to repair small cracks. This ability is limited and is only activated when direct contact with water takes place. The study mainly dealt with crack repair and sustainability. For the repair of cracks, autogenous healing mechanism is adopted. Autogenous healing can be accelerated by the addition of mineral additives. The mineral additive used for autogenous healing is metakaolin. The use of metakaolin in high performance concrete can improve both the strength and durability properties of concrete. The ability of autogenous self-healing will increase with higher cement content. The experiment was conducted on high performance concrete of M_{70} grade. 10% replacement of cement with metakaolin is considered for the study. Specimens with addition of metakaolin and without metakaolin were subjected to 0 day, 3 days and 7 days curing. This study mainly focused on the strength retained, type of crack formed, propagation of cracks and amount of crack healed. The compressive strength and flexural strength at 7 day, 28 day and 180 days were carried out. Flexural strength test carried on beam specimens of size 700 × 150 × 150 mm was subjected to two-point loading. It has been observed that the specimens with metakaolin gives increased compressive and flexural strength than those specimens without adding metakaolin. The early age cracks in concrete structure will get reduced due to the addition of metakaolin. Also the addition of metakaolin promoted autogenous self-healing.

Keywords Autogenous · Self-healing · Crack · Sustainability · Two-point loading

R. P. Soman · N. Philip (✉)
Saintgits College of Engineering, Pathamuttom, Kerala, India

F. Kanavaris et al. (eds.), *International RILEM Conference on Early-Age and Long-Term Cracking in RC Structures*, RILEM Bookseries 31,
https://doi.org/10.1007/978-3-030-72921-9_29

1 Introduction

1.1 General

Most of the concrete structures are prone to cracking. Cracks are formed because of comparatively low tensile strength. If micro-cracks are formed and once they reach the reinforcement, not only corrosion of steel takes place but also the entire failure of the structure occurs. Early age cracking in concrete may occur due to various factors like exposure environment, mix composition, hydration rate and curing conditions [1]. So the cracks should be healed as early as possible. In order to resolve the problems caused due to the early age cracks, self-healing mechanisms can be adopted. Self-healing techniques involves: autogenous self-healing, autonomous and polymer based mechanism. The autogenous self-healing proves to be the most sustainable among the other [2]. The performance of concrete can be improved by the replacement of Portland cement by the use of mineral admixtures such as fly ash, silica fume, metakaolin, blast furnace slag etc. [3].

The mineral admixture used for the study is metakaolin. One of the supplementary cementitious materials used in the production of high performance concrete is metakaolin. Metakaolin is a highly reactive pozzolan which enhances the mechanical properties of the concrete in short term and long term. The use of metakaolin increases the durability of the structure and also improves the microstructure [4]. Metakaolin is used as a partial replacement for cement, as it gives high strength to the structure and reduces the amount of cracks. Concrete with 10% replacement of cement with metakaolin is considered in the study, as it exhibits better strength than other replacements of metakaolin [5, 6]. The specimens were cast with M70 grade of concrete.

1.2 Objective

The main objectives of the study include:

- To determine the type of crack occurred
- To investigate the propagation of cracks at early age of concrete
- To find the flexural strength of concrete after crack formation
- To find the amount of crack healed.

1.3 Scope

The present study investigates the early age crack in M_{70} grade concrete. The study is restricted to 0 day, 3 days and 7 days curing of concrete. The study is limited to 7, 28 and 180 days strength of concrete.

Table 1 Chemical composition of cement and metakaolin

Chemical composition	Cement (%)	Metakaolin (%)
Silica (SiO_2)	17–25	50–60
Alumina (Al_2O_3)	3–8	30–40
Ferric oxide (Fe_2O_3)	0.5–6	0.5–5
Calcium oxide (CaO)	60–65	0–0.5
Magnesium oxide (MgO)	1–3	0–2
Potassium oxide (K_2O)	0–1	0.5–1.5
Sulfuric anhydride (SO_4)	1–3	1–3

2 Materials Used

Ordinary Portland cement of grade 53 conforming to IS 12269:1987 was used for the experimental program [7]. The specific gravity of the cement used in the study is 3.12. Fine aggregate confirming to zone II having specific gravity 2.63 was used [8]. Crushed stone metal of size 20 mm having a specific gravity of 2.74 was used [8]. Fresh potable water free from contaminants is used for mixing and curing. The mineral additive used is metakaolin. It is used for partial replacement of cement [9]. The specific gravity of metakaolin was 2.50. The super plasticizer used is Master Glenium sky 8233. It is polycarboxylic ether based chemical admixture used for high performance concrete having a relative density of 1.08. The chemical composition of cement and metakaolin is shown in Table 1.

3 Experimental Procedures

3.1 Mix Design

Mix design was carried out for M_{70} grade concrete by IS 10262:2019, having a mix proportion of 1:1.3:2.2. 10% replacement of cement with metakaolin is considered. The target strength for the grade is 79.9 N/mm^2 [10]. The concrete mix proportion of strength 70 N/mm^2 is shown in Table 2.

3.2 Beam Design

The reinforcement details for beam specimen includes, 8 mm diameter bars are placed as main bars. The stirrups are of 6 mm diameter. The stirrups were placed at 90 mm spacing [11].

Table 2 Mix proportion for M_{70}

Grade	M_{70}
Water/cement ratio	0.29
Mass of cement (kg/m^3)	508.997
Mass of fine aggregate (kg/m^3)	671.095
Mass of coarse aggregate (kg/m^3)	1140.74
Metakaolin (kg/m^3)	56.70
Superplasticizer (kg/m^3)	2.4

3.3 Compressive Strength Test

The compressive strength test is carried in accordance to IS: 516-1959. The cube specimens of size 150 × 150 × 150 mm diameter were cast and cured [12]. The specimens were subjected to 0 day, 3 days and 7 days curing. The 7, 28 and 180 day compressive strength test were carried out.

3.4 Flexural Strength Test

Flexural strength test is carried in accordance to IS code 516-1959. The specimens of size 700 × 150 × 150 mm were made [12]. The beam specimens are also subjected to 0 day, 3 days and 7 days curing respectively. The flexural strength was tested was carried out at 7, 28 and 180 days.

4 Test Results

4.1 Type of Crack

There are various types of cracks which occur in reinforced concrete structures such as flexural cracks, shear cracks, torsional cracks, corrosion cracks etc. The type of crack occurred in this study is the flexure crack. Flexure cracks are formed in the regions with maximum bending moment. In reinforced concrete beams for short term condition, the width of the flexural cracks stays narrow from the surface to the steel. However, in long term condition under continuous loading, the crack width may get increased [13].

Fig. 1 Propagation of crack

4.2 Propagation of Crack

Visual examination was performed during the flexural testing to know the crack pattern. It was found that the crack begins to form at the bottom and then the crack bends in the diagonal direction as it moves upwards with widening in width and propagation in length. Figure 1 shows the propagation of cracks.

4.3 Healing of Crack

The healing of crack is done through autogenous self-healing. In autogenous self-healing no external agent is required. The autogenous self-healing is carried out by adding metakaolin. The partial replacement of cement with metakaolin helps in crack control and increased compressive and flexural strength. The amount of crack healed is more for beams with metakaolin than without metakaolin. Up to 60% of crack can be healed by the addition of metakaolin. More cracks will get healed by the addition of metakaolin. For autogenous healing to takes place, the presence of water is important because healing consist of chemical reactions of compounds exposed on the cracked surfaces. These reactions involve the formation of calcium hydroxide or calcium carbonate and create new hydrates and the other material ultimately bridges the crack. Figure 2a, b shows the specimen without metakaolin and the healed specimen with the addition of metakaolin.

Fig. 2 **a** Beam specimen without metakaolin. **b** Beam specimen with metakaolin

4.4 Compressive Strength

The compressive strength of cube specimens was tested in the compressive strength testing machine. The 7th day, 28th day and 180th day compressive strength values of specimens are shown in Table 3, Table 4 and Table 5 respectively.

Table 3 7th day compressive strength of cube specimens

Compressive strength								
Days cured	7 day strength without metakaolin				7 day strength with metakaolin			
	Stress (MPa)	% variation	SD	Avg (MPa)	Stress (MPa)	% variation	SD	Avg (MPa)
0 day	19.68	0	0.93	18.7	20.18	15	1.97	23.59
	17.85	9			23.76	0		
	18.47	6			23.43	1.38		
3 day	25.17	5	0.99	25.55	28.87	1.5	0.24	29.03
	24.81	7			29.32	0		
	26.68	0			28.91	1.3		
7 day	29.69	3	0.52	30.28	34.59	1.6	0.75	34.48
	30.63	0			33.68	0		
	30.54	0.3			35.17	4.2		

Table 4 28th day compressive strength of cube specimens

Compressive strength								
Days cured	28 day strength without metakaolin				28 day strength with metakaolin			
	Stress (MPa)	% variation	SD	Avg (MPa)	Stress (MPa)	% variation	SD	Avg (MPa)
0 day	45.52	0	0.73	45.01	45.41	2.9	0.70	46.12
	44.17	2.9			46.81	0		
	45.35	0.3			46.13	1.4		
3 day	54.34	1.6	0.43	54.75	56.72	1.7	0.50	57.23
	54.65	1.1			57.72	0		
	55.27	0			57.25	0.8		
7 day	59.98	1.2	0.38	60.36	66.70	1.6	0.83	66.92
	60.75	0			67.84	0		
	60.37	0.6			66.22	2.3		
28 day	61.51	2	0.69	61.99	76.71	0.2	0.43	76.54
	61.69	1.7			76.87	0		
	62.79	0			76.05	1		

Table 5 180th day compressive strength of cube specimens

Compressive strength								
Days cured	180 day strength without metakaolin				180 day strength with metakaolin			
	Stress (MPa)	% variation	SD	Avg (MPa)	Stress (MPa)	% variation	SD	Avg (MPa)
0 day	45.98	0	0.74	45.47	49.36	3	0.70	46.12
	44.62	2.9			50.89	0		
	45.81	0.3			50.15	1.4		
3 day	54.89	1.6	0.47	55.31	61.65	1.7	0.50	57.23
	55.21	1.1			62.74	0		
	55.83	0			62.23	0.8		
7 day	60.59	1.2	0.39	60.98	72.51	1.6	0.83	66.92
	61.37	0			73.74	0		
	60.98	0.6			71.98	2.3		
28 day	62.14	2	0.69	62.63	83.39	0.2	0.86	82.54
	62.32	1.7			83.56	0		
	63.43	0			82.67	1		

- When compared to the specimens without adding metakaolin, the maximum percentage increase in 7 days compressive strength for specimens without undergoing curing was 20.72%, 11.98% with 3 days curing and 12.18% for 7 days curing.
- Similarly maximum percentage increase in 28 days compressive strength was observed as 2.4% for 0 day cured specimens, 4.3% for 3 days cured specimens, 9.8% for 7 days cured specimens and 19% for 28 days cured specimens.
- The maximum percentage increase in 180 days compressive strength was observed as 9.3% for 0 day cured specimens, 11.07% for 3 days cured specimens, 16.16% for 7 days cured specimens and 24.12% for 28 days cured specimens.

4.5 Flexural Strength

The flexural strength testing was carried out on the universal testing machine by the application of two-point loading. The maximum load that can be taken by the beam specimens were noted down. The modulus of rupture was calculated using the load in the equation. The 7 day, 28 day and 180 day flexural strength values of the beam specimens without metakaolin and with the addition of metakaolin are shown in Tables 6, 7 and 8.

- When compared to the specimens without adding metakaolin, the maximum percentage increase in 7th day flexural strength for specimens without undergoing curing was 29.87%, 9.37% with 3 days curing and 13.55% for 7 days curing.
- Similarly maximum percentage increase in 28th day flexural strength was observed as 7.89% for 0 day cured specimens, 11.63% for 3 days cured specimens, 5.14% for 7 days cured specimens and 13.05% for 28 days cured specimens.
- The maximum percentage increase in 180th day flexural strength was observed as 7.58% for 0 day cured specimens, 14.16% for 3 days cured specimens, 7.6% for 7 days cured specimens and 15.91% for 28 days cured specimens.

5 Conclusion

On the basis of the results obtained by conducting the experiment, the following conclusions have been drawn:

- The type of crack occurred in the structure is flexure crack.
- The crack propagated from the bottom and then bends in the diagonal direction as it moves upwards with widening in width and propagation in length.
- By the addition of metakaolin, autogenous self-healing has occurred and the formation of early age cracks got retarded.
- By replacing the cement with 10% metakaolin, the strength has got increased and also caused a reduction in the formation of early age cracks. Approximately 60% of the early age cracks formed got healed by the addition of metakaolin.

Table 6 7th day flexural strength of beam specimens

Flexural strength										
Days cured	7 day strength without metakaolin					7 day strength with metakaolin				
	Load (kN)	Stress (MPa)	% variation	SD	Avg (MPa)	Load (kN)	Stress (MPa)	% variation	SD	Avg (MPa)
0 day	36.03	5.76	0	0.23	5.4	50.89	7.69	2.1	0.1	7.7
	32.78	5.39	6.4			52.06	7.86	0		
	35.23	5.32	2.2			51.72	7.81	0.6		
3 day	56.68	8.56	9.9	0.65	8.7	60.10	9.08	0	0.6	9.6
	54.63	8.25	13			58.68	10.4	12		
	53.54	9.51	0			62.49	9.44	11		
7 day	64.87	9.80	12	0.76	10.2	68.99	12.26	2.6	0.2	11.8
	65.18	9.84	11.7			70.89	12.60	0		
	62.76	11.15	0			71.17	10.75	14		

Table 7 28th day flexural strength of beam specimens

Flexural strength										
Days cured	28 day strength without metakaolin					28 day strength with metakaolin				
	Load (kN)	Stress (MPa)	% variation	SD	Avg (MPa)	Load (kN)	Stress (MPa)	% variation	SD	Avg (MPa)
0 day	48.76	8.1	0	0.36	7.7	49.42	8.7	0	0.35	8.36
	50.38	7.6	6.1			50.01	8.4	3.4		
	49.51	7.4	8.6			50.61	8.0	8		
3 day	59.70	10.61	0	0.59	9.95	65.02	11.55	0.4	0.54	11.26
	61.32	9.45	10			66.53	10.64	8.2		
	62.20	9.81	7.5			65.28	11.60	0		
7 day	68.81	12.23	0	0.60	11.61	72.75	12.09	5.3	0.48	12.24
	69.33	11.03	9.8			71.89	12.78	0		
	68.67	11.59	5.2			73.16	11.85	7.2		
28 day	73.26	11.46	9.7	0.67	11.92	80.15	14.24	0.7	0.99	13.71
	72.60	11.61	8.5			80.69	14.34	0		
	71.46	12.70	0			81.76	12.57	12		

Table 8 180th day flexural strength of beam specimens

Flexural strength										
Days cured	180 day strength without metakaolin					180 day strength with metakaolin				
	Load (kN)	Stress (MPa)	% variation	SD	Avg (MPa)	Load (kN)	Stress (MPa)	% variation	SD	Avg (MPa)
0 day	49.26	8.7	0	0.55	8.16	53.72	9.5	0	0.65	8.83
	49.89	7.6	12			54.36	8.2	13		
	50.02	8.2	5.7			55.02	8.8	7.3		
3 day	60.31	9.11	13	1.03	10.06	70.68	12.56	0	0.7	11.72
	62.83	11.16	0			72.32	11.57	7.8		
	61.94	9.91	11			70.96	11.03	12		
7 day	69.51	12.35	0	0.57	11.91	79.08	12.72	7.8	0.83	12.89
	70.04	11.26	8			78.15	13.8	0		
	69.37	12.14	1.7			79.53	12.15	11		
28 day	74.01	11.57	11.2	0.79	12.47	87.12	15.48	0.7	1.22	14.83
	73.34	13.03	0			87.71	15.59	0		
	72.19	12.83	1.5			88.87	13.42	13		

- Metakaolin contributes strength mainly through the quick pozzolonic reaction at early ages. It has been observed that 7 day, 28 day and 180 day compressive strength and flexural strength of the specimens has increased due to the addition of metakaolin when compared to the specimens without adding metakaolin.

References

1. Safiuddin, M., Amrul, A.B.M.K., Woon, C., Sudharshan, N.R.: Early age cracking in concrete: causes, consequences, remedial measures and recommendations. Appl. Sci. **8**, 1–25 (2018)
2. Rajczakowska, M., Habermehl-Cwirzen, K., Hedlund, H., Cwirzen, A.: Autogenous self-healing: a better solution for concrete. J. Mater. Civ. Eng. **31**, 1–19 (2019)
3. Güneisi, E., Gesoglu, M., Mermerdas, K.: Improving strength, drying shrinkage, and pore structure of concrete using metakaolin. Mater. Struct. **41**, 937–949 (2008)
4. Brooks, J.J., Johari, M.A.M.: Effect of metakaolin on creep and shrinkage of concrete. Cement Concr. Compos. **23**(6), 495–502 (2001)
5. Muralinathan, P., Joshua, A.D., Sivakamasundari, S.: Study of high strength concrete using metakaolin at elevated temperatures. Int. J. Pure Appl. Math. **119**, 1267–1273 (2018)
6. Pavankumar, V., Krishnamurthy, G.R.: Experimental study on replacement of cement with metakaolin using m sand with steel fiber of M70 grade of concrete. Int. J. Sci. Res. **6**(8), 115–119 (2015)
7. IS 12269:1987: Specification for 43 Grade Ordinary Portland Cement. Bureau of Indian Standards, New Delhi
8. IS 383:1970: Specification for Coarse and Fine Aggregates from Natural Sources for Concrete. Bureau of Indian Standards, New Delhi
9. Sunny, A.J., Mohan, N.S., Sambhaji, L.K.: Effect of metakaolin on the properties of concrete. Int. J. Eng. Technol. **4**(7), 643–645 (2017)
10. IS 10262: Guidelines for Concrete Mix Design Proportioning. Bureau of Indian Standards, New Delhi (2019)
11. IS 456: Plain and Reinforced Concrete—Code of Practice. Bureau of Indian Standards, New Delhi (2000)
12. IS 516:195: Methods of Tests for Strength of Concrete. Bureau of Indian Standards, New Delhi
13. Krishna, T.T., Subathra, S., Meikandaan, T.P.: An experimental study of crack patterns on reinforced concrete beam. Int. Res. J. Eng. Technol. **5**, 3195–3202 (2018)

Bacteria Based Self-healing of Later-Age Cracks in Concrete

Lorena Skevi, Bianca Reeksting, Susanne Gebhard, and Kevin Paine

Abstract Monitoring and rehabilitation of cracks in concrete structures can be rather expensive, labour intensive and often presents technical difficulties. Nonetheless, it is vital for prolonging their lifespan. As an alternative to traditional repair techniques self-healing in cementitious materials has been proven a promising technology for crack remediation. The use of bacteria in self-healing relies on their ability to metabolically facilitate the precipitation of calcite, which can act as a crack sealant. Since cracking is likely to occur throughout the life of concrete it is important to investigate the long-term effectiveness of the method. In this study bacterial spores of the species *B. cohnii* were added in cement mortar mix, along with nutrients for the bacteria. Reference samples of plain mortar and control samples containing only nutrients were also prepared for comparison. Two sets of samples were prepared: one with samples that were cracked after 28 days of curing and one with samples cracked after 9 months. A single crack of 0.4–0.5 mm crack width was introduced to all samples. After cracking, samples were left to heal semi-submerged in water. Healing was examined at two temperatures, at 7.5 and 20 °C, for both sets of samples. Optical microscopy and water-flow tests were used for evaluating the healing. Results showed that substantial healing can be achieved in 9 months old samples containing bacteria, both at 7.5 and 20 °C, comparable to that of 28 days old samples. The long-term competence of bacteria-based self-healing technology for crack remediation in concrete is, therefore, demonstrated.

Keywords Self-healing · Bacteria · Cracking · Permeability · Microscopy

---

L. Skevi (✉) · K. Paine
BRE Centre for Innovative Construction Materials, University of Bath, Claverton Down, Bath B2 7AY, UK
e-mail: l.skevi@bath.ac.uk

B. Reeksting · S. Gebhard
Department of Biology & Biochemistry, University of Bath, Claverton Down, Bath B2 7AY, UK

F. Kanavaris et al. (eds.), *International RILEM Conference on Early-Age and Long-Term Cracking in RC Structures*, RILEM Bookseries 31,
https://doi.org/10.1007/978-3-030-72921-9_30

1 Introduction

Cracking is an inherent vice of concrete structures. Depending on their cause, size and place in the structure, cracks can seriously impair the material endangering the long-term serviceability of the structure. Ingression of salts such as chlorides and sulphates in concrete through the cracks can lead to spalling of the surface, carbonation, and corrosion of the reinforcement, even for cracks as small as 0.1 mm in width [1]. Monitoring and rehabilitation of concrete structures like bridges, dams, and tunnels where cracks are often not visible or easy to access, can be technically challenging and expensive. In Europe maintenance and repair cost of existing structures was estimated to reach 45% of the total construction budget in 2018 [2].

Self-healing concrete has been widely investigated over the past years as an alternative way of managing damage in cementitious materials. The mechanism includes the incorporation of healing agents in the composite, which are released upon cracking repairing the damage. Utilisation of bacteria as healing agents is based on their ability to promote mineralisation of calcite, which acts as crack sealant. They do that through their metabolic activities, mainly by increasing the concentration of inorganic carbon: CO_2, HCO_3^-, CO_3^{2-} [3] and by regulating the alkalinity of their surrounding environment [4]. Alkaliphilic bacterial spores that can endure the high-alkaline concrete environment are employed, and nutrients and calcium salts are also added in the composite for ensuring calcite formation. Spores are often encapsulated in a protective material for increasing their viability in the composite and for avoiding their contact with the nutrients before cracking.

Various approaches in bacteria-based self-healing concrete (BBSHC) have been investigated leading to effective crack healing of 0.2–0.3 mm [5], 0.4–0.5 mm [6, 7] and even 1 mm crack widths [8, 9]. Parameters regarding the materials used, i.e., bacteria, nutrients, and spore-carriers, have been researched in depth. However, less attention has been paid so far to the long-term efficiency of the method. Qian et al. [10] examined cracks generated at 7, 14, 28, 60 days and reported a reduction in the healing as the cracking age increased. Similar results were obtained by Luo et al. [11]. Since both studies used vegetative bacteria cells instead of spores the results can be related to the decrease of the cells' viability with time. Khaliq and Ehsan [12] cracked concrete samples at the age of 3, 7, 14 and 28 days and showed that the efficiency of the carrier material was influenced with time resulting in lower healing. Cracking of samples older than 28 days was studied in [6], where 56 days old samples were cracked, and crack widths up to 0.46 mm were healed completely after 100 days. Ersan et al. [13] was the first to study long-term cracks in 6 months old mortar samples. A crack closure close to 90% was observed for cracks up to 0.4 mm wide and 70% for crack widths between 0.4–0.5 mm. Zheng and Qian [14], also, reported significant healing up to 96% for 0.2–0.3 mm wide cracks in 6 months old samples.

This study investigates for the first time the long-term efficiency of BBSHC in 9 months old samples as compared to that of 28 days old samples. Furthermore, healing at lower temperature (7.5 °C) is examined along with healing at room temperature (20 °C) for samples of both ages, with a view to consider the application of BBSHC in more realistic environmental conditions.

2 Materials and Methods

2.1 Materials

Portland limestone cement, CEM II/A-L 32.5R, and standard sand complying with BS EN 196-1 were used for making cement mortar samples. The bacteria used in this study are an alkaliphilic, spore-forming strain of the species *Bacillus cohnii.* Spores of the bacteria were prepared in sporulation medium and were collected by centrifugation after 48 h. Aerated concrete granules (ACG) of 1–4 mm diameter were used as carriers for the spores. Yeast extract was used as organic food source for the bacteria and calcium nitrate was employed as a precursor. Both yeast extract and calcium nitrate were added directly in the mortar mix.

2.2 Spore Encapsulation

The encapsulation process followed the method described in [7]. In brief, the ACG were soaked in a water solution containing the spores and a concentration of 1.44 x 10^7 spores per gr of ACG was achieved. After impregnation, the granules were left to completely dry in the oven at 50 °C. Subsequently, they were coated with a layer of PVA, for avoiding an early germination of the spores when contacting the mixing water and the growth media, i.e., yeast extract and calcium nitrate.

2.3 Sample Preparation

Three different types of cement mortar samples were prepared in triplets: reference samples (REF) of plain mortar with no bacteria or nutrients, control samples (CTRL) containing only growth media and bacterial samples (BIO) containing both growth media and spores. Two sets of samples were prepared regarding the age of cracking. First set was cracked after 28 days of water curing. Samples of the other set were cured in water for 28 days and then stored in a conditioning room with controlled temperature and relative humidity at 20 °C and 50% respectively for 9 months before cracking. In each set, half of the samples were healed at 7.5 °C and the other half at 20 °C. Thus, a total of 36 samples, 12 of each type, were prepared.

Mixing of mortars was carried out in accordance with BS EN 196-1. The spore-bearing granules replaced 6% of the sand by volume. Yeast extract and calcium nitrate were diluted in the mixing water before added directly to the mortar mix. Tap water was used for preparing all samples.

In order to economize on spores and nutrients, BIO samples of 65 × 40 × 40 mm^3 were made in two layers: bottom layer contained the growth media and the spore loaded ACG, while the top half was filled with plain cement mortar (Fig. 1). Reference and control samples were prepared in the same way for consistency. Specimens were demoulded 48 h after casting and were cured in water at 20 °C for 26 days. Composition of all samples is given in Table 1. Only the mix design of the bottom layer is considered in this table, as standard mortar mix complying to BS EN 196-1 was prepared for the top layer of all specimens.

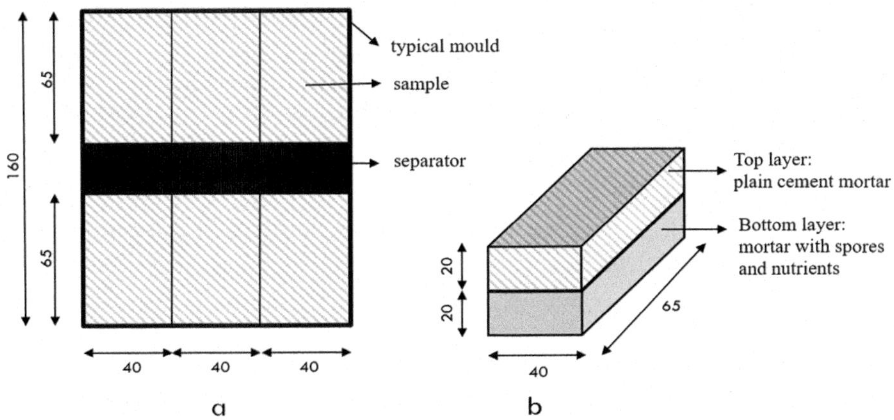

Fig. 1 Schematic presentation of BIO samples: **a** plan view of the specimens in the mould and **b** 3D view of one specimen. Dimensions in the figure are given in mm

Table 1 Mix design of the three mortar types used in this study. Quantities are given for the total number of samples (12) made per type

	Cement (g)	Water (g)	Sand (g)	ACG (g)	Bacterial spores	Yeast extract (g)	Calcium nitrate (g)
REF	368	184	1104	–	–	–	–
CTRL	368	184	1104	–	–	18.2	4
BIO	368	184	1040	21	3.02 x 10^8	18.2	4

2.4 Cracking

At the age of 28 days the first set of specimens were taken out of the curing tanks and the top half of them, i.e., the plain mortar, was wrapped with carbon fibre reinforced polymer strips before the cracking to avoid the complete split of the samples during-loading. Hence, only the bottom half of the sample was cracked. Before cracking, a small notch of approximately 1.5 mm depth was introduced in the middle of the bottom half of the sample. Samples were then cracked under a 3-point bending load so that a single-edge crack was created. The crack width was monitored by a crack mouth opening displacement gauge (CMOD) and a crack width of about 0.4–0.5 mm was achieved for all samples. The second set of samples were cracked after 9 months following the same procedure.

2.5 Healing Conditions and Evaluation

After cracking samples were left to heal in containers, semi-submerged in tap water. Separate containers were used for each specimen type to avoid any leak of growth media and spores in the water. Half of the samples were healed at room temperature (20 °C) and the other half were kept in a fridge with a fixed temperature at 7.5 °C.

The self-healing efficiency of the mortar samples was evaluated visually using a LEICA M205C light optical microscope. Crack width was measured immediately after cracking and then after 7, 28 and 56 days of healing using the image analysis software of the microscope. Percentage of healing (H_W) was calculated through Eq. (1). A triplet of samples was measured for each specimen type and the final value occurred from their average.

$$H_W = ((W_0 - W_t)/W_0) \times 100\,\% \tag{1}$$

where

W_0: crack width before healing, immediately after cracking.

W_t: crack width after t time of healing.

For assessing healing in terms of water tightness RILEM Tube Test Method 11.4 [15] was used. Only freshly cracked samples and samples healing for 28 days were subjected to this test. Water-flow coefficient was given by Eq. (2) and healing ratio (H_K) was calculated by Eq. (3).

$$K = ((a \times L)/(A \times t)) \times \ln(h_1/h_2) \tag{2}$$

where

a: the area of the Rilem tube (cm^2).

A: the area of the specimen that is in touch with the acrylic plate (cm^2).

L: the height of the specimen (cm).
t: time need for the water to flow from h1 to h2 (s).
h1 and h2: the initial and final water head respectively

$$H_K = ((K_0 - K_{28}) / K_0) \times 100\% \quad (3)$$

where
K_0: water-flow coefficient before healing, immediately after cracking.
W_{28}: water flow coefficient after 28 days of healing.

3 Results and Discussion

3.1 Crack Closure

Crack closure was estimated by comparing crack widths measured before and after certain time of healing. Results are shown in Fig. 2 for 28 days and 9 months old samples healed at 7.5 and 20 °C. A healing ratio up to 70% was reported for 9 months old samples containing bacteria after 56 days of healing at both temperatures. Healing ratio in reference and control samples on the other hand was limited to 30% and 40% respectively at 7.5 °C and slightly lower at 20 °C. It is, thus, concluded that healing temperature does not significantly affect the crack closure in mature samples. Contrarily, a notable decrease is observed in the healing ratio of 28 days old samples containing bacteria that were healed at 7.5 °C as compared to those healed at 20 °C. This difference does not appear in reference and control samples, showing that they are not significantly affected by the temperature. It should also be noted that healing ratio of control samples, containing only growth media, appears to be strongly influenced by the age of the sample. In 28 days old samples growth media components seem to greatly contribute to the healing, whilst in mature samples their contribution is much lower. Finally, as an overall trend for all samples it is noted that most of the healing is already achieved after 7 days and healing ratios increase only slightly from day 7 to day 56 of healing.

Images of the cracks were taken immediately after cracking and then weekly between 7 and 56 days of healing. The same areas of the cracks were captured every time. Figures 3 and 4 show images of the reference, control, and bacterial samples before and after 56 days of healing at 7.5 °C and 20 °C respectively. It should be noted that while the images show one sample of each type, healing ratio values represent measurements in a triplet of samples for each type. Nonetheless, in general, crack closure results are well depicted visually in these images.

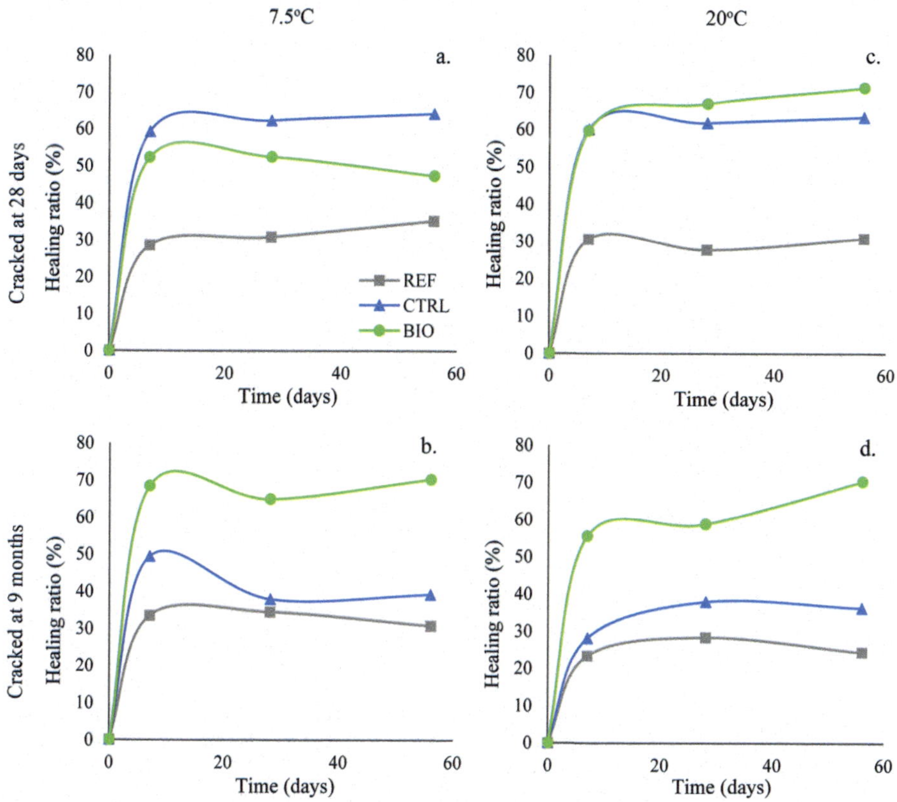

Fig. 2 Healing ratios in terms of crack closure of 28 days old (**a**, **b**) and 9 months old (**c**, **d**) samples at 7.5 °C (**a**, **c**) and 20 °C (**b**, **d**)

3.2 Water Tightness

Water-flow test was performed in freshly cracked samples and after 28 days of healing and the healing percentage for all samples is presented in Table 2. Results for 28 days old samples roughly agree with the crack closure healing results presented in the previous section (Fig. 2a, b). The significant contribution of the growth media to the healing of these samples is, thus, verified by the water flow results. In addition, BIO samples of 9 months old age seem to be more watertight than reference and control samples at both temperatures.

However, for older samples a strong effect of the temperature is noticed in water-flow results, which was not reported in crack closure healing (Fig. 2c, d). More specifically, 9 months old samples of all types are significantly less watertight at 7.5 °C than at 20 °C. This discrepancy between crack closure and water flow results can be explained by the fact samples contract when healing at low temperature resulting in higher crack closure even when the cracks are not covered with healing

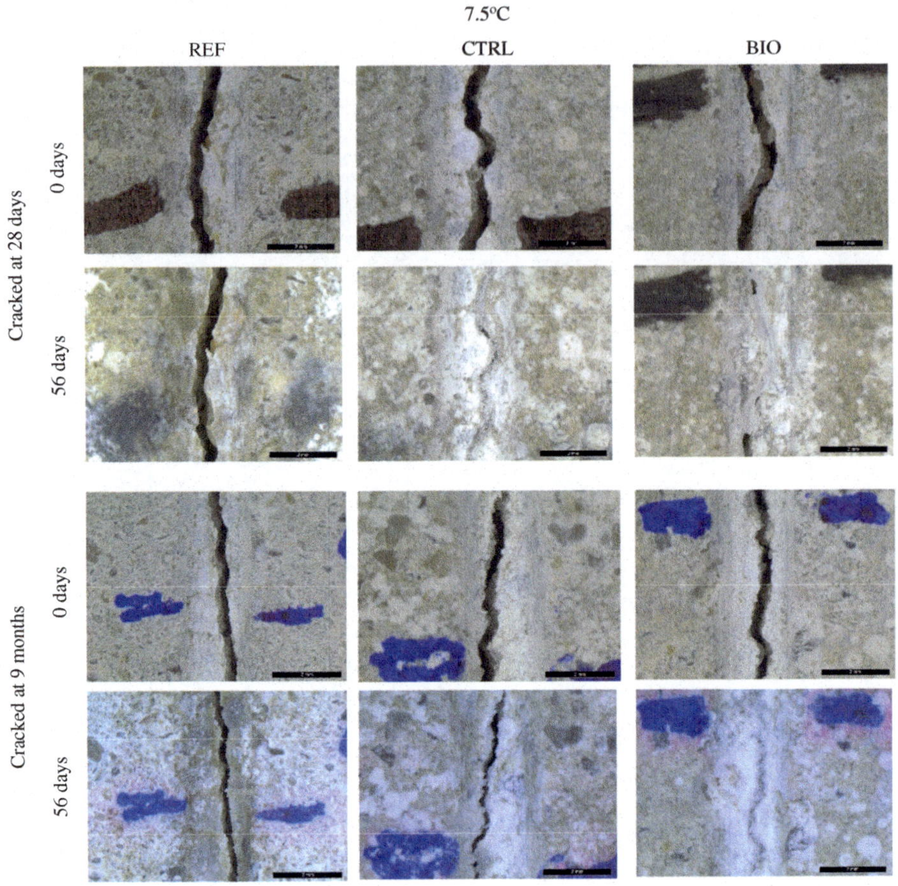

Fig. 3 Images of 28 days and 9 months old samples before and after healing at 7.5 °C

Table 2 Healing ratio in terms of water tightness of 28 days and 9 months old samples after 28 days of healing at 7.5 and 20 °C

		7.5 °C (%)	20 °C (%)
Cracked at 28 days	REF	38.16	26.85
	CTRL	80.32	72.63
	BIO	60.08	70.11
Cracked at 9 months	REF	7.77	39.74
	CTRL	26.97	65.64
	BIO	47.04	80.72

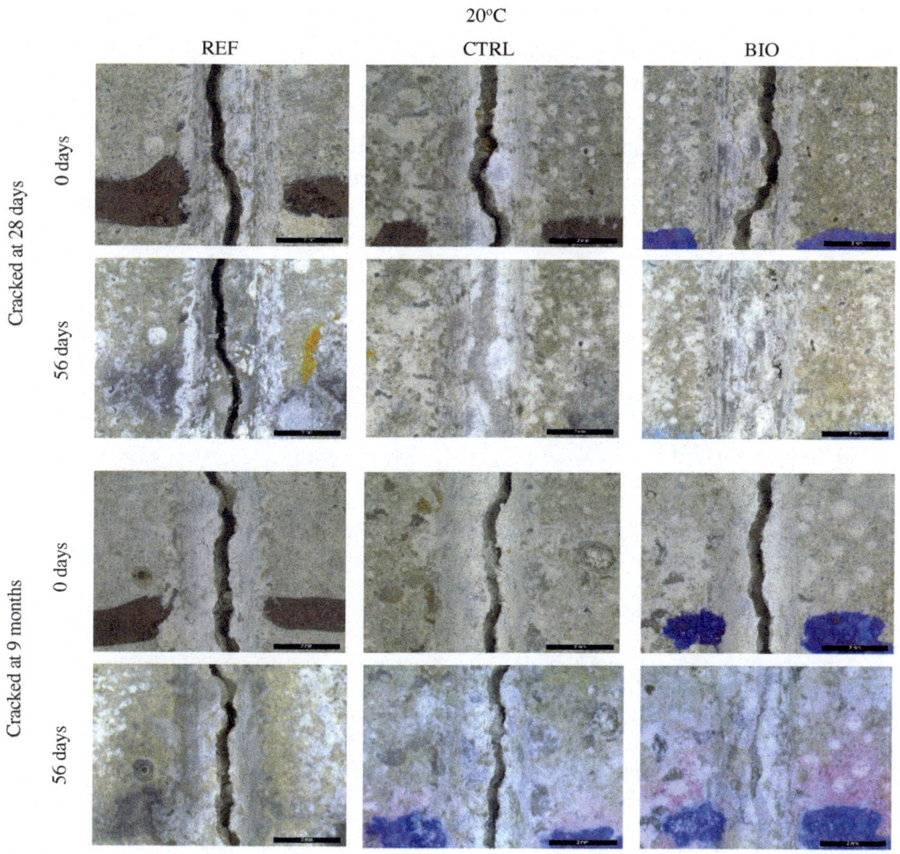

Fig. 4 Images of 28 days and 9 months old samples before and after healing at 20 °C

product. Thus, cracks are shrunk in size but not healed with calcite, allowing the water to flow through the samples. This phenomenon appears to affect 9 months old samples that had the smaller initial crack width, ~0.4 mm, while 28 days old samples, whose crack width was ~0.5 mm do not seem to be influenced. This can be confirmed by the images in Fig. 3, where healing at 7.5 °C for samples of both ages is shown. Finally, the reduction of healing for all types of older samples at low temperature shows that both autogenous and biogenic calcite yield was impaired.

4 Conclusions

This investigation demonstrated that bacteria-based self-healing in cementitious materials is efficient even for 9 months old samples. Healing of older samples at room temperature was found to be comparable to that of 28 days old samples, reaching a healing ratio of approximately 70–80% after 56 days. At lower temperature, though, less healing was reported for older samples of all types. Further research is carried by the authors towards that direction, investigating the use of psychrophilic bacteria for achieving self-healing at low temperatures.

Long-term performance of BBSHC opens the road for the application of self-healing technology in real concrete structures, thus, resulting in substantial maintenance cost reduction in construction industry.

References

1. Faye, P.N., Ye, Y., Diao, B.: Effects of crack width on chloride penetration and performance deterioration of RC columns with sustained eccentric compressive load. KSCE J. Civ. Eng. **22**(2), 637–646 (2018)
2. Abrahamsen, Y.: Civil engineering market. In: 80th Euroconstruct Summary Report, 77-97. Euroconstruct, Budapest (2015)
3. De Belie, N., Gruyaert, E., Al-Tabbaa, A., Antonaci, P., Baera, C., Bajare, D., et al.: A review of self-healing concrete for damage management of structures. Adv. Mater. Interfaces **5**(17), 1800074 (2018)
4. De Muynck, W., De Belie, N., Verstraete, W.: Microbial carbonate precipitation in construction materials: a review. Ecol. Eng. **36**(2), 118–136 (2010)
5. Tziviloglou, E., Pan, Z., Jonkers, H.M., Schlangen, E.: Bio-based self-healing mortar: an experimental and numerical study. J. Adv. Concrete Technol. **15**(9), 536–543 (2017)
6. Wiktor, V., Jonkers, H.M.: Quantification of crack-healing in novel bacteria-based self-healing concrete. Cement Concr. Compos. **33**(7), 763–770 (2011)
7. Tan, L., Reeksting, B., Ferrandiz-Mas, V., Heath, A., Gebhard, S., Paine, K.: Effect of carbonation on bacteria-based self-healing of cementitious composites. Constr. Build. Mater. **257**, 119501 (2020)
8. Wang, J.Y., Soens, H., Verstraete, W., De Belie, N.: Self-healing concrete by use of microencapsulated bacterial spores. Cem. Concr. Res. **56**, 139–152 (2014)
9. Jiang, L., Jia, G., Jiang, C., Li, Z.: Sugar-coated expanded perlite as a bacterial carrier for crack-healing concrete applications. Constr. Build. Mater. **232**, 117222 (2020)
10. Qian, C., Chen, H., Ren, L., Luo, M.: Self-healing of early age cracks in cement-based materials by mineralization of carbonic anhydrase microorganism. Front. Microbiol. **6**, 1225 (2015)
11. Luo, M., Qian, C., Li, R.: Factors affecting crack repairing capacity of bacteria-based self-healing concrete. Constr. Build. Mater. **87**, 1–7 (2015)
12. Khaliq, W., Ehsan, M.B.: Crack healing in concrete using various bio influenced self-healing techniques. Constr. Build. Mater. **102**, 349–357 (2016)
13. Ersan, Y.C., Gruyaert, E., Louis, G., Lors, C., De Belie, N., Boon, N.: Self-protected nitrate reducing culture for intrinsic repair of concrete cracks. Front. Microbiol. **6**, 1228 (2015)
14. Zheng, T., Qian, C.: Self-healing of later-age cracks in cement-based materials by encapsulation-based bacteria. J. Mater. Civ. Eng. **32**(11), 04020341 (2020)
15. Rilem TC 25-PEM: Recommended tests to measure the deterioration of stone and to assess the effectiveness of treatment methods. Mater. Struct. **13**(75), 175–253 (1980)

Effects of *Bacillus subtilis* on Crack Remediation in Thermally Degraded Limestone Calcined Clay Cement Mortars

Joseph Mwiti Marangu and Mark Bediako

Abstract This paper reports experimental findings on the effect of *Bacillus subtilis* on crack remediation in thermally degraded Limestone Calcined Clay Cement (LC3) mortars. Mortar prisms measuring 160 mm × 40 mm × 40 mm were cast using LC3 at water/cement (w/c) ratio of 0.5 and cured for 28 days. Half of the 28-day cured mortar prisms were thermally degraded by heating them at 1000 °C in a furnace to induce the cracks while the other half was used as a control. Both cracked and un-cracked mortar prisms were subjected to compressive strength, porosity and accelerated chloride ingress tests. Moreover, half of the cracked and un-cracked mortars were immersed in bacterial solution containing *Bacillus subtilis* while the other half was separately immersed in curing water until the 90th day. Compressive strength, porosity and chloride ingress tests were also repeated on the 90th day. Microstructural changes in cracked LC3 mortars were carried out using Scanning Electron Microscope (SEM) before and after immersion in bacterial solution. Results showed that at 28 days of curing, un-cracked mortars exhibited higher compressive strength, lower porosity and lower apparent chloride diffusion coefficients than cracked mortars. However, compressive strength, porosity and apparent chloride diffusion coefficients of both cracked and un-cracked mortars were equivalent after 90 days of curing in bacterial solution. SEM images showed visible micro-cracks after thermal treatment and healed cracks with calcite deposition after curing in bacterial solution. In conclusion, *Bacillus subtilis* was found to improve the crack healing capacity of thermally cracked mortars.

Keywords Bio-concrete · Cement · Concrete cracks · Compressive strength · Chloride ingress · Porosity

J. M. Marangu (✉)
Department of Physical Sciences, Meru University of Science & Technology, Meru, Kenya

M. Bediako
Advanced Material Science Division, CSIR-Building & Road Research Institute, Kumasi, Ghana

F. Kanavaris et al. (eds.), *International RILEM Conference on Early-Age and Long-Term Cracking in RC Structures*, RILEM Bookseries 31,
https://doi.org/10.1007/978-3-030-72921-9_31

1 Introduction

Concrete is one of the mostly widely used materials in building and construction. However, it is subject to degradation when used in day to day construction activities [1–3]. In its natural state, concrete is susceptible to cracks. Cracks in concrete structures occur due to various mechanisms such as shrinkage, alkali-silica reactions and mechanical compression among others [4–6]. The presence of cracks allows ingress of water, suphate ions, carbon dioxide, acids and other deleterious media into the hardened cement matrix. This leads to degradation, loss of aesthetics and ultimate reduction in service life of cement based structures [4]. This calls for costly constant repairs and maintenance [7].

The long term durability of cement based structures is crucial in order to ensure safe, durable and resilient built infrastructure globally. In the built environment, cracks occur in cement based structures when it is exposed to elevated temperatures [8]. High temperatures have a significant influence on the thermal deformation, cracking, spalling and compressive strength losses. The loss of strength due to cracks reduces the durability of cement structures [8]. Extensive efforts aimed at crack remediation in cement based structures are therefore being adopted to improve their service life. There are many commercial products for crack remediation such as structural epoxy, silica fume and epoxy mortars among others [9, 10]. However, the use of bacteria in remediating cracks has attracted some great attention since it is considered environmental friendly. To this end, a Microbially Induced Calcite Precipitation (MICP) has been reported as a potentially viable technique for remediation of concrete cracks. MICP process imparts special ability to cementitious materials to repair cracks autonomously [10–15]. This is mainly through calcite precipitation which results in self-healing of concrete cracks. Earlier studies using concrete demonstrated that *Bacillus subtilis* is a potential ureolytic bacteria for use in bioconcrete [11, 15–17]. It has been reported that the optimum cell concentration of *Bacillus subtilis* in MICP process was found to be 10^6 to 10^8 cells/ml [18, 19]. The present study investigated the effects of *Bacillus subtilis* on crack remediation in thermally degraded Limestone Calcined Clay Cement LC3) vis a vis Ordinary Portland Cement (OPC) mortars.

2 Materials and Methods

2.1 Materials

Materials were sampled from their respective places in Kenya. OPC (42.5 N/mm^2), limestone and standard sand were supplied by the East African Portland Cement Company. Fired Rejected Clay Bricks (FRCB)/calcined clay was obtained from a waste dump site in Kitui County. *Bacillus subtilis* was supplied by the Kenya Bureau of Standards.

2.2 Methods

The limestone and FRCB (Calcined Clay) were separately oven- dried at 105 °C for 24 h. The chemical composition of, dried limestone, FRCB, clinker and gypsum was obtained using X-Ray Florescence (XRF) technique. LC3 was prepared by inter-grinding of 50% clinker, 30% calcined clay, 15% limestone and 5% gypsum in a laboratory ball mill. The chemical composition of limestone, FRCB, clinker and gypsum used is given in Table 1 [20].

Mortar prisms measuring 160 mm × 40 mm × 40 mm were cast using LC3 and OPC at water/cement (w/c) ratio of 0.5 and immersed in curing water for 28 days. Mortar prisms were tested for compressive strength at the 2nd, 7th and 28th day of curing in accordance with accordance to KS EAS 148-1:2000.

To induce the cracks, half of the 28-day cured LC3 and OPC prisms were thermally degraded by heating them at 1000 °C in an electric Muffle furnace for 2 h while the other half was used as a control. Both cracked and un-cracked mortar prisms were subjected to compressive strength, porosity and accelerated chloride ingress tests. Porosity and accelerated chloride ingress tests were carried out following the procedures described in [20]. Cracked mortars were immersed in bacterial solution containing *Bacillus subtilis* with a cell concentration of 1×10^6 cells/ml while the un-cracked mortars were also separately immersed in curing water for 90 days. Compressive strength, porosity and accelerated chloride ingress tests were also repeated on the 90-day cured mortars. In addition, microstructural examination of polished mortar surfaces was conducted using the Scanning Electron Microscope (SEM) after thermal degradation and later after re-curing in bacteria solution using. Table 2 gives the characteristics of different binders used.

Table 1 Chemical composition of limestone, FRCB, clinker and gypsum [20]

Chemical composition	Clinker	FRCB	Limestone	Gypsum
SiO_2	21.78	58.13	1.85	2.44
Al_2O_3	4.55	15.55	0.98	6.23
Fe_2O_3	3.97	9.88	1.63	0.79
CaO	62.97	1.66	68.63	34.25
MgO	1.05	0.83	1.86	1.79
SO_3	2.07	0.04	0.022	40.82
K_2O	0.6	4.78	2.79	0.05
Na_2O	0.17	3.18	2.15	0.05
H_2O	2.33			
Cl^-	0.01			
Sum (SiO_2, Al_2O_3, Fe_2O_3)		83.56		
Loss on ignition (LOI)	0.96	2.92	10.26	12.99
Blaine fineness (m^2/kg)	339	445		

Table 2 Characteristics of different binders used

Binder type	Characteristics	
	Composition	Micro-cracks
LC3-C	50% clinker, 30% calcined clay, 15% limestone and 5% gypsum	Cracked
LC3-UC	50% clinker, 30% calcined clay, 15% limestone and 5% gypsum	Un-cracked
OPC-C	95% clinker and 5% gypsum	Cracked
OPC-UC	95% clinker and 5% gypsum	Un-cracked

3 Results and Discussions

3.1 Compressive Strength

The compressive strength results for cracked/un-cracked mortars after curing in water/bacterial solution are presented in Fig. 1.

At 28 days of curing, the compressive strength of un-cracked is higher than that of cracked LC3 and OPC mortars. This can be attributed to the fact that the cracks destroy the CSH network in the hydrated cement matrix hence decreased strength [21, 22].

At 90-days of bacterial curing, there is no significance difference in terms of compressive strength of un-cracked and cracked mortars for both LC3 and OPC cured in bacterial solution. This can be attributed to the deposition of calcite mainly from

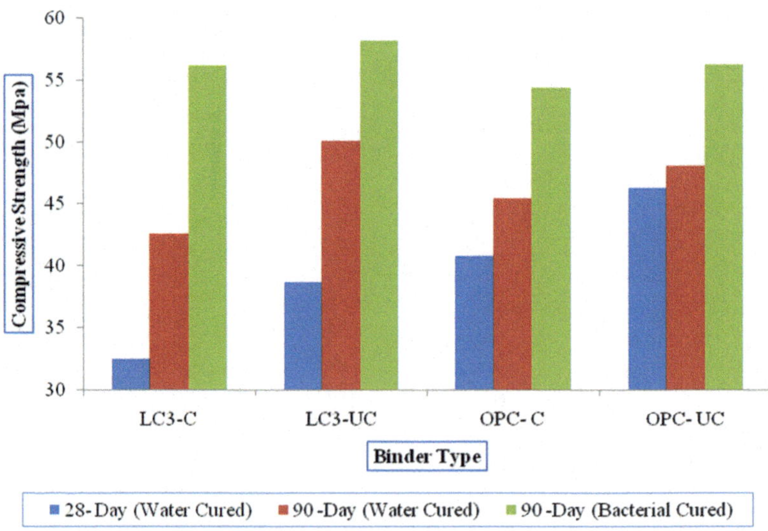

Fig. 1 Compressive strength results for cracked/un-cracked mortars cured in water/bacterial solution

the microbial activity of *Bacillus subtilis*. Ureolytic type of bacteria has been greatly used to modify the performance of cement based materials [11]. The bacteria have been used to improve the compressive strength of cement mortars by the precipitation of calcium carbonate. The bacteria produces urease, which catalyzes the hydrolysis of urea $[CO(NH_2)_2]$ into ammonium (NH_4^+) and carbonate (CO_3^{2-}). This occurs in a sequence of steps. At the beginning, urea is hydrolyzed into carbamate (NH_2COOH) and ammonia (NH_3) as shown in Eq. (1) [11].

$$[CO(NH_2)_2] + H_2O \rightarrow NH_2COOH + NH_3 \tag{1}$$

Consequently, the carbamate acid spontaneously hydrolyzes to form ammonia and carbonic acid as given in Eq. (2) [23].

$$NH_2COOH + H_2O \rightarrow NH_3 + H_2CO_3 \tag{2}$$

The carbonic acid partly dissociates to form bicarbonate ions as shown in Eq. (3).

$$H_2CO_3 \leftrightarrow HCO_3^- + H^+ \tag{3}$$

Similarly, ammonia formed in Eq. (2) subsequently dissociates to form ammonium and hydroxide ions as given in Eq. (4).

$$2NH_3 + 2H_2O \rightarrow 2NH_4^+ + 2OH^- \tag{4}$$

The reactions represented by Eqs. (3) and (4) results in the increase in the pH of the media causing the shifts in the bicarbonate equilibrium, resulting in the formation of carbonate ions as shown in Eq. (5) [23].

$$HCO_3^- + H^+ + 2NH_4^+ + 2OH^- \rightarrow CO_3^{2-} + 2NH_4^+ + H_2O \tag{5}$$

The bacteria have cell walls that are negatively charged. This cause the bacteria attract positively charged ions from the cementitious material environment matrix, including Ca^{2+} which are subsequently deposited on their cell surface [11, 23–26]. The Ca^{2+} which are drawn to the cell wall of the bacteria reacts with react with the CO_3^{2-} ions. This leads to the precipitation of $CaCO_3$ at the cell surface that serves as a nucleation site and subsequently seals the cracks hence improving the compressive strength of cement based materials [11, 23–26].

On the other hand, at 90 days, the improvement in compressive strength in cracked mortars cured in water can be attributed to the re-hydration of residual C_2S and C_3S. When re-curing of cracked mortars take place, more CSH is formed as shown in Eqs. (6) and (7) [27].

$$2C_3S + 6H \rightarrow C_3S_2H_3 + 3CH \tag{6}$$

$$2C_2S + 4H \rightarrow C_3S_2H_3 + CH \quad (7)$$

The additional CSH formed during re-hydration results in densification of hydrated cement mortars hence improved strength.

From the study, it was noted that regardless of the binder type used, cracked mortars cured in bacterial solution exhibited higher compressive strength than those cured in water. This can be ascribed to the densification of hydrated cement matrix as a result of calcite precipitation due to bacteria action [19].

3.2 *Porosity*

The porosity results for cracked and un-cracked mortars after curing in water and bacterial solution are presented in Fig. 2.

At 28 days of curing, it was observed that all (cracked and un-cracked) LC3 mortars exhibited lower porosity than OPC mortars. This can be attributed to the inclusion of calcined clays in LC3. Calcined clays that are rich in SiO_2 and Al_2O_3 when present in blended cement act as the nucleation sites for precipitation of hydration products such as CAH and CSH phases during the pozzolana reaction as shown in Eqs. (8) and (9) [20, 28, 29];

$$3CH + 2S + 3H \rightarrow C_3S_2H_3 \quad (8)$$

$$A + H + CH \rightarrow CAH \quad (9)$$

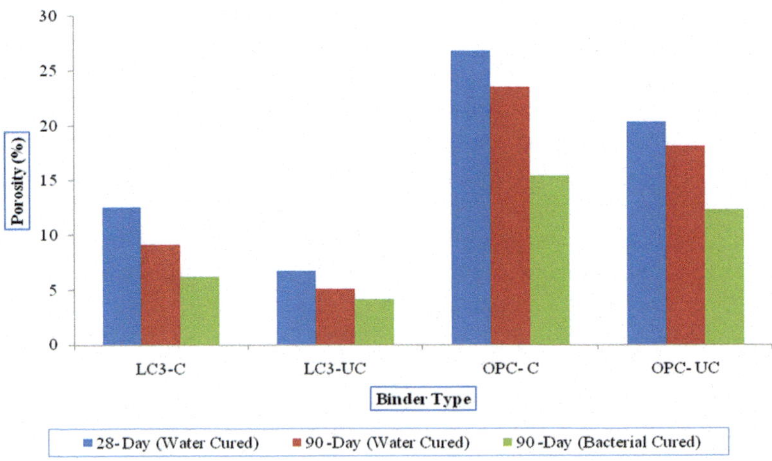

Fig. 2 Porosity results for cracked/un-cracked mortars cured in water/bacterial solution

The additional CSH and CAH formed result in pore refinement of hydrated cement matrix hence decreasing their porosity [30–33]. Moreover, the secondary CSH and CAH formed during the pozzolanic reaction in results in densification of LC3 mortars thus lowering their porosity. No pozzolanic reaction occurs in OPC mortars due to the absence of pozzolanic materials. In addition, the formation of additional CAH and CSH in fills the pores in hydrated cement matrix.

Moreover, at 28 days of curing, it was also observed that both un-cracked LC3 and OPC mortars exhibited lower porosity than cracked LC3 and OPC mortars. The presence of cracks increases the pores/voids in hydrated cements hence cracked LC3 and OPC mortars exhibit higher porosity than un-cracked LC3 and OPC mortars.

At 90-days, both cracked and un-cracked LC3 and OPC mortars showed lower porosity when cured in bacterial solution than in water. Calcite precipitation occurs in bacterial cured mortars resulting in pore refinement. On the other hand, re-hydration processes take place during re-curing of mortars in water. From the study it was found that pore refinement as a result of calcite precipitation is greater than deposition of hydration products due to re-hydration.

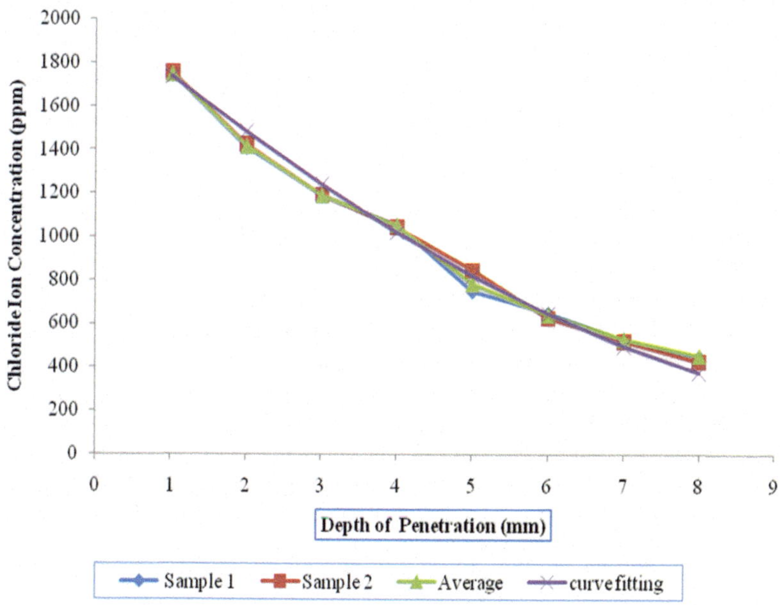

Fig. 3 Error fitting curve for LC3-UC mortar at 28-days of curing

Table 3 D_{app} values for LC3-UC mortar at 28-days of curing

	28-day (water cured)	90-day (water cured)	90-day (bacterial cured)
	D_{app} ($m^2\ s^{-1}$) × 10^{-12}	D_{app} ($m^2\ s^{-1}$) × 10^{-12}	D_{app} ($m^2\ s^{-1}$) × 10^{-12}
LC3-C	9.18	6.77	4.28
LC3-UC	6.93	5.82	3.87
OPC- C	20.2	17.4	13.55
OPC- UC	18.23	16.32	10.68

3.3 Chloride Ingress

The error fitting curve for LC3-UC mortar specimens is presented in Fig. 3. Similar error fitting curves were used to obtain apparent diffusion coefficients (D_{app}). The D_{app} values obtained from the error fitting curves of all the mortar specimens are also presented in Table 3.

At 28 days of curing, it was noted that all (cracked and un-cracked) LC3 mortars exhibited lower apparent diffusion coefficients than OPC mortars. This is because LC3 mortars were less porous than OPC mortars. Lower porosity effectively reduces the permeability of mortars hence minimal chloride penetration. OPC mortars exhibited higher porosity thus greater permeability of chlorides depicted by the apparent diffusion coefficients. The additional secondary hydration products arising from the pozzolanic reaction in LC3, fills the voids, which result in more dense hydrated cement microstructure and consequently reduce the permeability of cement based materials from a pore refining process point of view [20].

The lower porosity exhibited by mortars cured in bacterial solution than in water explains why at 90-days, both cracked and un-cracked LC3 and OPC mortars showed lower apparent diffusion coefficients when cured in bacterial solution than in water. Bacterial solution increases the self-healing capacity compared with water.

In addition, at 90-days, it was noted that regardless of the binder type used, cracked mortars cured in bacterial solution exhibited lower apparent diffusion coefficients than those cured in water. This could be explained by their differences in porosity. Cracked mortars exhibit greater voids which serve as pathways for the permeability of chlorides than un-cracked mortars.

3.4 Microstructural Analysis

Microstructural examination of LC3-C before and after re-curing in bacteria solution in given in Fig. 4a, b.

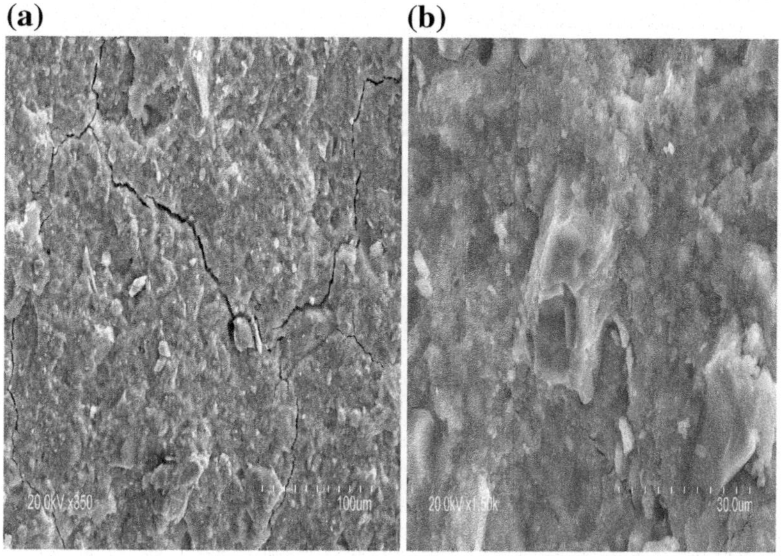

Fig. 4 (**a**) SEM images of LC3-C before re-curing in bacteria solution while (**b**) SEM images of LC3-C before and after re-curing in bacteria solution. The SEM images in (**a**) showed that visible cracks with no calcium carbonate deposit. On the contrary, (**b**) showed no cracks and the deposition of calcium carbonate was significant. The presence of cracks can be attributed to thermal degradation of the mortars while the microbial process induced by *Bacillus subtilis* accounts for the deposition of calcium carbonate which results in the healing of cracks

4 Conclusions

From the study, the following conclusions were made;

(a) This demonstrated that *Bacillus subtilis* solution has a higher potential for self-healing than water.

(b) Un-cracked mortars exhibited higher compressive strength, lower porosity and lower apparent chloride diffusion coefficients than cracked mortars at 28 days of curing. This is mainly because cracks destroy the CSH gels which are mainly responsible for strength of cementitious materials; create voids in the pore network of hydrated cement thus affecting porosity and the voids acts as pathways for increased permeability of aggressive media such as chlorides.

(c) At 90 days of curing, the compressive strength, porosity and apparent chloride diffusion coefficients of cracked mortars were equivalent to those of un-cracked mortars. This is mainly due to the microbially induced calcite precipitation process which occurs in concrete as a result of the microbial action of *Bacillus subtilis* in which there is precipitation of calcium carbonate that fill the cracks leading to their repair.

(d) Microstructural examination the mortars showed visible micro-cracks after thermal treatment which healed as a result of calcite deposition after curing in bacterial solution.
(e) *Bacillus subtilis* was found to improve the crack healing capacity of thermally cracked mortars than water.

Acknowledgements This research was financed from the National Innovation Award 2020 awarded by the Kenya National Innovation Agency (KENIA) in the built environment and Technologies. The East African Portland Cement Company (EAPCC) and Meru University of Science and Technology (MUST) are all duly acknowledged and highly appreciated for the provision of research facilities used in this project.

References

1. Marangu, J.M., Thiong'o, J.K., Wachira, J.M.: Review of carbonation resistance in hydrated cement based materials. J. Chem. **2019**, 8489671 (2019). https://doi.org/10.1155/2019/8489671
2. Marangu, J.M.: Effects of sulfuric acid attack on hydrated calcined clay–limestone cement mortars. J. Sustain. Cem. Based Mater. 1–15 (2020). https://doi.org/10.1080/21650373.2020.1810168
3. Gao, J., Yu, Z., Song, L., Wang, T., Wei, S.: Durability of concrete exposed to sulfate attack under flexural loading and drying-wetting cycles. Constr. Build. Mater. **39**, 33–38 (2013). https://doi.org/10.1016/j.conbuildmat.2012.05.033
4. Safiuddin, Md., Kaish, A., Woon, C.-O., Raman, S.: Early-age cracking in concrete: causes, consequences, remedial measures, and recommendations. Appl. Sci. **8**, 1730 (2018). https://doi.org/10.3390/app8101730
5. Fernandes, I., Broekmans, M.A.T.M.: Alkali–silica reactions: an overview. Part I. Metallogr. Microstruct. Anal. **2**, 257–267 (2013). https://doi.org/10.1007/s13632-013-0085-5
6. Sule, M., van Breugel, K.: The effect of reinforcement on early-age cracking due to autogenous shrinkage and thermal effects. Cem. Concr. Compos. **26**, 581–587 (2004). https://doi.org/10.1016/S0958-9465(03)00078-7
7. Wang, H.: 25—Life-cycle analysis of repair of concrete pavements. In: Pacheco-Torgal, F., Melchers, R.E., Shi, X., Belie, N.D., Tittelboom, K.V., Sáez, A. (eds.) Eco-Efficient Repair and Rehabilitation of Concrete Infrastructures, pp. 723–738. Woodhead Publishing (2018). ISBN 978-0-08-102181-1
8. Mwiti, M.J., Karanja, T.J., Muthengia, W.J.: Thermal resistivity of chemically activated calcined clays-based cements. In: Martirena, F., Favier, A., Scrivener, K. (eds.) Proceedings of the Calcined Clays for Sustainable Concrete, pp. 327–333. Springer Netherlands, Dordrecht (2018)
9. Czarnecki, L., Geryło, R., Kuczyński, K.: Concrete repair durability. Materials **13**, 4535 (2020). https://doi.org/10.3390/ma13204535
10. Jonkers, H.M., Schlangen, E.: Self-Healing of Cracked Concrete: A Bacterial Approach, pp. 6
11. Abo-El-Enein, S.A., Ali, A.H., Talkhan, F.N., Abdel-Gawwad, H.A.: Utilization of microbial induced calcite precipitation for sand consolidation and mortar crack remediation. HBRC J. **8**, 185–192 (2012). https://doi.org/10.1016/j.hbrcj.2013.02.001
12. De Belie, N.: Application of bacteria in concrete: a critical evaluation of the current status. RILEM Tech. Lett. **1**, 56 (2016). https://doi.org/10.21809/rilemtechlett.2016.14

13. Harkes, M.P., van Paassen, L.A., Booster, J.L., Whiffin, V.S., van Loosdrecht, M.C.M.: Fixation and distribution of bacterial activity in sand to induce carbonate precipitation for ground reinforcement. Ecol. Eng. **36**, 112–117 (2010). https://doi.org/10.1016/j.ecoleng.2009.01.004
14. Van Tittelboom, K., De Belie, N., De Muynck, W., Verstraete, W.: Use of bacteria to repair cracks in concrete. Cem. Concr. Res. **40**, 157–166 (2010). https://doi.org/10.1016/j.cemconres.2009.08.025
15. Wong, L.S.: Microbial cementation of ureolytic bacteria from the genus bacillus: a review of the bacterial application on cement-based materials for cleaner production. J. Clean. Prod. **93**, 5–17 (2015). https://doi.org/10.1016/j.jclepro.2015.01.019
16. Chaurasia, L., Verma, R.K., Bisht, V.: Microbial Carbonate Precipitation by Urease Producing Bacteria in Cementitious Materials, vol. 9
17. Joshi, S., Goyal, S., Mukherjee, A., Reddy, M.S.: Microbial healing of cracks in concrete: a review. J. Ind. Microbiol. Biotechnol. **44**, 1511–1525 (2017). https://doi.org/10.1007/s10295-017-1978-0
18. Okwadha, G.D.O., Li, J.: Optimum conditions for microbial carbonate precipitation. Chemosphere **81**, 1143–1148 (2010). https://doi.org/10.1016/j.chemosphere.2010.09.066
19. Mutitu, K.D., Munyao, M.O., Wachira, M.J., Mwirichia, R., Thiong'o, K.J., Marangu, M.J.: Effects of biocementation on some properties of cement-based materials incorporating *Bacillus Species* bacteria—a review. J. Sustain. Cem. Based Mater. **8**, 309–325 (2019). https://doi.org/10.1080/21650373.2019.1640141
20. Marangu, J.M.: Physico-chemical properties of kenyan made calcined clay-limestone cement (LC3). Case Stud. Constr. Mater. **12**, (2020). https://doi.org/10.1016/j.cscm.2020.e00333
21. Mihashi, H., de Leite, J.P.B.: State-of-the-art report on control of cracking in early age concrete. J. Adv. Concr. Technol. **2**, 141–154 (2004). https://doi.org/10.3151/jact.2.141
22. Szeląg, M.: Evaluation of cracking patterns in cement composites-from basics to advances: a review. Materials (Basel) **13**, 2490 (2020). https://doi.org/10.3390/ma13112490
23. Abo-El-Enein, S.A., Ali, A.H., Talkhan, F.N., Abdel-Gawwad, H.A.: Application of microbial biocementation to improve the physico-mechanical properties of cement mortar. HBRC J. **9**, 36–40 (2013). https://doi.org/10.1016/j.hbrcj.2012.10.004
24. Canakci, H., Sidik, W., Halil Kilic, I.: Effect of bacterial calcium carbonate precipitation on compressibility and shear strength of organic soil. Soils Found. **55**, 1211–1221 (2015). https://doi.org/10.1016/j.sandf.2015.09.020
25. Chidara, R., Nagulagama, R., Yadav, S.: Achievement of early compressive strength in concrete using *Sporosarcina Pasteurii* bacteria as an admixture. Adv. Civ. Eng. **2014**, 1–7 (2014). https://doi.org/10.1155/2014/435948
26. Khattra, S.K., Parmar, M., Phutela, U.G.: Study of strength variation of concrete using ureolytic bacteria. **3**, 4 (2016)
27. Neville, A.M.: Properties of Concrete, 4th and final ed., reprint ed.. Longman, Harlow, Essex (1997). ISBN 978-0-582-23070-5
28. Marangu, J.M.: Physico-chemical properties of kenyan made calcined clay-limestone cement (LC3). Case Stud. Const. Mater. **12**, (2020). https://doi.org/10.1016/j.cscm.2020.e00333
29. Marangu, J.M., Muturia M'thiruaine, C., Bediako, M.: Physicochemical properties of hydrated Portland cement blended with rice husk ash. J. Chem. **2020**, 5304745 (2020) https://doi.org/10.1155/2020/5304745
30. Lafhaj, Z., Goueygou, M., Djerbi, A., Kaczmarek, M.: Correlation between porosity, permeability and ultrasonic parameters of mortar with variable water/cement ratio and water content. Cem. Concr. Res. **36**, 625–633 (2006). https://doi.org/10.1016/j.cemconres.2005.11.009
31. Wu, B., Ye, G.: Development of porosity of cement paste blended with supplementary cementitious materials after carbonation. Constr. Build. Mater. **145**, 52–61 (2017). https://doi.org/10.1016/j.conbuildmat.2017.03.176
32. Chen, X., Wu, S., Zhou, J.: Influence of porosity on compressive and tensile strength of cement mortar. Constr. Build. Mater. **40**, 869–874 (2013). https://doi.org/10.1016/j.conbuildmat.2012.11.072

33. Chindaprasirt, P., Rukzon, S.: Strength, porosity and corrosion resistance of ternary blend Portland cement, rice husk ash and fly ash mortar. Constr. Build. Mater. **22**, 1601–1606 (2008). https://doi.org/10.1016/j.conbuildmat.2007.06.010

Effect of Binary-Use Mineral Admixtures for the Advanced Autogenous Self-healing Behavior of Fiber-Reinforced Cementitious Composites

Oğuzhan Öztürk and Marta Roig-Flores

Abstract This paper comprises the enhanced self-healing properties of fiber-reinforced cementitious composites by taking advantage of synergistic effect of using different mineral admixtures. To do this, single and binary use of metakaolin (MK) and zeolite (Z) were utilized in cement-based mixtures, and early age self-healing behavior of specimens was assessed based on sorptivity tests. The autogenous self-healing performance of each mixture was also evaluated via crack width assessments by using a video microscope. Results were compared with the sound and preloaded specimens having different crack width levels. In addition, mechanical properties of four different mixtures were evaluated by conducting experiments on 7 and 28 days. The multiple crack width closures and sorptivity coefficients of preloaded specimens indicated that an effective combination of MK and Z can trigger a higher level of self-healing in comparison with reference mixture. However, single-use of Z in mixtures was also promising for achieving autogenous self-healing attributes compared to mixtures containing only MK and to reference mixtures. Having said that, this behavior should also be addressed for different pre-cracking and prolonged curing age of each mixture.

Keywords Cement-based materials · Self-healing · Synergistic effect · Mineral admixtures

1 Introduction

Cracking in concrete structures are tightly related with a non-durable behavior, mainly because of an increase in the permeability, which allows the ingress of deleterious materials, such as aggressive ions in liquid and gaseous substances [1, 2].

O. Öztürk (✉)
Konya Technical University, Konya, Turkey
e-mail: oozturk@ktun.edu.tr

M. Roig-Flores
Universitat Jaume I, Castelló de la Plana, Spain
e-mail: roigma@uji.es

F. Kanavaris et al. (eds.), *International RILEM Conference on Early-Age and Long-Term Cracking in RC Structures*, RILEM Bookseries 31,
https://doi.org/10.1007/978-3-030-72921-9_32

Consequently, under certain circumstances, structural integrity of a cracked structure may fall into fast degradation, often needing resulting in necessity to repair or maintenance actions. However, repair/maintenance applications are costly and time-consuming processes. As an example of the economic impacts produced by repair actions, the American Society of Civil Engineers expects infrastructures deficits to be around $4.59 trillion for maintaining the structures lifespan by year 2025 [3]. In addition to the financial aspect of the problem, environmental impacts of the repair/maintenance programs may also reduce the feasibility of some applications [4]. In those structures with difficult access or with very expensive repair/maintenance, one solution would be designing concrete mixtures with self-healing properties.

Fiber Reinforced Concrete and High-Performance Fiber Reinforced Cementitious Composites (HPFRCC) made significant step forward in the controlling the crack formation of cement-based materials [5]. Among HPFRCCs, Strain-Hardening Cementitious Composites (SHCCs) are considered to be promising materials due to their high tensile strain capacity and multi-cracking behavior [6, 7]. These composites are able to form multiple closely spaced microcracks even under the highest strain capacity [8]. In addition to the strain-hardening behavior of SHCCs, the small micro-scale cracks that are produced have the potential to be healed by themselves, for some appropriate mixture designs. This capability is known as autogenous self-healing and is defined as a mechanism of recovery of the original properties of the uncracked element that does not require external intervention [9, 10]. The autogenous self-healing of cement-based materials is mainly related to formation of $CaCO_3$ and additional C-S-H gels [9, 10] and is mainly controlled by the matrix, crack width distribution and the availability of water. Although both formations are considered to be effective for providing crack healing, calcite precipitation ($CaCO_3$) has been reported to be more significant for the structures that are in contact with water or humidity [11]. For instance, some studies reported that concrete specimens immersed in water were able to produce effective crack closing for cracks up to 0.35 mm [12].

These three mechanisms involved in the penetrability of concrete are: permeability, diffusion and sorption. For concrete elements exposed at the same time partially to wet and dry states (in different areas of the same element), or undergoing wet and drying stages at different times in the same element, capillary suction forces are of interest since they may become the main transport mechanism taking place [13]. In the literature, capillary suction of cracked specimens has already been proposed as a method to evaluate self-healing [14, 15].

Studies related to autogenous self-healing revealed that incorporation of pozzolans such as slag, fly ash and silica fume improved the self-healing ability of cracked cementitious composites, mainly due to their delayed reactivity [16]. The effect of other pozzolans, such as Metakaolin or Zeolites, on autogenous healing has not been widely studied yet. Zeolites are aluminosilicates, and in concrete they can act as natural pozzolans, internal curing agents, and lightweight aggregates [17]. In the field of self-healing concrete, they have been used as carrier for bacterial agents due to their high surface area and porosity [18]. Usually, studies related to autogenous self-healing are mostly focused on one type of pozzolanic material at a time, even though synergistic effects of the use of binary or ternary systems of pozzolanic materials in cementitious systems are reported in the literature [19, 20]. These synergies are detected as an improvement in the pore structure of hardened sound concrete, and have been produced when combining fly ash with nanosilica [19] and fly ash with metakaolin [20]. Because of that, it may be hypothesized that this effect would be also very influential on the autogenous self-healing due to very close relation between sorptivity properties and pore structure.

This work presents a preliminary experimental investigation performed on SHCCs with metakaolin (MK) and zeolite (Z), added independently and combined to study the presence of synergistic effects. The self-healing response at early ages has been evaluated through the sorptivity and crack closing of sound and pre-damaged SHCCs specimens after healing for 28 days in water immersion. Additionally, the mechanical properties of the mixes has been characterized.

2 Research Program

2.1 Materials and Mixture Proportions

The cement was a commercially available Portland cement (PC) similar to ASTM Type I. Silica sand with maximum aggregate size of 700 μm, tap water, polyamid fibers (PA) having a length of 12 mm, diameter of 550 μm, specific gravity of 1.14, tensile strength of 900 MPa and polycarboxylate-based high range water reducing admixture (HRWRA) were used in the mixtures. In addition, mineral admixtures (MA) of different chemical compositions were incorporated into mixtures with by substitution of PC. Metakaolin (MK) with specific gravity of 2.5, surface area of 17 m^2/g and zeolite (Z) with specific gravity of 2.1 and 40 m^2/g were used. The

Table 1 Chemical and physical properties of cementitious materials

Chemical composition (%)	PC	MK	Z
SiO_2	20.80	45.4	71.60
Al_2O_3	5.55	37.8	11.30
Fe_2O_3	3.35	1.6	1.39
MgO	2.49	–	0.40
CaO	61.40	0.2	2.27
Na_2O	0.19	–	0.86
K_2O	0.77	0.3	3.67
Loss on ignition	2.20		–
Physical properties			
Specific gravity	3.06	2.5	2.1
Blaine fineness (m^2/kg)	325	17000	44000

Table 2 Proportions of the four SHCC mixtures

Mixture ID	PC	Sand/PC[a]	MK/PC[a]	Z/PC[a]	MK + Z/PCa	W/CM[a]	SP[b]	PA Fibers[b]
SHCC/Ref.	1	1	–	–	–	0.35	4.3	22
SHCC/MK	1	1	0.7	–	–	0.35	5.9	22
SHCC/Z	1	1	–	0.7	–	0.35	6.2	22
SHCC/MK + Z	1	1	–	–	0.35 + 0.35	0.35	7.5	22

[a]By mass; [b]In kg/m^3

physical and chemical compositions of MK and Z are displayed in Table 1, together with PC.

The water to cementitious materials ratio (W/CM) and mineral admixture to Portland cement ratio (MA/PC [MA accounts for the total weight of MK + Z]) of mixtures were constant at 0.35 and 0.7 by mass, respectively. PA fibers were used at 2% by volume of total each mixture. Given varying particle sizes of different MA, it was not possible to keep HRWRA amounts constant for four mixtures. Therefore, to evaluate consistency of different mixtures, mini slump tests were conducted and the amount of HRWRA was taken into consideration based on similar flow deformation levels of around 15 cm. The mixture proportioning of four different SHCC mixtures is reported in Table 2.

2.2 *Production of Specimens and Curing Conditions*

PC, MK, Z and silica sand were mixed in a conventional mortar mixer together at 100 rpm for a minute. Then total water and SP were slowly added into the dry blend and mixing was kept on for a minute at 100 rpm followed by an additional 2 min at 200 rpm. The PA fibers were incorporated with as the last step and mixing at 200 rpm continued for further 3 more minutes. The produced fresh mixtures were poured into cubic molds of side 50 mm, and prismatic molds by $40 \times 40 \times 160\ mm^3$. After 24 h curing in the molds at 50 ± 5% RH, 23 ± 2 °C, specimens were demolded and placed into plastic bags isolated from the outside environment to complete a total of 7 days at 95 ± 5% RH and 23 ± 2 °C. Basic mechanical properties of each mixture were evaluated through compressive and flexural strength tests (including deformation levels at yield and fracture). Compressive strength tests were conducted based on ASTM 109 [21] by using six different cubic specimens from each mixture. On the other hand, flexural tests were applied by using three different prism specimens with the loading rate of 0.04 kN/mm and deformation and strength values were averaged for 7- and 28-day old specimens.

2.3 *Pre-cracking and Testing for Self-healing Evaluation*

The prismatic specimens, of size 40 × 40 × 160 mm^3 (width × depth × length) with different SHCC mixtures were used for the sorptivity evaluation. After 7 days of curing in plastic bags, three specimens from each mixture were loaded up until failure under 3-point bending tests of unnotched specimens (Fig. 1) in order to characterize their ultimate load by means of evaluating the deflections at the yielding point (first-crack deflection) and fracture. Afterwards, three specimens of each mix were preloaded up to 80% of their deformation capacity. Unnotched specimens were used in order to avoid the influence of the notch on the water sorptivity tests. The zeolite-bearing specimens (SHCC/Z, SHCC/MK + Z) exhibited higher ductile fracture with a higher number of microcracks than MK- and reference mixtures, which had only 1–2 cracks per specimen. The crack size obtained through this method ranged from 40 to 200 microns, in different crack distributions.

Sorptivity experiments (Fig. 2) were performed by using three pre-cracked specimens having silicon coated edges except pre-cracked side in conformity with the procedures given in ASTM C 1585 [22]. Three separate sound specimens from each mixture were also similarly used in the tests for comparison with pre-cracked elements. After isolation in plastics bags and pre-cracking of specimens at 7 days, specimens healed immersed in water for 7 (7 + 7) and 28 (7 + 28) days. Prior to sorptivity testing, specimens were dried at 50 ± 2 °C until the weights of specimens were stabilized with the accuracy of 0.01 g. Changes of the mass during the capillary suction of specimens was recorded at 1, 2, 3, 4, 6, 8, 12, 16, 20, 25, 36, 49, 64, 81, 120 and 360 min of intervals. The rate of absorption (mm) was calculated by averaging each specimen using the change of mass (g) divided by cross section area of tested surface (40 × 160 mm^2) and the density of water and then was graphed versus square root of time ($min^{1/2}$). Also, each crack from the same specimens used in sorptivity tests was tracked by using a video microscope.

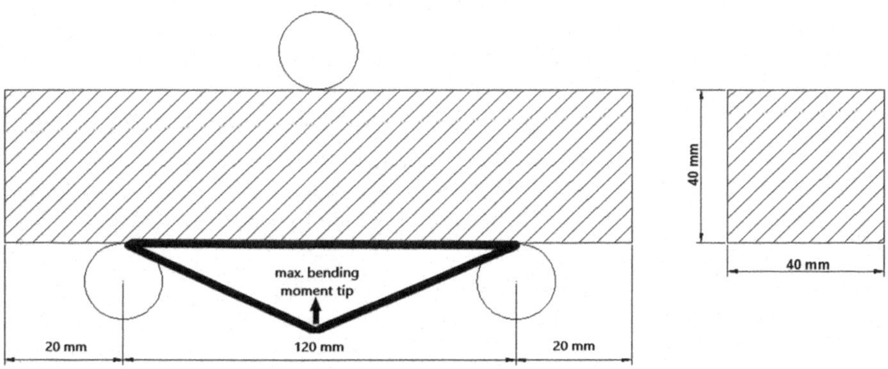

Fig. 1 A typical pre-loading of specimen and stress concentration

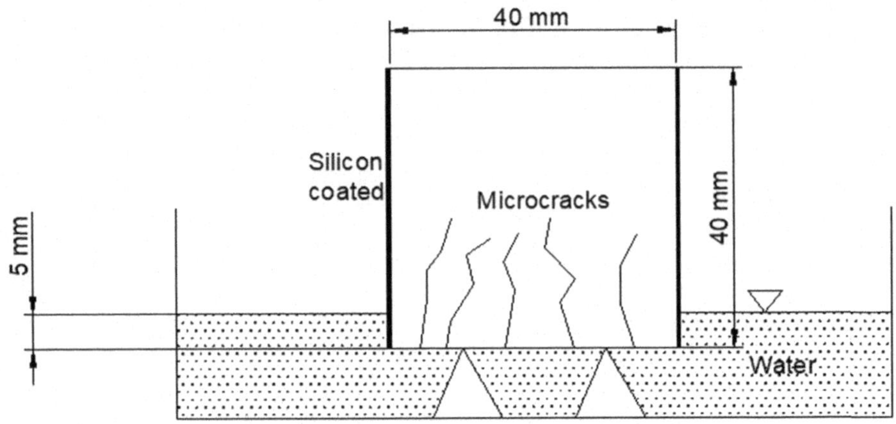

Fig. 2 The scheme of sorptivity tests

3 Results and Discussion

3.1 Mechanical Properties

Compressive strength of the specimens is displayed in Fig. 3. The compressive strength of 7- and 28- day-old SHCC/Ref., SHCC/MK, SHCC/Z, SHCC/MK + Z were found to be (39.1–54.1), (28.9–55.5), (17.1–30.1) and (24.6–46.5) MPa, respectively. With further curing up to 28 days, compressive strength of all specimens increased irrespective of mixture type, as expected. However, the extent of increment was varying in the mixtures developed with single- and binary-use MA. For example, increment rates of compressive strength were recorded 38.3%, 92%, 76%, and 89% between 7 and 28 days for the SHCC/Ref., SHCC/MK, SHCC/Z, SHCC/MK + Z mixtures, respectively. The higher strength development of MA-bearing mixtures was due to pozzolanic ingredient of specimens which highly contributed to the strength acquisition particularly after 7 days. However, this trend should be also compared with the older specimens beyond 28 days.

The lowest compressive strength results were obtained in the Z-bearing mixtures both at 7 and 28 days. The results were in accordance with the studies available in the literature [23, 24]. The lower compressive strength of SHCC/Z can be caused by a very slow pozzolanic behavior of the zeolite, which in some cases have displayed pozzolanic activity even after 2–3 years [23, 24]. On the other hand, it is worth noting that although zeolite is a pozzolanic material, it may produce rheology problems due to its high surface area, and lead to higher porosity compared to other pozzolans [25]. On the other side, the compressive strength of MK-bearing specimens was comparable to SHCC/Ref. at 28 days (Fig. 3). This can be related with the chemical composition of MK having higher reactive aluminosilicates than the zeolite used. The average compressive strength of the SHCC/MK specimens was the only mixture that

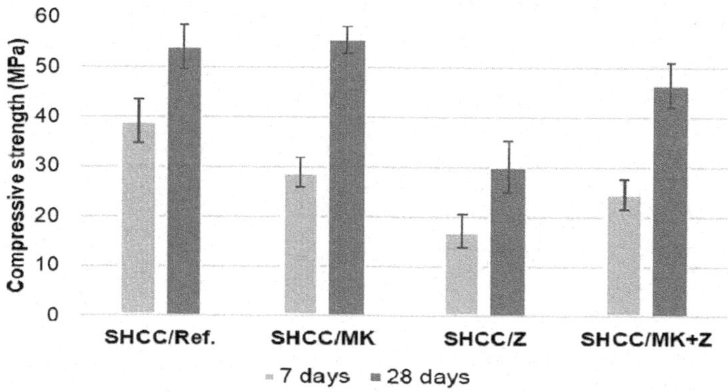

Fig. 3 Compressive strength of mixtures at 7 and 28 days (average and standard deviation)

showed up higher compressive strength than reference mixtures at 28 days (Fig. 3). For the mix with the binary-use of MA (SHCC/MK + Z), although higher increments were recorded at 7 and 28 days in comparison with the SHCC/Z, zeolite seemed to be also very regulator on the mechanical properties of mixtures, especially pronounced for early age (7 days). With the incorporation of Z and MK, greater particle size distribution [26] and high-early pozzolanic behavior of MK may have improved the packing density of binary-use MA and resulted in higher compressive strength than SHCC/Z. However, it should be noted that compressive strength all mixtures were satisfying for general-purpose concrete applications both at 7 and 28 days (>30 MPa).

The flexural strength of the specimens is displayed in Fig. 4. The flexural strength of 7- and 28- day-old SHCC/Ref., SHCC/MK, SHCC/Z, SHCC/MK + Z were in the range of (8.12–8.60), (7.18–8.74), (2.98–5.22) and (4.28–7.00) MPa, respectively. The flexural strength displays a similar trend to the results obtained in compressive strength. The differences in the flexural strength were not as high as in compressive tests. This behavior is generally ascribed to higher delicate properties of materials

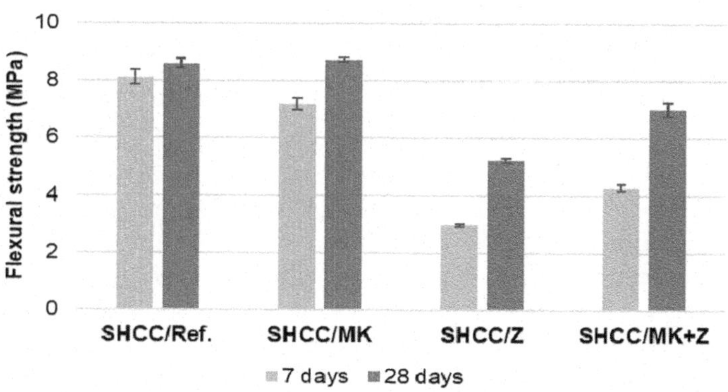

Fig. 4 Flexural strength of mixtures at 7 and 28 days (average and standard deviation)

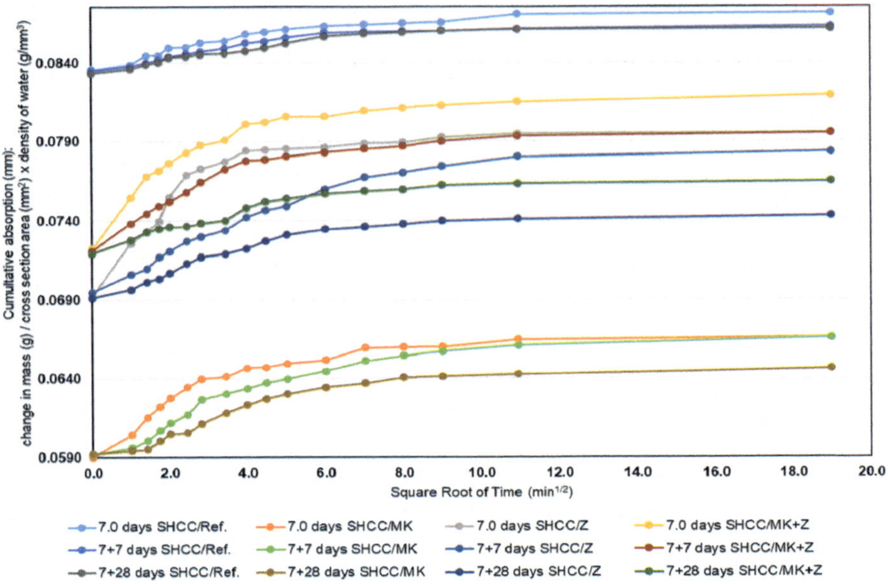

Fig. 5 Average cumulative sorption curves obtained for each mix for sound specimens (mm/min$^{1/2}$)

such as tensile first cracking strength, ultimate tensile strength and strain capacity [27].

3.2 *Tests for Autogenous Self-healing of Cracks*

Figures 5 and 6 show the cumulative absorption rate versus square root of time (mm/min$^{1/2}$) for sound and preloaded SHCC mixtures. Each curve represents the average curve obtained from the three specimens tested per group. In Figs. 5 and 6, cumulative sorptivity rates of sound 7-day-old SHCC/Ref., SHCC/MK, SHCC/Z, SHCC/MK + Z mixtures are (0.0836–0.0871 mm/min$^{1/2}$ [4.2%]), (0590–0.066 mm/min$^{1/2}$ [12.8%]), (0.0692–0.0795 mm/min$^{1/2}$ [14.9%]), (0.0722–0.0819 mm/min$^{1/2}$ [4.2%]), respectively while preloaded 7-day-old specimens showed average cumulative sorptivity rates of (0.0791–0.0838 mm/min$^{1/2}$ [6.0%]), (0.0684–0.0775 mm/min$^{1/2}$ [13.3%]), (0.0661–0.0773 mm/min$^{1/2}$ [16.9%]), (0.0689–0.0817 mm/min$^{1/2}$ [18.5%]) in same order. Results indicate significant differences between sound and preloaded specimens. Sound specimens exhibited lower cumulative sorptivity rate compared to preloaded specimens irrespective of mixture type. Generally, SHCC/Ref. specimens were more resistant to capillary suction, which is thought to be produced by a more complete hydration state, that lead to higher amount of disconnected pores. All the pre-cracked specimens had lower absorption when increasing the healing time (Fig. 6), due to the additional pathways

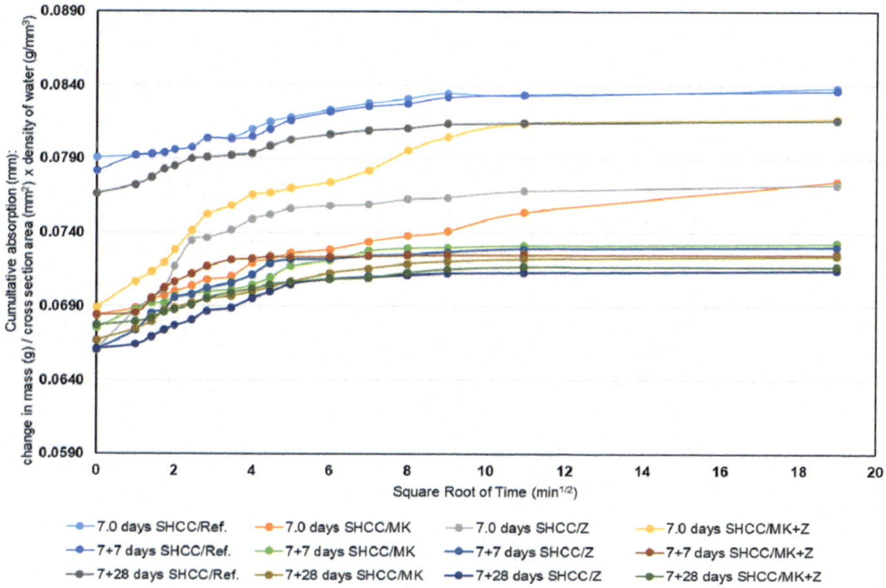

Fig. 6 Average cumulative sorption curves obtained for each mix for sound specimens (mm/min$^{1/2}$)

produced by the presence of cracks, which are able to suction water through capillary actions [27]. After healing (7 + 7 and 7 + 28 days), a similar trend can be observed between the reference and MA-bearing sound specimens.

Figure 7 shows the average sorptivity index of all the groups tested. Before healing under water immersion (7 + 0), cracked specimens had higher sorptivity index than their equivalent sound specimen. Reference specimens always displayed higher sorptivity for cracked specimens. However, all specimens with pozzolan display a decrease of the sorptivity index with the age, with lower sorptivity for cracked specimens in some cases. This result may have been produced by the pozzolanic reactions densifying the matrix or by the autogenous healing of the cracks. Since sound specimens followed the same conditions than cracked specimens, they also experienced the densification due to the pozzolanic reactions. Thus, the improvements produced in cracked conditions may have been produced by the enhanced autogenous self-healing.

Table 3 shows the number of cracks produced in each cracked specimen and the number of healed cracks at the different healing times. Higher number cracks of 7-day old specimens (SHCC/Z + SHCC/MK + Z) was saturated faster and provided more efficient autogenous self-healing probably due to the smaller cracks. Especially the mixture SHCC/MK + Z had all the cracks with 100% of crack closing after 28 days healing under water immersion. Figure 8 shows pictures of the crack width evolution. As depicted in Fig. 8, SHCC/Z and SHCC/MK + Z mixtures underwent a higher crack number and higher self-healing capability than SHCC/MK and SHCC/Ref. mixtures. This higher self-healing capability can be mainly attributed to a higher

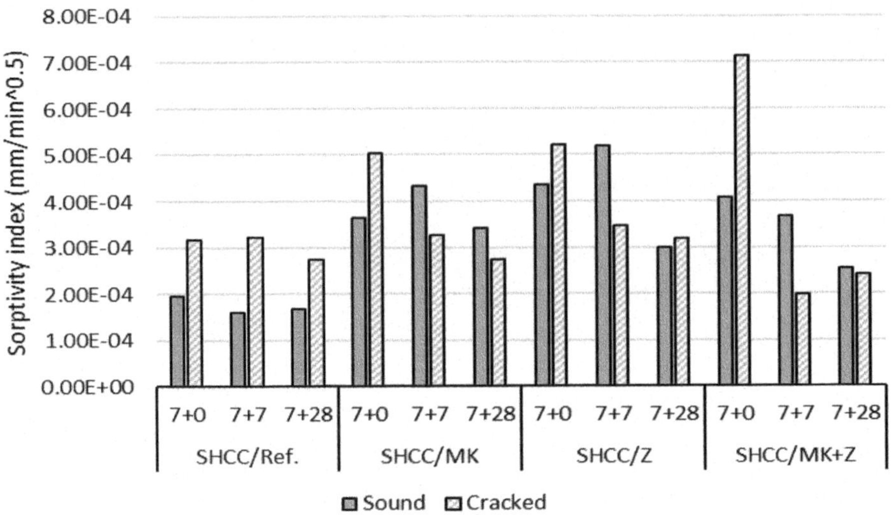

Fig. 7 Sorptivity index for all the mixtures in sound and pre-cracked conditions

Table 3 Number of cracks after preloading (7 + 0 days) and number of cracks that complete (100%) autogenous self-healing after each curing age

Mixture ID	# of specimen	7 + 0 days	7 + 7 days	7 + 28 days
		# of cracks	# of cracks	# of cracks
SHCC/Ref.	1	2	–	–
	2	1	–	–
	3	2	–	–
SHCC/MK	1	1	1	1
	2	1	–	–
	3	1	–	–
SHCC/Z	1	9	7	7
	2	5	4	4
	3	3	2	2
SHCC/MK + Z	1	2	2	2
	2	5	5	5
	3	4	4	4

volume of $CaCO_3$ production at a given crack width. The higher number of cracks produced in the mixtures with zeolites could have been caused by their less brittle matrix or by an improvement of the fiber-matrix bond due to a better quality ITZ. The delayed reaction due to its lower pozzolanity as well as the tight cracks produced would be favorable for the cracks to be totally filled with precipitates. Alterations of the fiber-matrix bond mechanisms produced by a lower pozzolanity fly ash compared

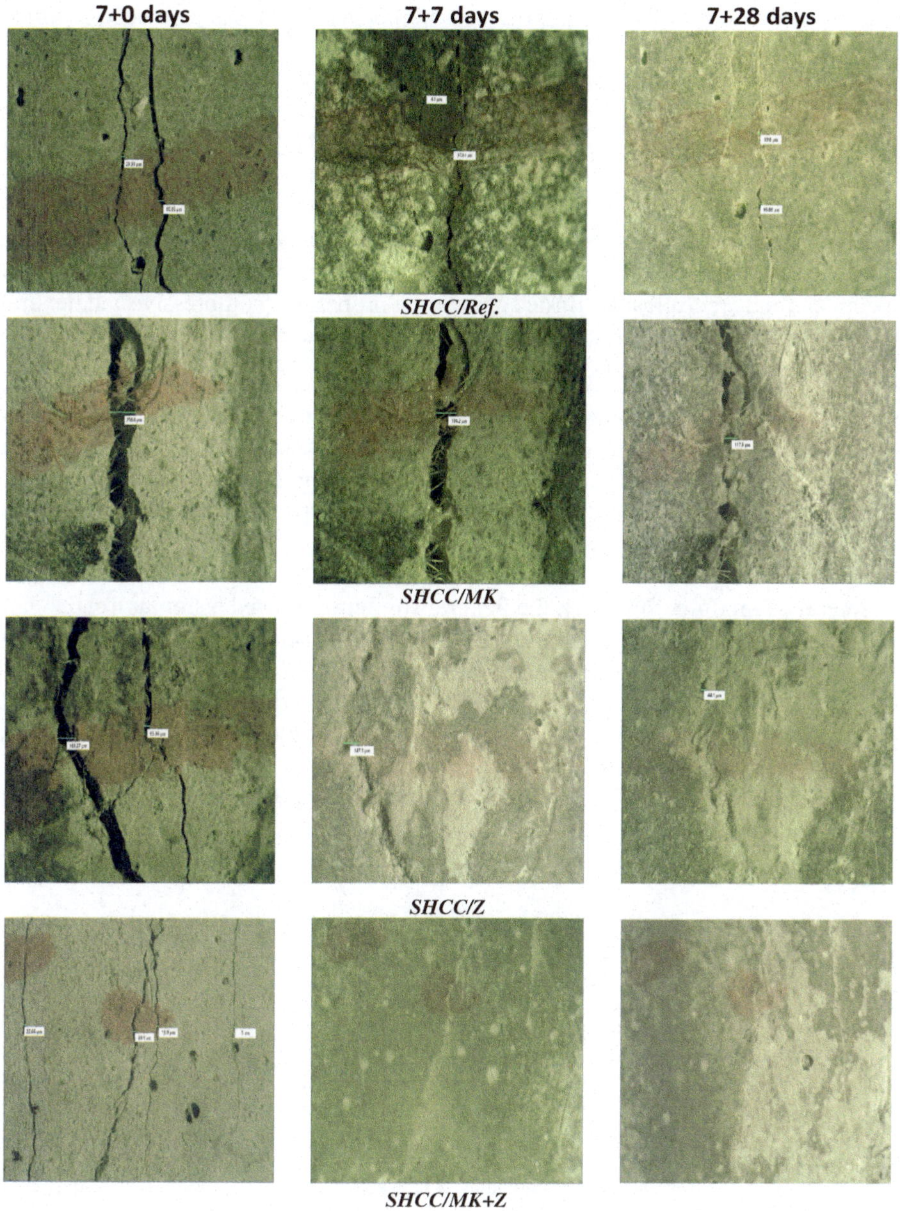

Fig. 8 Visual crack closing of the cracks for the four mixtures after 7 and 28 days healing

to other types of mineral admixtures with higher reactivity has also been reported previously [28].

4 Conclusions

This work studied autogenous self-healing of SHCCs with single and binary use of metakaolin and zeolite, through sorptivity properties of sound and pre-damaged specimens and crack closing. Basic engineering properties (compressive and flexural strength) of the four different mixtures were also evaluated. The study reached the following conclusions:

- The compressive strength of all mixtures was able to exhibit more than 30 MPa for at 28 days. However, the acquisition of compressive strength varied between 7 and 28 days for the mixtures developed with single- and binary-use of MA. SHCC/Z mixture showed the lowest compressive strength in comparison with the other mixtures both at 7 and 28 days. At 28 days, compressive strength of SHCC/MK mixture was the only mixture that was found to be higher than the reference mixture. Similarly, the flexural strength followed a similar trend.
- During preloading, mixtures with different mineral admixtures showed different cracking behavior. Zeolite-bearing specimens (SHCC/Z, SHCC/MK + Z) exhibited higher number of cracks than SHCC/MK and SHCC/Ref. SHCC/MK + Z was the only mixture that showed complete self-healing in terms of visual crack closing (after only 7 days) for all microcracks.
- The sorptivity index evaluated in sound and cracked specimens at the different healing ages, shows that the pozzolanic additions improved autogenous healing.

References

1. Aldea, C.-M., Shah, S.P., Karr, A.: Permeability of cracked concrete. Mater. Struct. **32** 370–376 (1999)
2. Wang, K., Jansen, D.C., Shah, S.P.: Permeability study of cracked concrete. Cement Concr. Res. **27**(3), 381–393 (1997)
3. American Society of Civil Engineers (ASCE): Infrastructure report card. Technical report, USA (2017)
4. Wittmann, F.H.: Separation of assignments: a new approach towards more durable reinforced concrete structures. In: Proceedings of the Fifth Workshop on Material Properties and Design, Durable Reinforced Concrete Structures, Aedificatio Publishers, pp. 151–160 (1998)
5. Naaman, A.E., Reinhardt, H. W.: Setting the stage: toward performance-based classification of FRC composites. In: High Performance Fiber Reinforced Cement Composites (HPFRCC-4), Published by RILEM S.A.R.L. (2003)
6. Sahmaran, M., Li, M., Li, V.C.: Transport properties of engineered cementitious composites under chloride exposure. ACI Mater. J. **104**(6), 604–611 (2007)
7. Li, V.C.: ECC—tailored composites through micromechanical modeling. In: Fiber Reinforced Concrete: Present and the Future Conference, pp. 64–97. CSCE Press, Montreal (1998)

8. Li, V.C.: On engineered cementitious composites (ECC): a review of the material and its applications. J. Adv. Concr. Technol. **1**(3), 215–230 (2003)
9. Tittelboom, K.V., Belie, N.D.: Self-healing in cementitious materials—a review. Materials **6**, 2182–2217 (2013)
10. Belie, N.D., Gruyaert, E., Al-Tabbaa, A., Antonaci, P., Baera, C., Bajare, D., Darquennes, A., Davies, R., Ferrera, L., Jefferson, T., Litina, C., Miljevic, B., Otlewska, A., Ranogajec, J., Roig-Flores, M., Paine, K., Lukowski, P., Serna, P., Tulliani, J.M., Vucetic, S., Wang, J., Jonkers, H.M.: A review of self-healing concrete for damage management of structures **5**, 17 (2018)
11. Lepech, M.D., Li, V.C.: Water permeability of engineered cementitious composites. Cement Concr. Compos. **31**(10), 744–753 (2009)
12. Roig-Flores, M., Serna, P.: Concrete early-age crack closing by autogenous healing. Sustainability **12**(11), 4476 (2020)
13. Martys, N.S., Ferraris, C.F.: Capillary transport in mortars and concrete. Cem. Concr. Res. **27**(5), 747–760 (1997)
14. Qureshi, T.S., Kanellopulos, A., Al-Tabbaa, A.: Encapsulation of expansive powder minerals within a concentric glass capsule system for self-healing concrete **121**, 629–643 (2016)
15. Feiteira, J., Gruyaert, E., Belie, N.D.: Self-healing of moving cracks in concrete by means of encapsulated polymer precursors. Constr. Build. Mater. **102**, 671–678 (2016)
16. Öztürk, O., Yıldırım, G., Keskin, U.S., Siad, H., Şahmaran, M.: Nano-tailored multi-functional cementitious composites. Composites Part B **182**, 107670 (2020)
17. Tran, Y.T., Lee, J., Kumar, P., Kim, K.H., Lee, S.S.: Natural zeolite and its application in concrete composite production. Composites Part B **165**, 354–364
18. Bhaskar, S., Hossain, K.M.A., Lachemi, M., Wolfaardt, G., Kroukamp, M.O.: Effect of self-healing on strength and durability of zeolite-immobilized bacterial cementitious mortar composites. Cement Concr. Compos. **82**, 22–33 (2017)
19. Zhang, X., Liu, M., Du, X., Yang, X., Zhou, Z.: Synergistic effect of nano silica and fly ash on the cement-based materials. Mater. Rep. **31**(24), 50–55 (2017)
20. Sujjavanich, S., Suwanvitaya, P., Chaysuwan, D., Heness, G.: Synergistic effect of metakaolin and fly ash on the properties of concrete. Constr. Build. Mater. **155**, 830–837 (2017)
21. ASTM C109/C109M—20b Standard test method for compressive strength of hydraulic cement mortars (Using 2-in. or [50 mm] Cube Specimens)
22. ASTM C1585—20 Standard test method for measurement of rate of absorption of water by hydraulic-cement concretes
23. Ramezanianpour, A.A., Mousavi, R., Kalhori, M., Sobhani, J., Najimi, M.: Micro and macro level properties of natural zeolite contained concretes. Constr. Build. Mater. **101**, 347–358 (2015)
24. Nagrockiene, D., Girskas, G.: Research into the properties of concrete modified with natural zeolite addition. Constr. Build. Mater. **113**, 964–969 (2016)
25. Influence of pumice and zeolite on compressive strength, transport properties and resistance to chloride penetration of high strength self-compacting concretes. Constr. Build. Mater. **151**(1), 292–311 (2017)
26. Radlinski, M., Olek, J.: Investigation into the synergistic effects in ternary cementitious systems containing portland cement, fly ash and silica fume. Cem. Concr. Compos. **34**(4), 451–459 (2012)
27. Ahmed, A., Mohamed, L., Gurkan, Y., Mustafa, Ş: Effect of self-healing on the different transport properties of cementitious composites. J. Adv. Concr. Technol. **1**(3), 112–123 (2015)
28. Li, V., Wang, S.: Engineered cementitious composites with high-volume fly ash. ACI Mater. J. **104**(3), 233–241 (2007)

Printed by Printforce, the Netherlands